Xingni Zhou, Zhiyuan Ren, Yanzhuo Ma, Kai Fan, Xiang Ji
Data structures and algorithms analysis – New perspectives

Also of interest

Data Structures and Algorithms Analysis – New Perspectives,
Volume 1: Data Structures Based on Linear Relations
Xingni Zhou, Zhiyuan Ren, Yanzhuo Ma, Kai Fan, Xiang Ji, 2020
ISBN 978-3-11-059557-4, e-ISBN (PDF) 978-3-11-059558-1,
e-ISBN (EPUB) 978-3-11-059318-1

Big Data Analytics Methods, Analytics Techniques in Data Mining,
Deep Learning and Natural Language Processing
Peter Ghavami, 2020
ISBN 978-1-5474-1795-7, e-ISBN (PDF) 978-1-5474-0156-7,
e-ISBN (EPUB) 978-1-5474-0158-1

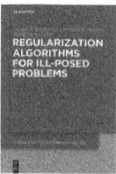

Regularization Algorithms for Ill-Posed Problems
Anatoly B.Bakushinsky, Mikhail M. Kokurin, Mikhail Yu.Kokurin, 2018
ISBN 978-3-11-055630-8, e-ISBN (PDF) 978-3-11-055735-0
e-ISBN (EPUB) 978-3-11-055638-4

Machine Learning and Visual Perception
Baochang Zhang, 2020
ISBN 978-3-11-059553-6, e-ISBN (PDF) 978-3-11-059556-7
e-ISBN (EPUB) 978-3-11-059322-8

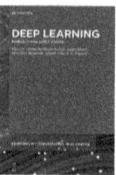

Deep Learning, Research and Applications
Siddhartha Bhattacharyya, Vaclav Snasel, Aboul Ella Hassanien,
Satadal Saha, B. K. Tripathy, (Eds.), 2020
ISBN 978-3-11-067079-0, e-ISBN (PDF) 978-3-11-067090-5
e-ISBN (EPUB) 978-3-11-067092-9

Xingni Zhou, Zhiyuan Ren, Yanzhuo Ma,
Kai Fan, Xiang Ji

Data structures and algorithms analysis – New perspectives

Volume 2: Data structures based on nonlinear relations
and data processing methods

DE GRUYTER   Science Press
Beijing

Authors

Xingni Zhou
School of Telecommunications Engineering
Xidian University
Xi'an, China
xnzhou@xidian.edu.cn

Zhiyuan Ren
School of Telecommunications Engineering
Xidian University
Xi'an, China
zyren@xidian.edu.cn

Xiang Ji
kloeckner.i GmbH
Berlin, Germany
hi@xiangji.me

Yanzhuo Ma
School of Telecommunications Engineering
Xidian University
Xi'an, China
yzma@mail.xidian.edu.cn

Prof. Kai Fan
School of Cyber Engineering
Xidian University
Xi'an, China
kfan@mail.xidian.edu.cn

ISBN 978-3-11-067605-1
e-ISBN (PDF) 978-3-11-067607-5
e-ISBN (EPUB) 978-3-11-067616-7

Library of Congress Control Number: 2020930334

Bibliographic information published by the Deutsche Nationalbibliothek
The Deutsche Nationalbibliothek lists this publication in the Deutsche Nationalbibliografie;
detailed bibliographic data are available on the Internet at http://dnb.dnb.de.

© 2020 China Science Publishing & Media Ltd. and Walter de Gruyter GmbH, Berlin/Boston
Cover image: Prill/iStock/Getty Images Plus
Typesetting: Integra Software Services Pvt. Ltd.
Printing and binding: CPI books GmbH, Leck

www.degruyter.com

Contents

Preface

Looking at old problems with new perspectives, requires creative thinking —— Einstein

Most of the contents have been mentioned in the Preface of Volume 1 of this book. Therefore, they are not repeated here.

Zhou Xingni
xnzhou@xidian.edu. cn
Spring 2019 in Chang'an

https://doi.org/10.1515/9783110676075-203

1 Nonlinear structure with layered logical relations between nodes – tree

Main Contents
- The definition of tree and basic terminologies
- The concept of binary tree and its storage structure
- The traversal of binary tree
- Huffman tree and its applications
- The concept, storage structure and basic operations of lists

Learning Aims
- Comprehend the transition from linear structure to nonlinear structure of the logical structure of data
- Comprehend linear structures that contain child structure
- Understand the functionality of linked storage structure in expressing nonlinear data structures
- Comprehend the concept, storage method and basic operations on trees
- Understand the concept, structural characteristics and storage/representation methods of lists

1.1 Tree in actual problems

1.1.1 Introductory example to tree 1 – tree-like structures in our daily life

If we enter a huge library, and want to find a reference book for data structures, how can we find it facing several floors, multiple rooms, rows of bookshelves and tens of thousands of books? The first thing before we search for it should be to understand the classification and layout rules of books in the library. After we know the representational methods of the classification of books, we can easily find the needed books in the library, and it will be very convenient for the librarian to uniformly manage the books. Figure 1.1 shows partial classifications about computer according to the library classification method of China.

Observe the classification structure of Fig. 1.1. It looks very much like a reversed tree in nature, that is, the point at the top is the "root," while the points at the bottom ends are "leaves." Therefore, we call it as tree structure, and abbreviate it as "tree."

Tree-like structures are very common in our daily life. Examples include genealogy records, organizational structures and match results of a tournament. They are all structures produced after a certain classification of information.

https://doi.org/10.1515/9783110676075-001

Fig. 1.1: Classification structure of books.

1.1.2 Introductory example of tree 2 – the directory structure of computer

Everyone who has used computers is familiar with the directory structure of files. This is also a classification structure, as shown in Fig. 1.2. In this directory hierarchy, we can see that it is a leveled (layered) structure. For the computer to manage such directories, what are the potential problems involved?

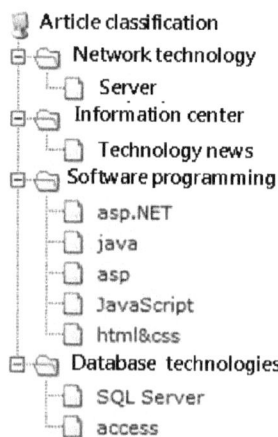

Fig. 1.2: Computer file directories.

According to our general concepts about directory management of computers, the operations include searching, adding files and deleting files. To realize such operations, the first thing to do is to store the folder information and the connections between layers into the machine, only then can it process them according to the functionality needs. Before the storage, we must perform further abstraction on its structure. After we abstract the folders into nodes with a uniform structure, the relations between various nodes are represented as shown in Fig. 1.3.

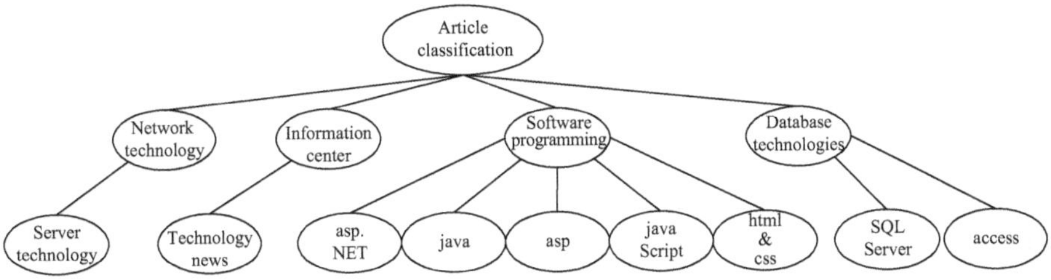

Fig. 1.3: The abstraction of directory structure.

Since the relation between nodes is not only one of predecessors and successors, tree structure is a nonlinear structure.

1.1.3 Introductory example of tree 3 – tree-structured website

The connection between various pages within a website also forms a web-like structure. Such a web-like structure can take on multiple forms. The tree-like web structure with clearly defined layers is shown in Fig. 1.4. Website structure considered ideal from the perspective of search engine optimization is a tree structure, where each page is only directly linked to the page one layer above. In this manner, users can quickly navigate to channels and articles they are interested in. The search engine can also understand the structure and layers of the website better and thus better crawl the contents, which is conducive to increasing the ranking of the website and bringing in more visitors, eventually augmenting the sales or marketing effect of the website.

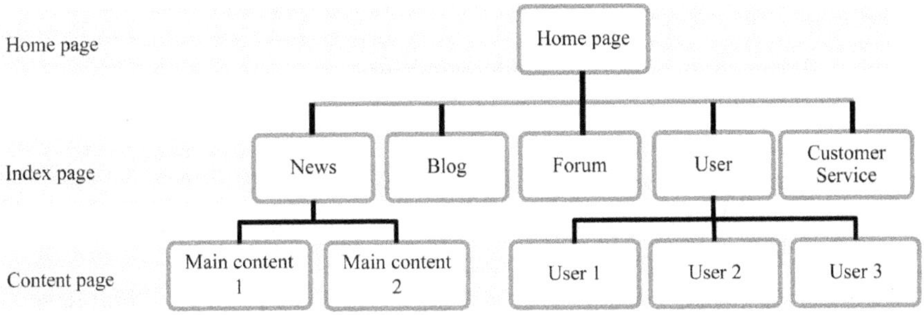

Fig. 1.4: Website with tree structure.

There are also a lot of application of tree structures in computer science. For example, when compiling source programs, the main job is to perform syntactical analysis on the program, in which the analysis of mathematical expressions is an important step. The result can constitute an expression tree as shown in Fig. 1.5. Corresponding mathematical expressions can be obtained by different search strategies. In database systems, tree structure is also an important form of information organization. Besides, there are decision trees used for the classification and prediction in data mining.

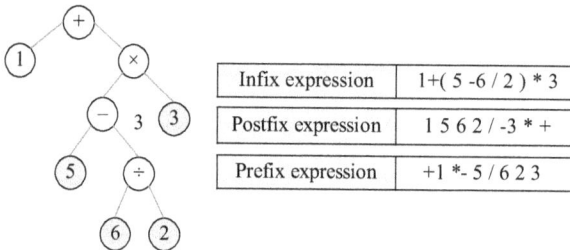

Infix expression	1+(5 -6 / 2) * 3
Postfix expression	1 5 6 2 / -3 * +
Prefix expression	+1 *- 5 / 6 2 3

Fig. 1.5: Expression tree.

Generally, tree is very widely applied in various fields of computer science. All problems with layered relation can be described using trees.

1.2 Logical structure of tree

From the previous introductory examples, we can see that when the relation between data elements is not one-to-one correspondence, but one-to-many correspondence. After the correspondence relation gets more complicated, linear structure does not suffice to conveniently describe such complex scenarios. We need other methods that fit the features of tree structure to describe such nonlinear structures with layered relations.

1.2.1 The definition and basic terminologies of tree

1.2.1.1 The definition of tree
Tree is a finite collection with n nodes, among which:
1. One specific node is called root node or root. It only has immediate successors, but has no immediate predecessors.
2. Data elements other than the root node are divided into m ($m \leq 0$) nonintersecting collections $T_1, T_2,..., T_m$, among which each collection T_i ($1 \leq i \leq m$) itself is also a tree, called a subtree of the root.

When the collection of trees is empty, $n = 0$. In this case, it is called empty tree. An empty tree has no nodes.

Tree is a recursive structure – in the definition of tree we use the definition of tree again: a nonempty tree constitutes several subtrees, and child trees can constitute several even smaller subtrees.

Tree has the following characteristics, as shown in Fig. 1.6.

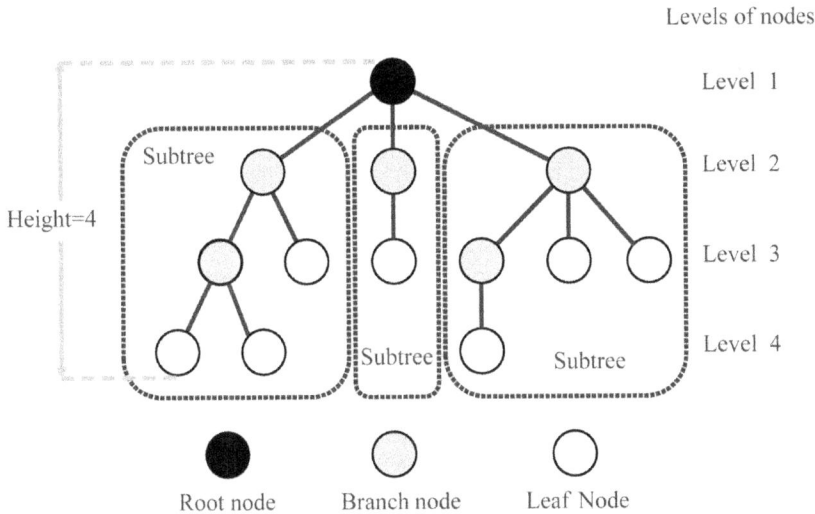

Fig. 1.6: Illustration of tree.

- Each node has zero or multiple child nodes. A node without any child node is called "leaf node."
- A node without parent node is called "root node." We specify its layer as the first layer.
- Each nonroot node has only one parent node.

1.2.1.2 The graphical representation of tree

There are multiple representation methods of tree's logical structure. Besides the normally seen tree-form representation, there are also Venn's diagram representation, indentation representation and list representation. Figure 1.7 is the same tree represented with various representation methods.

Tree-form representation method is the most basic representation method, which uses one reversed tree to represent a tree structure.

Venn diagram representation method uses sets and subset relations to describe a tree structure.

Tree representational method

Indentation representational method

Venn diagram representational method

(A(B(E,F),C,D(G)))

·(List representational method)

·Embedded brackets representational method

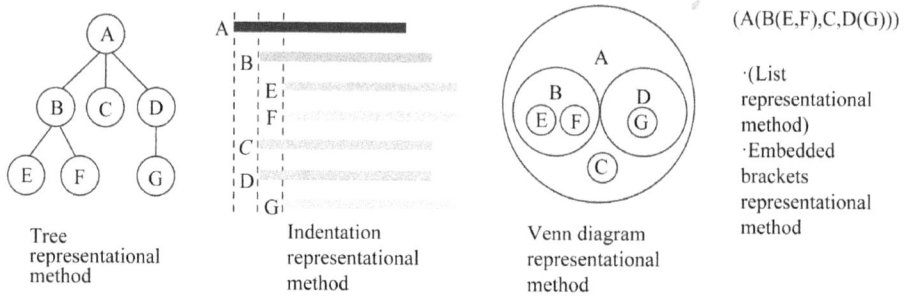

Fig. 1.7: Representation methods of tree.

Indentation representation method uses the relative positions of line segments to describe the tree structure. For example, in the directory hierarchy in Fig. 1.2, the directory of all books uses such a form to represent the tree structure.

List representation method is also called bracket representation method. It writes the root node to the left of the outer bracket and all the other nodes within the outer bracket, and uses commas to describe the tree structure.

1.2.1.3 Related terminologies for tree

Based on the tree representation method in Fig. 1.7, we can describe various terminologies of tree as shown in Fig. 1.8.

Ancestor
A is an ancestor of FG
B is an ancestor of KL

Level
A's level : 1
M's level : 4

Parent
H's parent : D
L's parent : F

Degree
A's degree :
M's degree : 0
Tree's degree : 3

Child
A's child : B, C, D
B's child : E, F

Depth
The depth of the tree : 4

Sibling
BCD are siblings
FG are cousins

Leaf
E,K,L,G
H,I,M

Fig. 1.8: Terminologies of tree structure.

1. Node of tree: contains one data element and can simultaneously record several branching information pointing to subtrees.
2. Child and parent: the root of a subtree of a certain node in the tree is called the child of the node. Correspondingly, this node is called the parent of the child. The children from the same parent are called siblings.
3. Ancestor and descendant
- Path
  If there is a sequence of nodes $k_1, k_2, \ldots, k_i, \ldots, k_j$ in the tree which makes k_i the parent of k_{i+1} $(1 \le i < j)$, then we call this sequence of nodes and the branches on it a path from k_i to k_j.

  The length of the path refers to the number of edges (i.e., line segments connecting two nodes), which is equal to $j - 1$.

  Note: If a sequence of nodes is a path, then in the tree representation of a tree, this sequence of nodes passes through each edge in the path in a "top-down" manner.

  From the root node of the tree to any node in the tree, there exists one unique path.
- Ancestor and descendant
  If there is a path from the node k to k_s in the tree, then we call k an ancestor of k_s and k_s a descendant of k.

  The set of ancestors of a node consists of all the nodes on the path from the root node to this node. The set of descendants of a node consists of all the nodes in the subtrees that have this node as the root.

  Convention: the ancestors and descendants of a node usually do not include the node k itself.
4. The level of nodes and the height of trees
   The level of node is counted beginning from the root: the level of the root is 1, while for the rest of the nodes the level is the level of its parent plus one. The nodes whose parents are at the same level are cousins.

   The maximum level value in a tree is called the height or depth of the tree.

   Note that some conventions also define the level of the root as 0.
5. Degree of node
   The number of subtrees present in a node in the tree is called the degree of this node. The degree of a tree refers to the maximum degree among all nodes. For example, in Fig. 1.7 the degree of the tree is 3. A node with degree 0 is called leaf node or terminal node. A node with a nonzero degree is called a branch node or a nonterminal node. Branch nodes except for the root node are called internal nodes. The root node is also called the starting node.

6. Ordered tree and unordered tree

If we see all the subtrees of all nodes in the tree as ordered from left to right (i.e., they cannot be interchanged), then we call this tree an ordered tree. Otherwise, we call it an unordered tree.

Note: If we do not specify it, our discussions are normally on ordered trees.

7. Forest

A forest is a collection of m ($m \geq 0$) nonintersecting trees.

Tree and forest are related. When we delete the root of a tree, we obtain a forest. On the contrary, if we add a node to become the root, then a forest becomes a tree.

> **!** **Think and Discuss** In the tree structures we are familiar with what are the ordered trees and nonordered trees?
>
> **Discussion:** The difference between ordered tree and unordered tree is on whether there is order requirement on the subtrees. Examples of ones with order requirement are genealogy records, book index and others. An example of unordered tree would be the folder structure of computers.

1.2.1.4 The logical characteristics of tree structures

The logical characteristics of tree structures can be described by the relations between the parent nodes and child nodes.

1. Any node in the tree can have zero or multiple immediate successor (i.e., child) nodes, but can have at most only one immediate predecessor (i.e., parent) node.
2. Only the root node in a tree does not have any predecessor. It is the beginning node. Leaf nodes do not have any successors; they are the terminal nodes.
3. The relation between ancestor and descendant is an extension on the relation between parent and child. It defines the vertical ordering of nodes in a tree.
4. In ordered trees, there is an ordering between a set of sibling nodes. If we extend this relation by specifying that if k_1 and k_2 are siblings, and k_1 is to the left of k_2, then any descendant of k_1 is also to the left of any descendant of k_2, then we have defined the horizontal ordering of all nodes in the tree.

1.2.2 Definition of operations on trees

There are normally the following kinds of operations based on the logical structure of tree:

1. Construction: construct a tree and initialize it.

2. Lookup: it is a lookup on root node, parent node, child node, leaf node, node with a given value and so on.
3. Insertion: insert a node at the given location.
4. Deletion: delete a node at the given location.
5. Traversal: following a certain search route, visit each node of the tree once and only once. The operations on the nodes being visited depend on the concrete problem.
6. Depth calculation: calculate the depth (height) of the tree.

1.3 The storage structure of tree

The storage of a tree structure should still follow the two basic principles of storage: "store the values as well as the connections" and "be able to store as well as retrieve." Since tree is a nonlinear structure, the storage of the connections between the nodes is much more complicated than that under a linear relation; it is the key to storing tree structures.

Think and Discuss What are the principles of storing the connections between the nodes in a tree structure?
Discussion: According to the features of a tree structure, there are only two types of possible relations between each node and a node directly connected to it: 1. parent; 2. child, as shown in Fig. 1.9. For a node, there can only be one parent, but there can be 0 to n children. The design principle of the storage structure of the tree structure is to realize the direct/indirect storage of the parent/child relation. The verification standard for a well-designed storage structure is to be able to find these two relations of a node in the storage structure. As long as we can do so, such a storage structure is feasible. We can call it "parent–child verification principle."

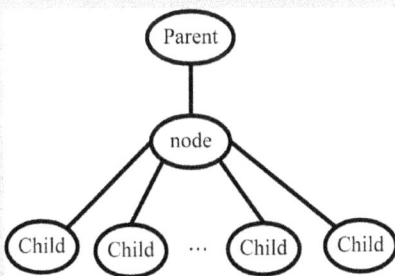

Fig. 1.9: Features of tree structure.

People have designed various storage structures to represent a tree. There are two main storage methods: continuous storage by arrays and discrete linked storage.

In the descriptions of this section, N represents the number of nodes of a tree, while D represents the depth (height) of a tree.

1.3.1 The continuous storage method of trees

1.3.1.1 Continuous storage of trees – parent–child representation method

According to the "store the values as well as connections" storage principle, we store the values of a tree with an array. The subscripts of the array correspond to the numberings of the nodes. Then, we can denote the relations between each node and its parent/children, as shown in Fig. 1.10 (–1 denotes that such a relation does not exist). The description of the data structure is shown in Fig. 1.11.

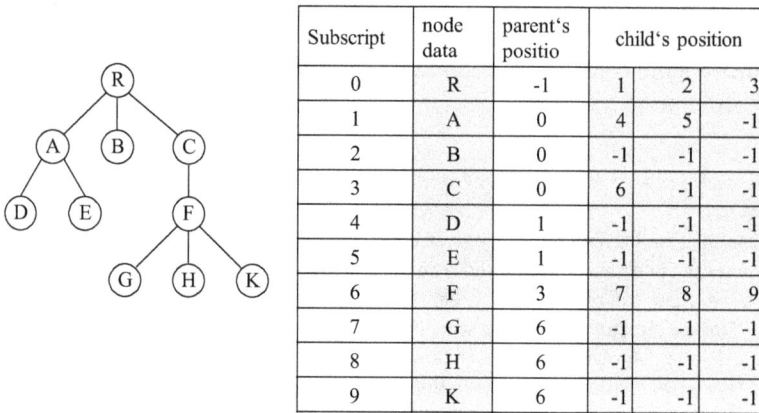

Subscript	node data	parent's positio	child's position		
0	R	-1	1	2	3
1	A	0	4	5	-1
2	B	0	-1	-1	-1
3	C	0	6	-1	-1
4	D	1	-1	-1	-1
5	E	1	-1	-1	-1
6	F	3	7	8	9
7	G	6	-1	-1	-1
8	H	6	-1	-1	-1
9	K	6	-1	-1	-1

Fig. 1.10: Continuous storage of tree – parent–child representation method.

1.3.1.2 Continuous storage of trees – parent representation method

A feature of tree structures is that the parent of each node is unique. We can obtain the information of the child indirectly from the parent relation. For example, in Fig. 1.11, if the parent of D, E is A, then the children of A is D, E. Therefore, we can obtain the information on both the parent and the children by only storing the positions of the parent. Consequently, the continuous storage method of trees can be simplified to Fig. 1.12.

1.3.1.3 Continuous storage of trees – children representation method

By the same logic, we can directly obtain the information of parent from the child relations. The continuous storage method of trees in Fig. 1.10 can be simplified to Fig. 1.13.

Structure design

data	parent	child
Node data	parent position	child position

typedef stuct

{ datatype data;   // Data field

  int parent;   // Parent position

  int child[D];   // Child position

} PCtree;

PCtree T[N];

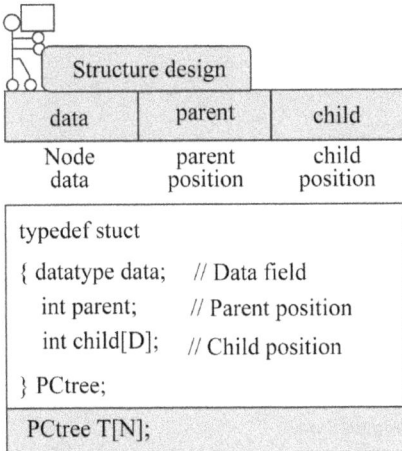

Fig. 1.11: Continuous storage of tree – description of data structure of parent–child representation method.

Subscript	node data	position
0	R	1
1	A	4
2	B	−1
3	C	6
4	D	−1
5	E	−1
6	F	7
7	G	−1
8	H	−1
9	K	−1

Structure design

data	parent
Node data	parent position

typedef struct

{ datatype data;// Data field

  int parent[D];// parent position

} Ptree;

Ptree T[N];

Fig. 1.12: Continuous storage of trees: parent representation method.

1.3.2 Linked storage method of trees

We can store the information of tree structure in the mold of linear linked lists. In a linear list, there is exactly one predecessor and one successor for a node. Therefore, the chain formed by linked storage of linear list is a straight line. On the contrary, a node in a tree structure has one predecessor, but multiple successors, and thus the successor part would contain multiple outgoing links, as depicted in Fig. 1.14. There, the length of array in the node represents the number of children of this node. It can be described by the word "degree" for a node.

Structure design

data	child

Node data	child position

typedef struct
{ datatype data;// Data field
  int child[D]; // Child position
} Ctree;

Ctree T[N];

Subscript	node data	child position		
0	R	1	2	3
1	A	4	5	-1
2	B	-1	-1	-1
3	C	6	-1	-1
4	D	-1	-1	-1
5	E	-1	-1	-1
6	F	7	8	9
7	G	-1	-1	-1
8	H	-1	-1	-1
9	K	-1	-1	-1

Fig. 1.13: Continuous storage of trees: child representation method.

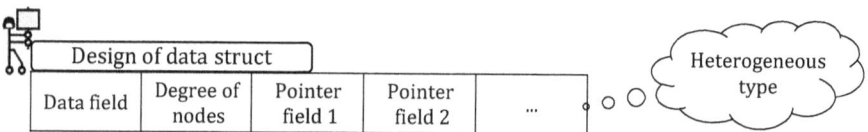

Fig. 1.14: Analysis of linked storage of trees.

Examining the "Parent–child verification principle," we can find the children of each node conveniently via the indication of the pointer fields, though we must start the search for its parent from the root node.

The structural design of nodes in linked tree is shown in Fig. 1.15. The number of link fields of each node is equal to the degree of each node. If two nodes differ in degrees, then they have different structures as well. We call this "heterogeneous type."

Design of data struct

Data field	Degree of nodes	Pointer field 1	Pointer field 2	...

Heterogeneous type

Fig. 1.15: Linked node structure of tree – heterogeneous structure.

Although the pointer fields of "heterogeneous type" is allocated on demand, its shortcomings are also apparent. First, its structural description is difficult. Also, certain operations on it are also difficult, such as node lookup: we cannot iterate all the subtrees of a node with the same strategy; another example is node insertion: since we did not reserve any additional space for pointer field, the operation is hard to proceed.

An improvement on the "heterogeneous type" is to change it into a "homogeneous type," as shown in Fig. 1.16. In a "homogeneous type," the number of pointer fields is the degree of the tree. The description of "homogeneous type" is shown in Fig. 1.17.

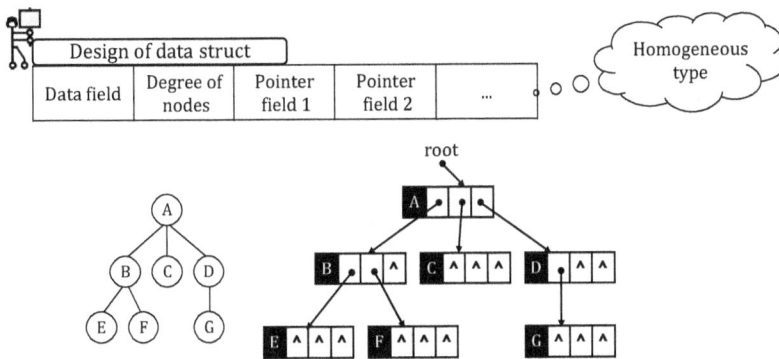

Fig. 1.16: Linked node structure of tree – homogeneous type.

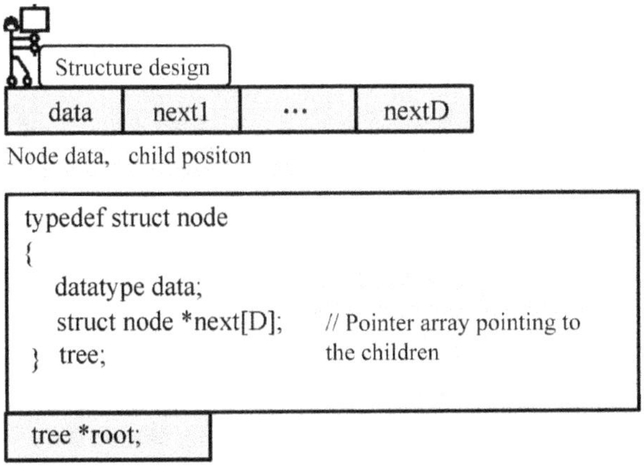

Fig. 1.17: Data type description of linked node structure of tree – homogeneous type.

Think and Discuss

1. What are the characteristics of a homogeneous type?
Discussion: Homogeneous structure eliminates the shortcomings of heterogeneous structure. The unification of structures makes management simple. However, if most nodes have a degree smaller than that of the tree, then some pointer fields will be empty, which causes a waste of storage space.

2. What kind of tree has the least amount of empty pointer fields when stored with homogeneous structure?
Discussion: The number of pointer fields is related to the degree of the node. The number of empty pointer fields is related to the total number of nodes of the tree. Suppose there are n nodes in the tree in total, then for a tree with degree d, if stored with homogeneous nodes, then:
- the total number of pointer fields in the whole linked list: $n \times d$;
- the number of useful pointer fields: $n - 1$ (note: the address of the root node does not occupy a pointer field);
- number of empty pointer fields: $n \times d - (n - 1) = n(d - 1) + 1$.

From the above analysis, we can see that the number of empty pointer fields is related to both the number of nodes n and the degree of the tree d. n is a number related to the tree itself, while d is a number related to the structure of the tree. We try to focus on d instead of n; thus, we would like to remove the n in expression (5.1). We can adopt a different way of representation, which gives us the proportion of empty pointer fields within all the pointer fields R, as shown in expression (5.2), and then obtain the value of R at the limit, as shown in expression (5.3). In this way, we have eliminated the influence of n:

$$R = \frac{\text{Number of empty pointer fields}}{\text{Total number of pointer fields in the whole linked list}} = \frac{n(d-1)+1}{nd} \tag{5.2}$$

$$\lim_{n \to \infty} R = \lim_{n \to \infty} \frac{n(d-1)+1}{nd} = 1 - \frac{1}{d} \tag{5.3}$$

In expression (5.3), we can see that when d gets smaller, R gets smaller as well. $d = 0$ is impossible since a tree cannot have degree 0; when $d = 1$, the degree of the tree is 1, and the structure of the tree morphs into a linear structure, which doesnot really correspond to the requirements of the question. Therefore, the ultimate conclusion is that in general, when $d = 2$, the number of empty linked fields would be the smallest.
   A tree of degree 2 is called a binary tree.

There are also the following methods for the linked storage of trees.

1.3.2.1 Linked storage of tree – child–sibling representation method
See Fig. 1.18. This storage method uses a linked storage structure. When storing each node of the tree, besides including the value of the field of this node, we also set two pointer fields, which point to the first child node of this node and the right sibling of this node, respectively, that is, we record at most the information of two

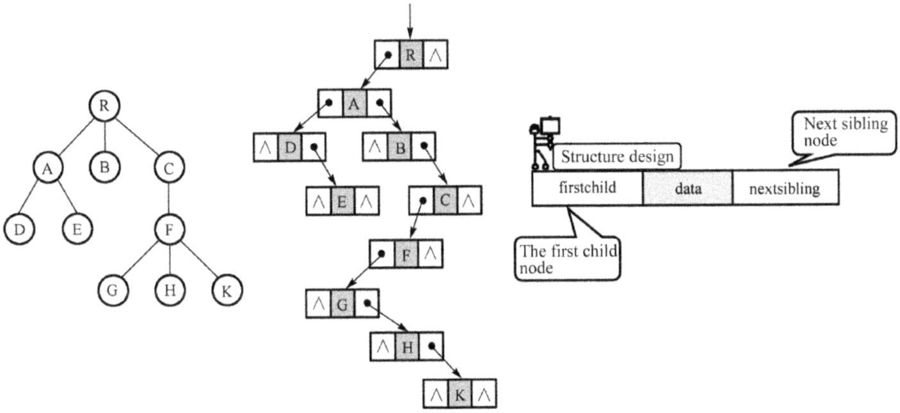

Fig. 1.18: Tree storage – child–sibling representation method.

nodes. In this way, we can store the information with a uniform binary linked list format. Therefore, this method is also usually called binary tree representation method.

In this storage structure, it is easy to realize operations such as finding the children of a node. It is more difficult to find the parent of a node. For example, if we want to visit the third child of the node F, K, we need only to first find the first child node G from the firstchild field, and, following the nextsibling fields of the children by two steps, we can find the third child of F.

1.3.2.2 Linked storage of trees – children linked list representation method

This method is a combination of continuous storage and linked storage of the tree. We arrange all the children nodes of a node together in a linear-list format, and we use singly linked list as the storage structure. Then, for n nodes in the tree, we will have n children linked lists (the children linked lists of leaves will be empty lists). Also, n head pointers constitute another linear list. To facilitate lookup, we can also adapt a linear storage structure, as shown in Fig. 1.19. We have seen this method of "representation method using singly linked lists with a vector of row pointers" a lot of times; for example, the linked storage is used for sparse matrices in Chapter 4. For this storage method, it is very convenient to look for the children of a node, but to look for the parent of a node, we will need to iterate the whole structure. The data type description of the structure is shown in Fig. 1.20.

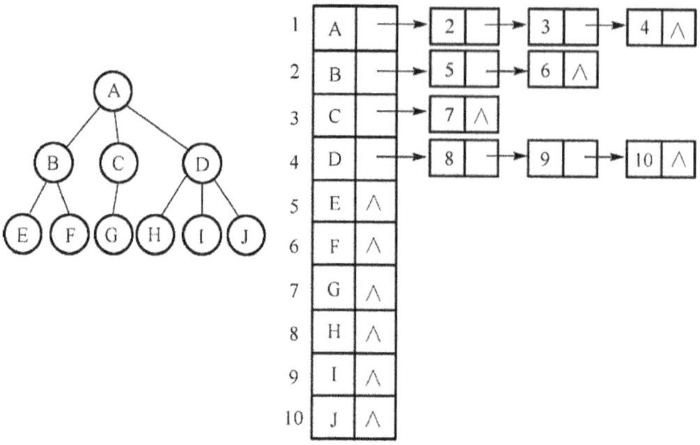

Fig. 1.19: Tree storage – children linked list representation method.

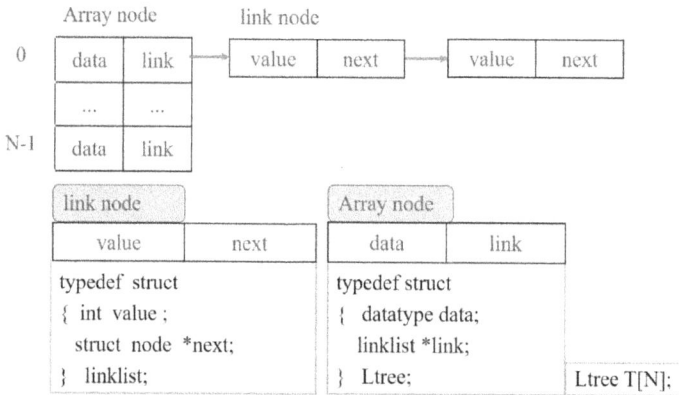

Fig. 1.20: Data type description of tree storage – children linked list representation method.

1.4 The logical structure of binary trees

According to the discussion on homogeneous tree structures above, a binary tree has relatively high storage efficiency, when we adopt the linked storage method. Binary tree is also the simplest form of a tree. Therefore, we focus on the features, storage and operations on a binary tree in the following sections.

> **!**
>
> **Think and Discuss** If we study and discuss binary trees as the typical structure, can we still apply the corresponding results on generic trees?
> **Discussion:** If we can find methods to convert a generic tree to a binary tree, and to convert a binary tree back to the original generic tree, then we have perfectly solved this problem.
> **Knowledge ABC** The conversion between tree and binary tree

1. Tree to binary tree
The conversion process is shown in Fig. 1.21.

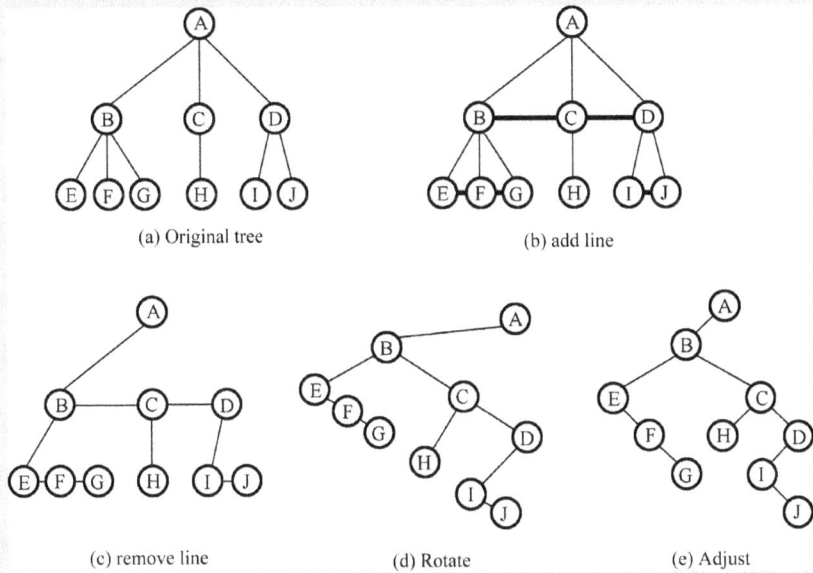

(a) Original tree

(b) add line

(c) remove line

(d) Rotate

(e) Adjust

Fig. 1.21: Conversion from tree to binary tree.

Add line: add lines to connect all the adjacent sibling nodes.
   Remove line: for each nonterminal node, except for its leftmost child, delete the lines between this node and all other child nodes.
   Rotate: using the root node as an axis, rotate rightward 45°.
   Adjust: adjust the resulting tree to a normal binary tree.

Think and Discuss What kinds of changes took place on the connections between all the nodes during the conversion process from tree to binary tree?
Discussion: During the process of adding line, direct connections between the nodes and their siblings are added. During the process of removing line, we removed all the connections except for the one with the first child; however, we can still indirectly obtain information on all the children via the sibling relations of the first child. This follows the same principle as the "linked storage of tree – child–sibling representation method" mentioned earlier.

2. Conversion from forest to binary tree
The conversion process is shown in Fig. 1.22.
   Conversion: convert each tree in the forest individually into a binary tree.
   Connection: starting from the last binary tree, use the root node of the latter binary tree as the right child of the root node of the previous binary tree, until all the binary trees are connected. Then we adjust the result into a binary tree.

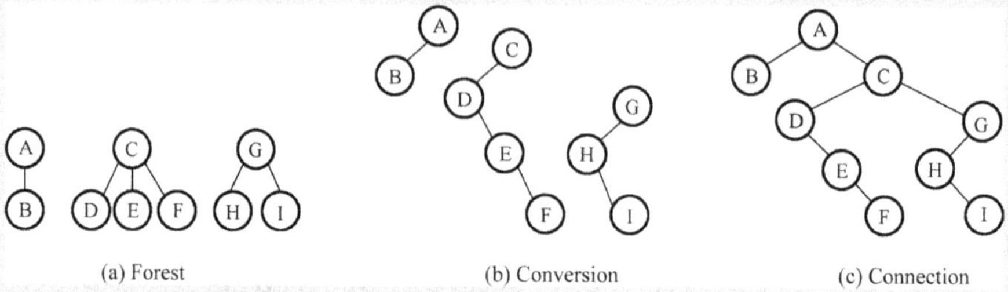

(a) Forest   (b) Conversion   (c) Connection

Fig. 1.22: Conversion from forest to binary tree.

3. Restoration from binary tree to tree
The conversion process is shown in Fig. 1.23.

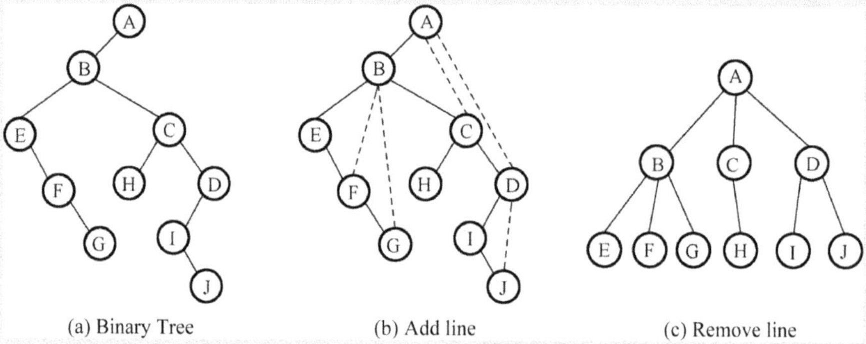

(a) Binary Tree   (b) Add line   (c) Remove line

Fig. 1.23: Restoration of binary tree into tree.

Add line: if some node x is the left child of its parent y, then we connect all the right descendants of x with y.

Remove line: delete all the lines between parent nodes and all their right children nodes in the original binary tree.

Adjust: adjust the tree or forest obtained via the previous two steps, so that their layers are clearly arranged.

Think and Discuss What are the changes in the connections between different nodes during the process of restoring a tree into a binary tree?

Discussion: The right descendants of the left child of a node x were all originally the siblings of this left child. For example, F and G in Fig. 1.23 are both descendants of E, while E, F, G were all originally children of B. Therefore, adding line is to restore the relations between a node and its original children; removing line is to remove the connections between the original siblings. In this way, we can restore the original tree structure.

4. The storage relation between tree and binary tree

In fact, the storage structure established by a tree using child–sibling representation method is exactly the same as the binary linked list storage structure of the corresponding binary tree. We just give different names and meanings to the two pointer fields. Figure 1.24 straightforwardly shows the correspondence and mutual conversion between tree and binary tree. The tree structure in Fig. 1.24(a) corresponds to Fig. 1.24(b), and the form of the corresponding binary linked list is Fig. 1.24(d). Figure 1.24(c) and (e) shows different explanations of Fig. 1.24(d). For example, in Fig. 1.24(c), node C is the right sibling of node B, while in Fig. 1.24(e), node C is the right sibling of node B. Therefore, the related algorithms on binary linked lists can be easily converted to algorithms on child–sibling linked lists of trees.

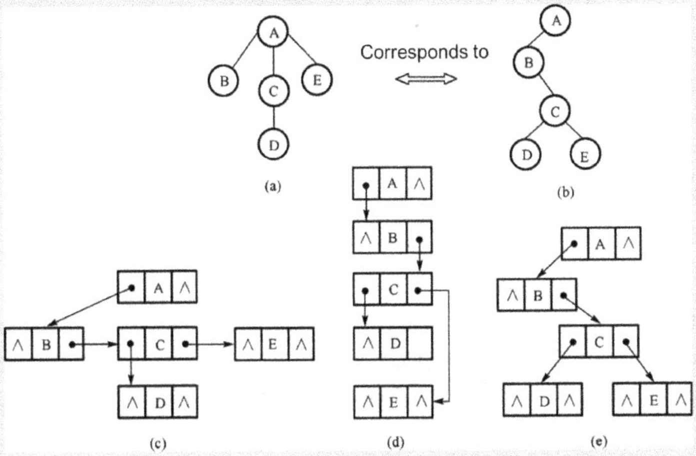

Fig. 1.24: The relation between tree and binary tree.

1.4.1 The concept of binary tree

1.4.1.1 Definition of binary tree

A binary tree is a finite collection of $n(n \geq 0)$. It can either be an empty tree ($n = 0$), or be constituted of a root node and two nonintersecting binary trees called left subtree and right subtree.

Note: A binary tree is an ordered tree whose each node has at most two subtrees. The subtrees of a binary tree are usually called left subtree and right subtree. The ordering of left subtree and right subtree cannot be switched.

Think and Discuss What is the difference between binary tree and tree?
Discussion: Although there are many similarities between binary tree and tree, there are two main differences between tree and binary tree:

!

1. There is no restriction on the maximum degree for nodes in a tree, while the maximal degree for nodes of a binary tree is 2.
2. In general, there is no left/right distinction for the nodes in a tree, while nodes of a binary tree are divided into left node and right node, as shown in Fig. 1.25.

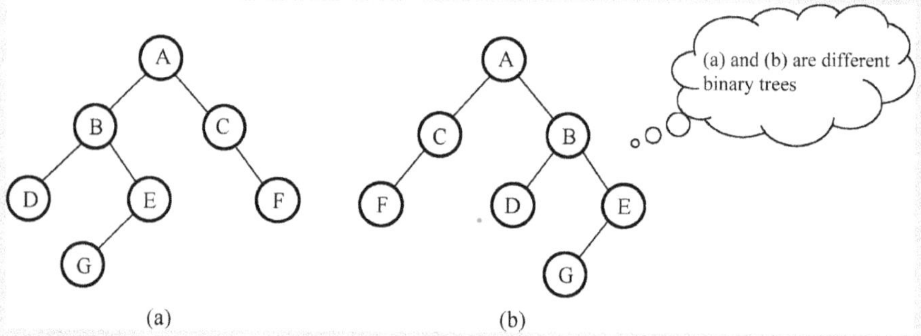

(a)   (b)

Fig. 1.25: Binary tree with left/right distinction.

1.4.1.2 Different forms of binary trees

There are different forms of binary trees. According to normal scenarios of problem-solving and the classification principles on special cases, we can summarize five fundamental forms and two special forms of binary trees, to facilitate discussions on binary trees.

1. The basic forms of binary trees

Logically, there are five basic forms of binary trees:

A binary tree can be an empty set; a root can have only left subtree or only right subtree; both nonempty subtrees; or both left and right subtrees are empty. See Fig. 1.26. The five basic forms of binary tree.

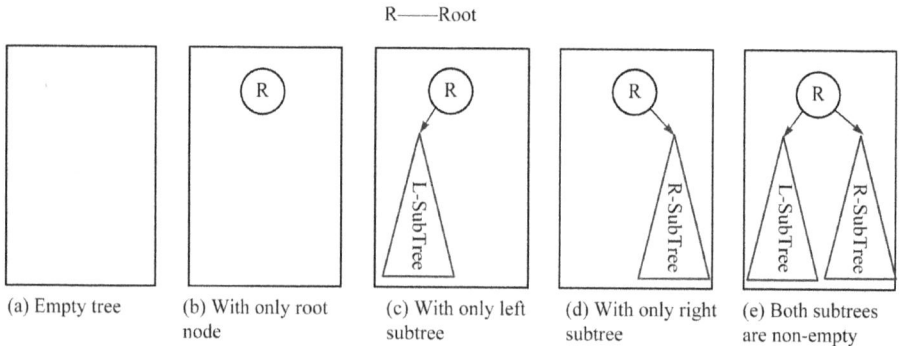

(a) Empty tree   (b) With only root node   (c) With only left subtree   (d) With only right subtree   (e) Both subtrees are non-empty

Fig. 1.26: Basic forms of binary trees.

2. Special forms of binary trees
   - Full binary tree
     A binary tree with depth k and $2^k - 1$ nodes is called a full binary tree. The characteristics of a full binary tree are that the number of nodes at each layer reaches the maximum number of nodes allowed for this layer. There is no node with degree 1. A sample is shown in Fig. 1.27(a).
   - Complete binary tree
     If we number the nodes of a binary tree with depth k, starting from the root, in a top-down, left-to-right manner, and obtain an identical sequence of numbering on the corresponding nodes as the numbering of a full binary tree, then we call this binary tree a complete binary tree. Otherwise, it is a noncomplete binary tree. Samples are shown in Fig. 1.27(b) and (c).

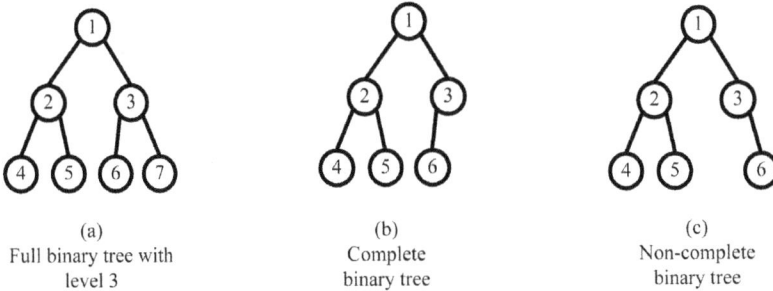

(a)
Full binary tree with
level 3

(b)
Complete
binary tree

(c)
Non-complete
binary tree

Fig. 1.27: Samples of binary trees.

1.4.2 Basic properties of binary trees

The following are the properties for the number of nodes, number of layers, depth, number of leaves and numbering of nodes of a binary tree.

Property 1 The ith layer of a binary tree has at most 2^{i-1} nodes ($i \geq 1$).

Property 2 A binary tree with depth h has at most $2^h - 1$ nodes ($h \geq 1$).

Property 3 For any binary tree, if it has n_0 leaf nodes, n_2 nodes with degree 2, then there must be a relation: $n_0 = n_2 + 1$.

Property 4 A complete binary tree with n nodes has depth $[\log_2 n] + 1$. (The square brackets indicate flooring.)

Property 5 If we number a complete binary tree with n nodes in a top-down, left-to-right manner with 1 to n, then for any node with numbering i in the complete binary tree:

If $i = 1$, then this node is the root of the binary tree. It has no parent. Otherwise, the node with numbering $[i/2]$ is its parent node.

If $2i > n$, then this node has no left child. Otherwise, the node with numbering $2i$ is its left child node.

If $2i + 1 > n$, then this node has no right child. Otherwise, the node with numbering $2i + 1$ is its right child node.

Example 1.1 Calculation of number of various types of nodes on a binary tree

If a complete binary tree has $n = 1{,}450$ nodes, then how many nodes with depth 1, nodes with depth 2, leaf nodes are there? How many left children and right children are there? What is the height of the tree?

Solution:

The height of the tree $h = [\log_2 n] + 1 = 11$.

$\because$ It is a complete binary tree

$\therefore$ Layers 1–10 are totally full, $k = 10$

The number of leaf nodes at the bottommost level $= n - (2^k - 1) = 1{,}450 - 1{,}023 = 427$

The number of nodes with leaves at level $k = [(427 + 1)/2] = 214$

The number of nodes at level $k = 2^{k-1} = 512$

The number of leaves at level $k = 512 - 214 = 298$

$\therefore$ Number of total leaves $n_0 = 427 + 298 = 725$

The number of nodes with degree $2 n_2 = n_0 - 1 = 724$

The number of nodes with degree $1 n_1 = n - 1 - 2n_2 = 1{,}450 - 1 - 2 \times 724 = 1$

The number of nodes with left child = number of nodes with degree 2 + number of nodes with degree 1 = 725

The number of nodes with right child = number of nodes with degree 2 = 724

1.4.3 Definition of operations on binary tree

According to the definition of the logical structure of binary trees, the following are the basic operations on the binary tree.

1. Construction: construct a binary tree.
2. Lookup: find the root node, parent node, child node, leaf node and so on.
3. Insertion: insert a node at the given position.
4. Deletion: delete a node at the given position.
5. Traversal: following a certain search route, visit each node in the binary tree once and only once in order.
6. Depth calculation: calculate the height of the binary tree.

1.5 The storage structure of binary tree and its implementation

Previously, we have discussed the storage plans for generic trees. Binary trees are simpler in form than generic trees; therefore, its storage form has corresponding characteristics. Next, we discuss the storage plans for binary trees in correspondence with those for generic trees.

1.5.1 The sequential structure of binary trees

Figure 1.28(a) represents a complete binary tree with the method shown in Fig. 1.9. According to the property 5 of binary trees, the relation between the nodes of a complete binary tree is shown in Fig. 1.28(b). Referencing the two figures, discuss whether we can improve the storage plan.

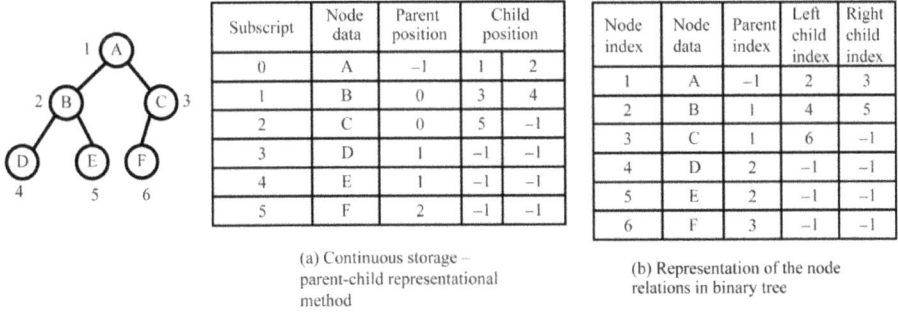

Subscript	Node data	Parent position	Child position	
0	A	−1	1	2
1	B	0	3	4
2	C	0	5	−1
3	D	1	−1	−1
4	E	1	−1	−1
5	F	2	−1	−1

Node index	Node data	Parent index	Left child index	Right child index
1	A	−1	2	3
2	B	1	4	5
3	C	1	6	−1
4	D	2	−1	−1
5	E	2	−1	−1
6	F	3	−1	−1

(a) Continuous storage – parent-child representational method

(b) Representation of the node relations in binary tree

Fig. 1.28: Discussion on the continuous storage method of complete binary trees.

1.5.1.1 Analysis on the relation between nodes
Because we already know the numbering of a node, we can deduce the numberings of its related nodes. Therefore, we only need to store the numbering of the node, and there is no need to store the position of its parent and children. This would save more than half of the space compared with parent–children representational method.

1.5.1.2 Analysis on the position of the node
Due to the continuity in the numberings of nodes, we can use them as the subscripts of the storage vector. Since C language specifies that array subscripts must start from 0, we can choose to not use unit 0, and store the root node at the position with subscript 1. Then, the numberings of the tree and the positions of the subscripts can correspond one by one.

The sequential structure of a full binary tree or a complete binary tree can be implemented using a group of contiguous memory units. The nodes of the complete binary tree can be stored sequentially according to the order of the numberings. The storage positions already imply the relation between the nodes in the tree, which satisfy property 5 of binary trees, as shown in Fig. 1.29. If a certain node has a unit numbering of I, then the position of its parent is $i/2$; the position of its left child is $2 \times i$ and the position of its right child is $2 \times i + 1$; the position of its left sibling is $i - 1$ and the position of its right sibling is $i + 1$.

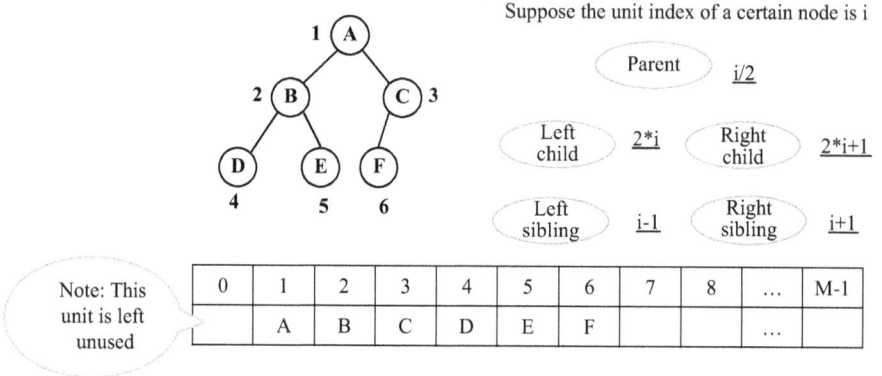

Suppose the unit index of a certain node is i

Parent	$i/2$

Left child	$2*i$	Right child	$2*i+1$

Left sibling	$i-1$	Right sibling	$i+1$

Note: This unit is left unused

0	1	2	3	4	5	6	7	8	...	M-1
	A	B	C	D	E	F			...	

Fig. 1.29: The continuous storage of a complete binary tree.

If $i = 3$ is the position of node C, then the position of parent A of C is $[3/2] = 1$. (Note we need to floor the result of the division.)

Think and Discuss Does the storage plan for complete binary trees apply for the storage of normal binary trees?

Discussion: For a normal binary tree, if we fill up the nodes that are lacking for the formation of a complete binary tree, then we can number the tree according to the method used for numbering complete binary trees. The "empty" nodes in the tree will have their values marked as Φ, as shown in Fig. 1.30. In this way, the nodes of the tree can be stored into continuous storage units sequentially. The sequence of the nodes reflects the logical relations between the nodes.

An exceptional case for binary trees is a degenerate binary tree with only the right branch. Its form and storage is shown in Fig. 1.31.

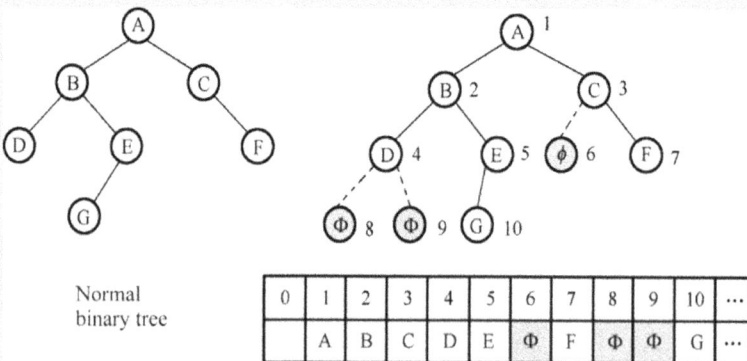

Normal binary tree

0	1	2	3	4	5	6	7	8	9	10	...
	A	B	C	D	E	Φ	F	Φ	Φ	G	...

Fig. 1.30: Storage method for normal binary trees.

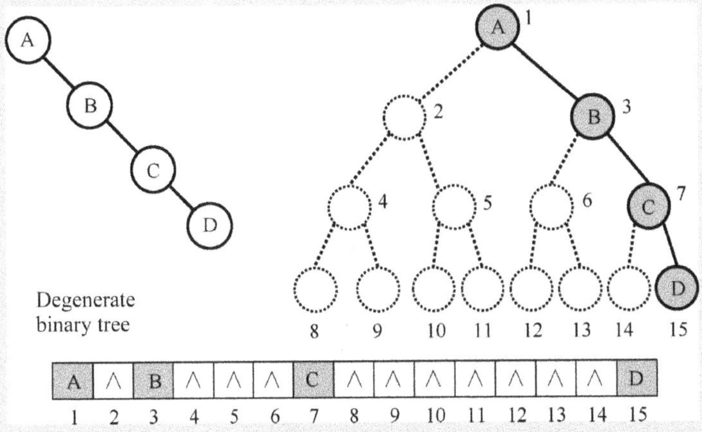

Fig. 1.31: Storage method for degenerate binary trees.

If a binary tree is not in complete binary tree form, then in order to maintain the relations between the nodes, we must leave a lot of elements as empty. This would cause waste of space. If we want to distribute memory space according to the actual number of nodes, we can use linked list data structure.

1.5.2 The linked storage structure of binary tree – binary linked list

Each node of the binary tree has two pointer fields that point to the corresponding branch. We call it binary linked list, as shown in Fig. 1.32.

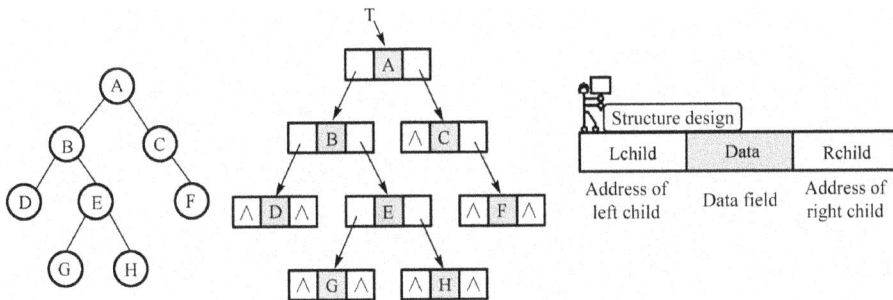

Fig. 1.32: Representation of binary linked list.

The description of the data structure of binary linked list:

```
typedef struct node
{
    datatype data;
    struct node *lchild,*rchild;
} BinTreeNode;
```

1.5.3 Constructing dynamic binary linked lists

To construct a binary linked list, we can number the nodes sequentially according to the level ordering of a complete binary tree. We enter the node information sequentially in order to construct the binary linked list. For a noncomplete binary tree, we need to add several virtual nodes to make it a complete binary tree. According to the differences in the forms of input of information on logical structure of binary trees, there are multiple algorithms for the construction of the linked storage structure of binary trees.

1.5.3.1 Method one of creating binary linked list with layered input method

Dynamically create all the nodes of the tree. The addresses of the nodes are to be sequentially filled into array Q according to the numberings of the nodes, and then we fill the left and right children pointer fields of each node sequentially. For example, in Fig. 1.33, the node A (subscript $i = 1$) has a left child B (position $= 2i$), right child C (position $= 2i + 1$); see Fig. 1.34. Because we process the nodes in Q in a head-to-tail sequence, array Q is actually also a queue.

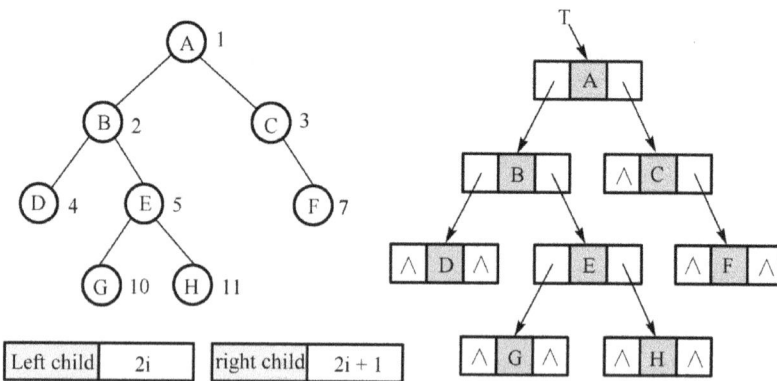

Fig. 1.33: Binary tree and the corresponding binary linked list.

We will illustrate the exact method for constructing the binary linked list with the step 1 in Fig. 1.34:

- Pop the element A of queue Q. The subscript of A $i = 1$.
- Fill in the node address of the left child of A (at subscript $2i$) into the left pointer field of node A.
- Fill in the node address of the right child of A (at subscript $2i + 1$) into the right pointer field of node A.
- Repeat the above dequeuing steps until the queue is empty, then we can establish the binary linked list.

The queue of tree nodes Q[16] (Depth $k = 4$, maximum number of nodes $2^k - 1 = 15$)

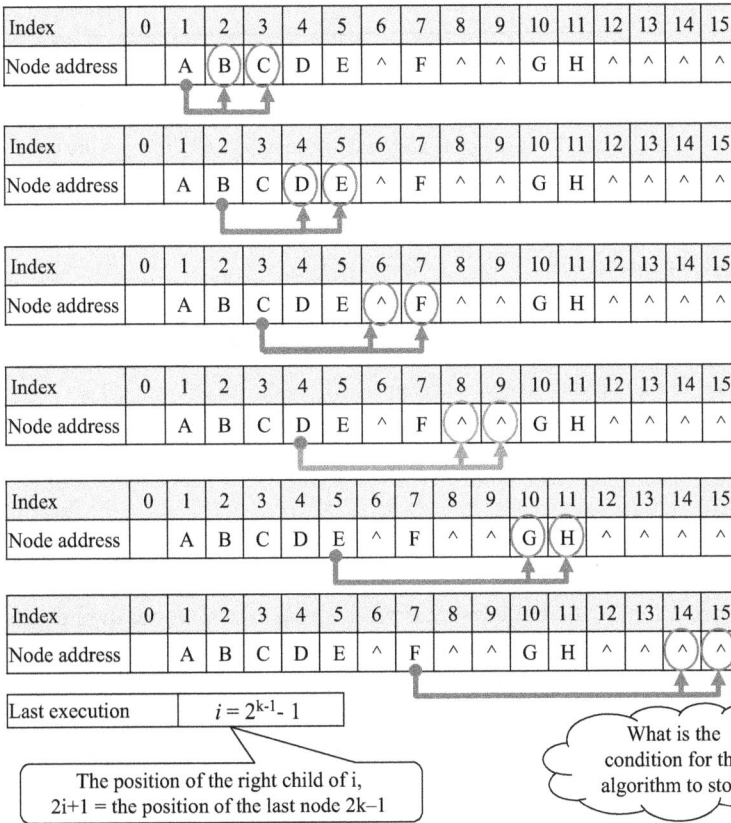

Index	0	1	2	3	4	5	6	7	8	9	10	11	12	13	14	15
Node address		A	B	C	D	E	^	F	^	^	G	H	^	^	^	^

Index	0	1	2	3	4	5	6	7	8	9	10	11	12	13	14	15
Node address		A	B	C	D	E	^	F	^	^	G	H	^	^	^	^

Index	0	1	2	3	4	5	6	7	8	9	10	11	12	13	14	15
Node address		A	B	C	D	E	^	F	^	^	G	H	^	^	^	^

Index	0	1	2	3	4	5	6	7	8	9	10	11	12	13	14	15
Node address		A	B	C	D	E	^	F	^	^	G	H	^	^	^	^

Index	0	1	2	3	4	5	6	7	8	9	10	11	12	13	14	15
Node address		A	B	C	D	E	^	F	^	^	G	H	^	^	^	^

Index	0	1	2	3	4	5	6	7	8	9	10	11	12	13	14	15
Node address		A	B	C	D	E	^	F	^	^	G	H	^	^	^	^

Last execution	$i = 2^{k-1} - 1$

The position of the right child of i, $2i+1$ = the position of the last node $2k-1$

What is the condition for the algorithm to stop?

Fig. 1.34: Method 1 to establish binary linked list.

> **!** **Think and Discuss** What is the condition for the algorithm to stop in the layered input method 1 to create binary linked list?
> **Discussion:** For a leaf node, it is certain that both its left child and right child will be empty. Therefore, there is no need to perform search for left/right child again. If we assume the depth of the binary tree to be k, then the kth level will definitely all be leaves. Therefore, we only need to process the left and right children of nodes at level $k - 1$. The subscript of the last node of the $k - 1$th level in array Q is the number of nodes from root to $k - 1$th level, $2^{k-1} - 1$. Therefore, the boundary condition for stopping the algorithm is subscript of the array element $i < 2^{k-1}$.

The pseudocode description of the algorithm is shown in Table 1.1.

Table 1.1: Method 1 for construction of binary linked list with leveled inputs.

Detailed description of the pseudocode
Assume the depth of the binary tree is k, ll
Generate all the nodes of the tree, fill in the node addresses sequentially according to the node numberings into the queue array Q[]
Subscript of head element i = 1
When the subscript of queue Q i < 2^{k-1} Dequeue element i of queue Q
Fill in the address of the left child of i (with subscript 2i) into the left pointer field of i
Fill in the address of the right child of i (with subscript 2i + 1) into the left pointer field of i
Return the address of the root node

1.5.3.2 Method 2 of creating binary linked list with leveled inputs

1. Algorithm description

Input node information. If the node being input is not a virtual node, then create a new node. If the new node is the first node, then set it as the root node; otherwise, link the new node as child onto its parent node. Repeat the above process, until the ending sign "#" is input.

Similar to Fig. 1.33, the second method to construct a binary linked list is shown in Fig. 1.35.

In step 1 of Fig. 1.35, A is the root node, B is enqueued and its parent is A. According to the subscript of B, we know it is the left child of A. We link node B into the left child field of node A. In step 2, node C is enqueued, and its parent is A. According to the subscript of node C, we know it is the right child of A. We link

Parent	i/2

Index	0	1	2	3	4	5	6	7	8	9	10	11	12	13	14	15
Node address		(A)	B													

Index	0	1	2	3	4	5	6	7	8	9	10	11	12	13	14	15
Node address		(A)	B	C												

Index	0	1	2	3	4	5	6	7	8	9	10	11	12	13	14	15
Node address		A	(B)	C	D											

Index	0	1	2	3	4	5	6	7	8	9	10	11	12	13	14	15
Node address		A	(B)	C	D	E										

Index	0	1	2	3	4	5	6	7	8	9	10	11	12	13	14	15
Node address		A	B	(C)	D	E	^	F								

Index	0	1	2	3	4	5	6	7	8	9	10	11	12	13	14	15
Node address		A	B	C	D	(E)	^	F	^	^	G					

Index	0	1	2	3	4	5	6	7	8	9	10	11	12	13	14	15
Node address		A	B	C	D	(E)	^	F	^	^	G	H				

Fig. 1.35: Method 2 of creating a binary linked list.

node C into the right child field of node A. Since both the left child and right child of A have been processed, A is dequeued.

The pseudocode description of the algorithm is given in Table 1.2.

Table 1.2: Method 2 of creating binary linked list with layered input method.

Detailed description of the pseudocode
Suppose the depth of the binary tree is k, then the length of queue Q can be set to be the numbering of the last node of the tree according to the numberings of complete binary tree.
When the input node information is not end indicator #'
If the node entered is not a virtual node, then create a new node and enqueue
If the new node is the first node, then make it the root node
Otherwise, link the new node to its parent node as a child node
If the new node is the right child of the parent node, then dequeue the parent node
Return the address of the root node

2. Description of function framework and functionalities: see Table 1.3.

Table 1.3: Function framework for creation of binary linked list with layered input.

Functionality description	Input	Output
Create binary tree CreateTree	None	Address of the root node
Function name	Parameter	Function type

3. Program implementation

Sequentially input the values of the nodes according to the numbering rules for the nodes of a complete binary tree to build a binary linked list:

```
/*================================================================
Functionality: Creation of binary linked list in layered input method
Function input: None
Function output: Root node of the binary linked list
Keyboard input: Input the node values according to the numbering rules of a
complete binary tree. An empty node is input as @
=============================================================*/
BinTreeNode *Q[16];
// Queue Q is used to store the addresses of the nodes of the tree
BinTreeNode *CreatBTree()
```

```
{
    char ch;
    int front=1,rear=0;
    BinTreeNode *root = NULL, *s;
    ch=getchar();
    while(ch!='#')      // End indicator
    {
        s=NULL;
        if (ch!='@')    // Empty node
        {
            s=(BinTreeNode *)malloc(sizeof(BinTreeNode)); // Generation of new node
            s->data=ch;
            s->lchild=NULL;
            s->rchild=NULL;
        }
        Q[++rear]=s;              // Enqueuing of the node
        if (rear==1) root=s;      // Record the root node
        else
        {
            if (s && Q[front])
            {
                if (rear%2==0) Q[front]->lchild=s;      // Enqueuing of the left child
                else Q[front]->rchild=s;                // Enqueuing of the right child
            }
            if (rear%2==1) front++;
            // The new node is the right child of the parent,
            // then pop the parent node
        }
        ch=getchar();
    }
    return root;
}
```

1.6 Search operation on nodes of binary tree – traversal of tree

In some applications of the binary tree, we usually want to search for nodes with certain features in the tree, or perform a certain operation on all the nodes in the tree.

1.6.1 Introductory example 1 of traversal – simple model of self-check of electrical devices

All electrical devices are composed of several parts. For example, computer hardware is composed of a series of components including motherboard/card type and nonmotherboard/card type [1] .We can construct a binary tree composed of computer hardware according to such a classification of computer, as shown in Fig. 1.36. Apparently, the precondition for the device to work normally is that every component works normally. It makes sense to start bottom-up to check whether every component and its parts as well as the whole device is running normally.

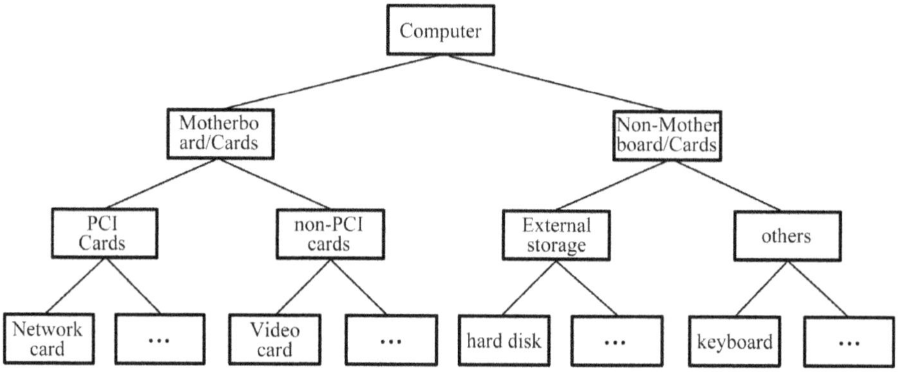

Fig. 1.36: Computer hardware system represented with binary tree.

Steps for the model of self-check of devices:
1. Left subtree: Network card normal -> PCI cards normal -> video card etc. normal -> non-PCI cards normal -> Motherboard/cards normal
2. Right subtree: Hard disk etc. normal -> Secondary storage normal -> Keyboard etc. normal -> Miscellaneous normal -> Nonmotherboard/cards normal
3. Whole tree: Motherboard/cards normal -> Nonmotherboard/cards normal -> Computer normal

During the check, if there is any error information, then an error should be reported. Otherwise the device will be shown as normal. In the checking process above, the root node of the tree is the last to be visited, no matter for the whole tree or left/right subtrees. Normally, we call this kind of traversal "post order traversal."

1.6.2 Introductory example 2 of traversal – management of goods in e-commerce

An online food shop organizes its foods according to the type, color and species. The classification diagram is tree-structured, as shown in Fig. 1.37.

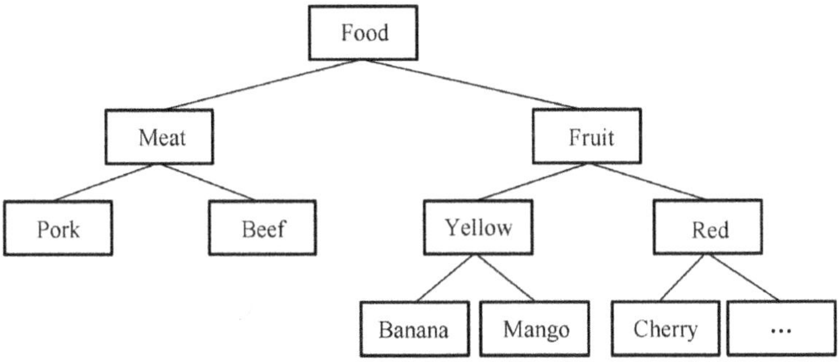

Fig. 1.37: Management of goods in food shop.

If we want to automatically read all the information in the tree structure with a program and print out the names of all goods, how should we do it so that it is both clear and without omission? We can first analyze the features of the data by a table, as shown in Fig. 1.38.

Food			
Meat	Pork	Beef	
Fruit	Yellow	Banana	Mango
	Red	Cherry	

Fig. 1.38: Table of goods.

What is the regularity in printing the nodes in the table? The order as seen in the table of goods, top-down, is:

Root node: Food
Left subtree: Meat→Pork/Beef
Right subtree: Fruit→ (Left subtree) Yellow→Banana/Mango
        Fruit→ (Right subtree) Red→Cherry

The order of printing is such that the root node will first be visited, whether for the whole tree or for the left/right subtree. Traversing the nodes of a tree in such an order can be described as "pre order traversal."

1.6.3 Introductory example 3 of traversal – the efficient searching strategy for nodes in the tree

To search for the keyword key among a group of data, there is a "binary search" strategy which is quick and efficient. Hereby we analyze the idea of this algorithm. Its feature is to evenly divide a group of ordered data into two parts, and compare with the number in the middle. This process is repeated until the result is obtained. This can be viewed as a recursive process, which fits the structural features of tree structure. Therefore, we can try to describe this process of data storage and lookup with tree.

Suppose there are 11 integers 1–11. The correspondence between each division each time and the central point is shown in Fig. 1.39. If we set the value of the corresponding central point each time as the root of the tree, we can draw the binary search tree structure in Fig. 1.40. The values in the left subtree of the root node are all smaller than the root, while the values in the right subtree are all larger than the root.

Subscript	0	1	2	3	4	5	6	7	8	9	10
Value	1	2	3	4	5	6	7	8	9	10	11
Middle point	3	4	2	3	4	①	3	4	2	3	4

Fig. 1.39: The central points during binary search.

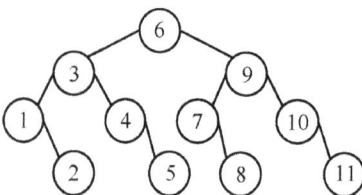

Fig. 1.40: Binary search tree.

Let the value to be searched for be x, then the possible scenarios of comparison during the search for nodes are shown in Fig. 1.41.

	x=key	Search is successful
Non empty node	x<key	Continue the search on the left subtree
	x>key	Continue the search on the right subtree

Fig. 1.41: Methods of search on ordered binary search.

Think and Discuss How does binary search tree embody the ascending order of the array?
Discussion: When we order the node values in ascending order, we can observe that the ordering for any subtree follows the pattern of "left subtree first, root node next, right subtree last." For example:

For the tree with root 1: Left subtree: (empty) -> Root: 1 -> Right subtree: (2) -> Result is 1, 2.

For the tree with root 3: Left subtree: (1, 2) -> Root: 3 -> Right subtree: (4, 5) -> Result is 1, 2, 3, 4, 5.

For the tree with root 9: Left subtree: (7, 8) -> Root: 9 -> Right subtree: (10, 11) -> Result is 7, 8, 9, 10, 11.

The traversal of tree nodes with such a pattern can be called in-order traversal. Here the "in" refers to the fact that the root is situated in the middle of the traversal order (in the middle of its children).

From the above three introductory examples, we can see that the search operation on nodes is easy for linear structures, but it is not so for binary trees, as shown in Fig. 1.42. Since binary tree is a nonlinear structure, each node might have two subtrees, that is, there is more than one successor to a node, thereby arises the problem on how to effectively visit such a nonlinear structure as tree, that is, how to visit

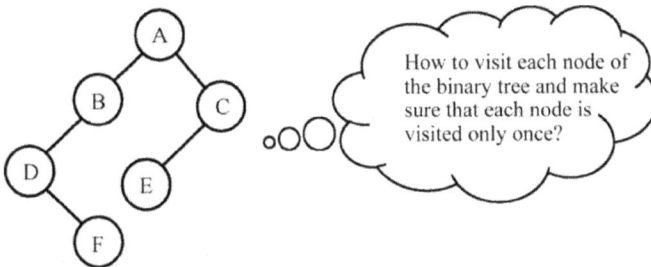

How to visit each node of the binary tree and make sure that each node is visited only once?

Fig. 1.42: Traversal of binary trees.

each node of the tree according to a certain search route, so that each node is visited once and only once. This kind of problem on effectively visiting nonlinear structure is called "traversal" in data structures.

For the traversal of binary trees, we need to find a pattern, so that the nodes in the binary tree can be listed in one linear sequence according to the order of visit, that is, the result of the traversal would be a linear sequence. One might say that traversal operation is to obtain a corresponding linear sequence from a nonlinear structure, which is the linearization of nonlinear structures.

In the traversal, "visit" has a broad meaning. It can mean all kinds of operations on a node, for example, modify or output the information stored in a node. Traversal is the most basic operation on various kinds of data structures. Many operations such as insertion, deletion, modification, lookup and ordering are all realized on the basis of traversal.

According to the differences in search strategies, the traversal of binary trees can normally be divided into traversals with breadth-first-search and traversals with depth-first-search. In the following, we discuss their implementation methods.

1.6.4 The breadth-first-search traversal of trees

Breadth-first-search is a traversal strategy on connected graphs. Because its central idea is to start from one vertex and traverse the broad zone of immediately adjacent nodes, thus the name.

The breadth-first-search on a binary tree is also called traversal by level. It starts from the first level (root node) of the binary tree, and traverses it top-down, level by level. In the same level, nodes are visited one by one from left to right.

The implementation of traversal is related to the storage structure. In the following, we discuss the breadth-first-search of trees in sequential storage and linked storage structure, respectively.

> **ℹ Knowledge ABC** Connected graph
> Connected graph is a structure in which there is a path between each pair of nodes. Tree is also a type of connected graph.

1.6.4.1 The breadth-first-search traversal of trees based on sequential storage structures

The sequential storage structure and related traversal of tree is shown in Fig. 1.43. Because sequential storage expresses the nonlinear structural information of trees with a linear sequence, the traversal of trees on sequential storage structure results exactly in correct ordering of stored nodes.

Subscript	0	1	2	3	4	5	6	7	8	9
Node		A	B	C	D	^	E	^	^	F

Traversal result:	A B C D E F

Fig. 1.43: Traversal of the sequential storage structure of tree.

1.6.4.2 The breadth-first-search traversal of trees based on linked storage structures

The breadth-first-search operation on trees starts from the root node, and then using this node as a clue, sequentially visit the nodes directly connected to it (children) sequences. And then proceed to the breadth-first-search of the next step. What kind of node should we use as the clue to begin with?

In Fig. 1.44, the first clue for search is node A. We search for the immediately adjacent nodes of A in the linked storage structure, and there are two nodes: B and C. Therefore, the sequence of nodes that we can directly visit is ABC. The next time, new search clue starts from node B, that is, we use the first node that has not been used as the clue among the sequence of visited nodes, and repeat the operation, until we have finished visiting all the nodes.

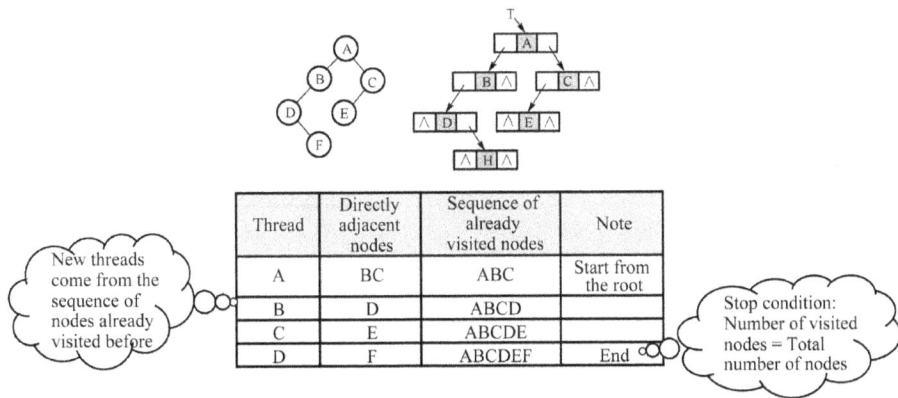

Thread	Directly adjacent nodes	Sequence of already visited nodes	Note
A	BC	ABC	Start from the root
B	D	ABCD	
C	E	ABCDE	
D	F	ABCDEF	End

Fig. 1.44: Traversal of tree on linked storage structure.

The above process of operation is a process of clue node being continuously moved forward in the "sequence of visited nodes" and newly visited nodes being continuously added in the "sequence of visited nodes." Therefore, the mode of operation on "sequence of visited nodes" is similar to the processing on queues. After we have analyzed the mode of data processing of depth-first-search traversal, we can directly apply the operations of queue to implement it.

1. Algorithm description
    - Initialize a queue and enqueue the root node.
    - Dequeue a queue element to obtain a node. Visit this node.
    - If this node's left child is not empty, then enqueue the left child of this node.
    - If this node's right child is not empty, then enqueue the right child of this node.
    - Repeat steps 2–4 until the queue is empty.
2. Description of data structure
    –Description of a binary tree node structure

```
typedef struct node
{
    datatype data;
    struct node *lchild,*rchild;
} BinTreeNode;
BinTreeNode *Q[MAXSIZE]; // Set array Q to be the queue, the element type of the
                            queue is the node datatype of binary linked list
```

3. Design of function framework: see Table 1.4.

Table 1.4: Function framework for breadth-first-search traverse of trees based on linked storage structure.

Functionality	Input	Output
Traverse the binary tree by level LevelOrder	Address of the root node of the binary tree	None
Function name	Parameters	Function type

4. Program implementation

```
/*=================================================================
Functionality: Traverse the binary tree by level, print the sequence of nodes
traversed
Function input: Address of the root node of the binary linked list BinTreeNode
*Ptr
Function output: None
=================================================================*/
```

```
void LevelOrder (BinTreeNode *Ptr)
{
  BinTreeNode *s;
  rear=1; front=0;      // Initialize the circular queue
  Q[rear]= Ptr;         // Enqueue the root node
  if ( Ptr!=NULL )      // The root node is non-empty
  {
    while ( front != rear )
    // The queue is non-empty, carry out the following operations
    {
      front= (front+1)% MAXSIZE;
      Ptr=Q[front];     // Dequeue the head element
      printf (" %c ", Ptr→data);     // Visit the node just dequeued
      if (Ptr→lchild!=NULL)     // Enqueue the left child of Ptr
      {
        rear=(rear+1) % MAXSIZE;
        Q[rear]= Ptr→lchild;
      }
      if (Ptr→rchild!=NULL)     // Enqueue the right child of Ptr
      {
        rear=(rear+1) % MAXSIZE;
        Q[rear]= Ptr→rchild;
      }
    }
  }
}
```

1.6.5 The depth-first-search traversal of trees

1.6.5.1 Depth-first-search traversal method
From the definition of a binary tree, we can note that a binary tree is constituted of three parts: its root node, left child of the root node and right child of the root node, as shown in Fig. 1.45. Therefore, we can also decompose the traversal of binary tree into three "subtasks":
1. Visit the root node
2. Traverse the left subtree (i.e., visit all the nodes in the left subtree in order)
3. Traverse the right subtree (i.e., visit all the nodes in the right subtree in order)

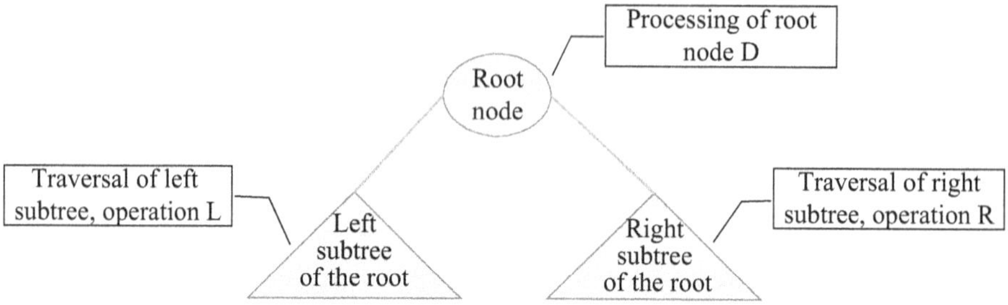

Fig. 1.45: The basic composition of binary tree and names of traversal operations.

Since both the left and right subtrees are binary trees (can also be empty binary trees), the traversal on them can continue to be decomposed according to the above method, until each subtree is an empty binary tree. Thereby, we can see that the relative order of carrying out the three subtasks mentioned above decides the order of the traversal. If we represent these three subtasks with D, L, R, then there are six possible orders: DLR, LDR, LRD, DRL, RDL and RLD. Normally, the restriction is "left to right," that is, subtask (2) should be finished before subtask (3). In this way, only the first three orders remain. The traversals according to these three orders are called, respectively:

DLR – pre-order traversal

LDR – in-order traversal

LRD – post-order traversal

The definitions of the three traversal methods are defined as follows:

- Pre-order traversal: visit the root node D first, and then traverse the left subtree and right subtree of D, respectively, according to the strategy of pre-order traversal. Visit the root node first, that is, whenever the subtree to be traversed is encountered, its root node is always visited first.
- In-order traversal: first, traverse the left subtree of root D according to the strategy of in-order traversal, then visit the root node D, lastly visit the right subtree of root D according to the strategy of in-order traversal. The root node is visited in the middle.
- Post-order traversal: first, traverse the left and right subtrees of root D with post-order traversal, and then visit the root node D. The root node is visited at last.

Note: The meaning of "pre, in, post" refer to the moment of visit on the root node D compared with the visit on the subtrees of D.

If the binary tree to be traversed is empty, then empty (no) operation will be carried out.

1. Example of DLR (pre-order) traversal

As shown in Fig. 1.46, each node encountered will be processed according to the DLR strategy.

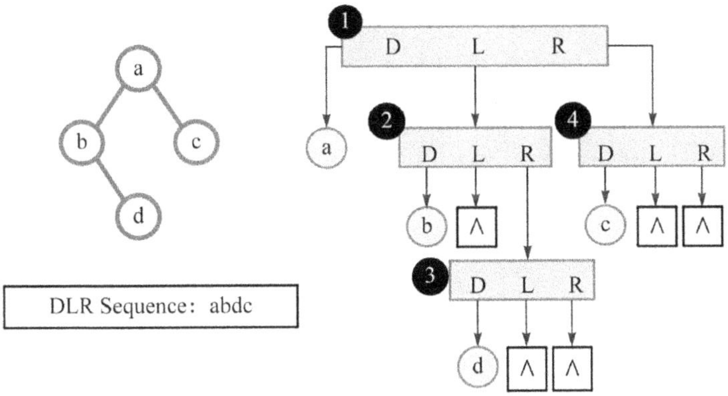

Fig. 1.46: DLR (pre-order) traversal.

In the rectangle numbered with 1:

D – D represents the root node. At this time, the root is a. s is represented with D(a).

L – L represents the left subtree of root node D. The left subtree of a has two nodes, b and d, represented with L(bd).

R – R represents the right subtree of the root node D. The right subtree of a has a node c, represented with R(c).

In the rectangle numbered with 2:

D(b), L(empty), R(d) – this means when b is the root node, its left subtree is empty, while its right subtree has node d.

In the rectangle numbered with 3:

D(d), L(empty), R(empty) – this means when d is the root node, its left subtree is empty, its right subtree is empty.

In the rectangle numbered with 4:

D(c), L(empty), R(empty) – this means when c is the root node, its left subtree is empty, its right subtree is empty.

From the above traversal process we can see that the root node is one node, while the left/right subtrees of the root node can have multiple nodes or no nodes. Whenever we traverse left subtree or right subtree, we follow the DLR strategy, which should correspond to one rectangle in the figure.

2. Example of LDR (in-order) traversal

As shown in Fig. 1.47, each node encountered should be processed by LDR strategy.

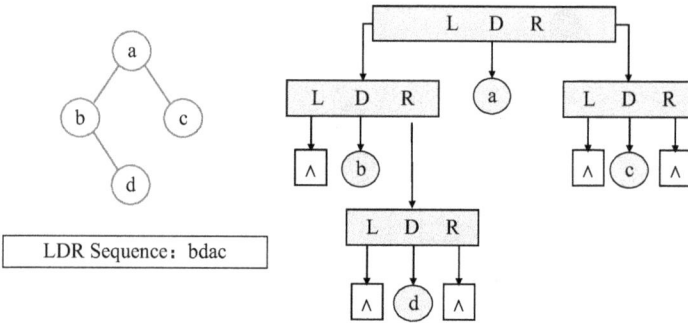

Fig. 1.47: LDR (in-order) traversal.

3. Example of LRD (post-order) traversal

As shown in Fig. 1.48, each node encountered is processed with LRD strategy.

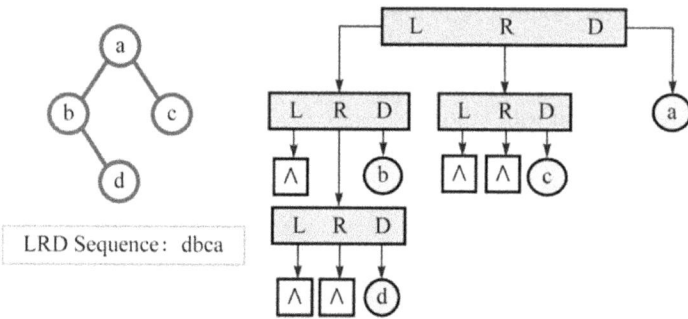

Fig. 1.48: LRD (post-order) traversal.

Apparently, the difference between the above three traversal methods is that the moment at which the subtask "visit root node" is carried out is different: carrying out this subtask in the beginning (in the middle, at last) corresponds to pre-order (post-order, in-order) traversal. When we traverse a binary tree according to a certain traversal method, then we get the visiting sequence of all the nodes in this binary tree.

Example 1.2 Example of traversal

Give the sequence of nodes obtained via pre-order, in-order and post-order traversals on the tree structure in Fig. 1.49.

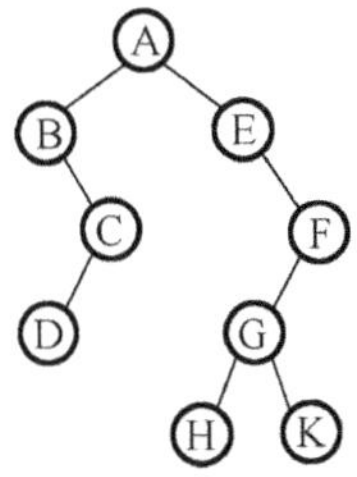

Fig. 1.49: Tree structure 1.

– The process of pre-order traversal is shown in Fig. 1.50. The concretely circled part is the part to be traversed at each step, while the dashed circles indicate that the parts within the circles cannot yield results of traversal directly. The meaning of the dashed circles is the same in the illustrations for in-order and post-order iterations. The description is found in Fig. 1.51.

Step 1: Pre-order traversal, visit the root A of the tree first.

Step 2: We need to traverse the left subtree L of root A. We cannot directly obtain the result on the part circled by dashed lines.

Step 3: Visit the root of L, B.

Step 4: Visit the left subtree of L, which is empty; then we visit the right subtree of L, which is the content in the dashed circle.

The steps after this can be similarly deduced. They all follow the order of DLR, which is visit the root node D of the current tree first, then visit the left subtree L of D, then visit the right subtree R of D.

The sequence obtained via pre-order traversal constitutes three parts: scenario 1, scenario 2′ and scenario 6′. See the gray cells in Fig. 1.51.

Root of the Tree – A's Left Subtree – A's Right Subtree: A BCD EFGHK.

– The process of in-order traversal is shown in Fig. 1.52. The parts in concrete circles are the parts to be traversed at each step. Description is shown in Fig. 1.53.

The in-order traversal sequence constitutes three parts: scenario 1′, scenario 5 and scenario 6′:

A's Left Subtree – Root of the Tree – A's Right Subtree: BDC A EHGKF.

– The process of in-order traversal is shown in Fig. 1.54. The parts in concrete circles are the parts to be traversed at each step. Description is shown in Fig. 1.55.

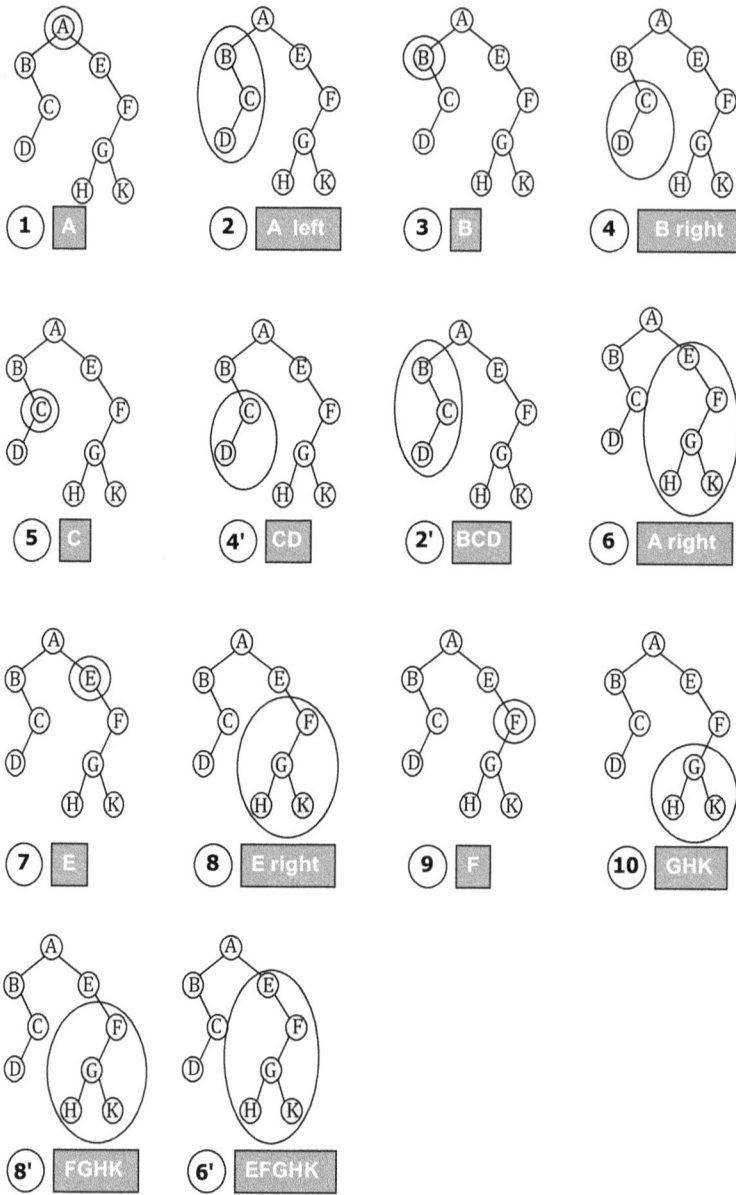

Fig. 1.50: Process of pre-order traversal.

Scenario	In the circle	DLR Sequence	Scenario	In the circle	DLR Sequence
1	Root of the tree	A	6	A's right subtree	Continue elaboration
2	A's left subtree	Continue elaboration	7	The root of A's right subtree	E
3	The root of A's left subtree	B	8	E's right subtree	Continue elaboration
4	B's right subtree	Continue elaboration	9	The root of E's right subtree	F
5	The root of B's right subtree	C	10	F's left subtree	GHK
4'	B's right subtree	CD	8'	E's right subtree	FGHK
2'	A's left subtree	BCD	6'	A's right subtree	EFGHK

Fig. 1.51: Illustration of the pre-order traversal process.

The post-order traversal sequence constitutes three parts: scenario 1', scenario 4' and scenario 7:

Left Subtree of A – Right Subtree of A – Root of Tree: DCB HKGFE A.
Summary: From the analysis of various kinds of traversal processes of a tree, we can see that traversal is a recursive process. When it is not possible to directly obtain the result of traversal on a big tree, it is decomposed into root and left/right subtrees, on which the traversal is carried out on the scale of trees that have one less level, until we get the result of the traversal. Then we return to the previous level to get the result of the traversal for the tree of the corresponding level. When we return in this way level by level, we obtain the final result.

1.6.5.2 The recursive algorithm for depth-first-search traversal
1. The recursive algorithm for pre-order traversal
   - Design of function framework: see Table 1.5.
   - Pseudocode description of the algorithm: see Table 1.6.
   - Description of data structure of binary linked list

```
typedef struct node
{
    datatype data;
    struct node *lchild, *rchild;
} BinTreeNode;
```

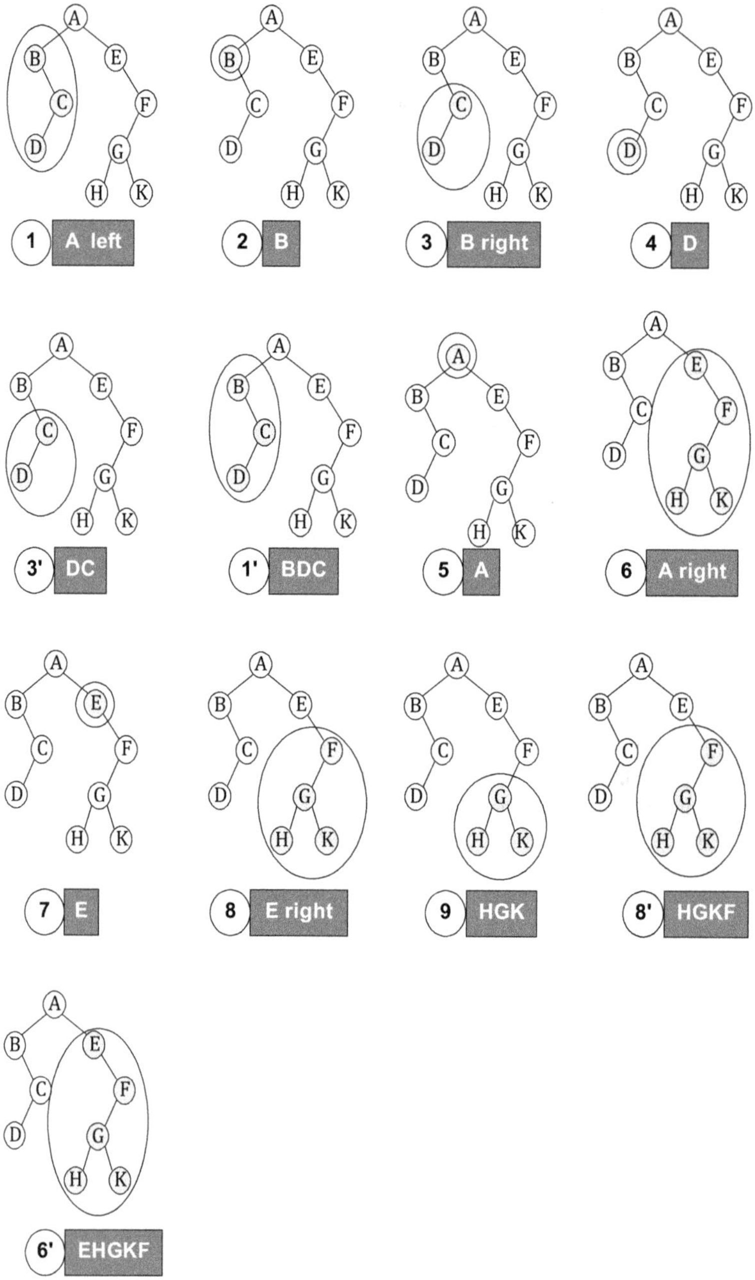

Fig. 1.52: Process of in-order traversal.

Scenario	In the circle	LDR Sequence	Scenario	In the circle	LDR Sequence
1	A's left subtree	Continue elaboration	6	A's right subtree	Continue elaboration
2	A's left subtree's root	B	7	A's right subtree's root	E
3	B's right subtree	Continue elaboration	8	E's right subtree	Continue elaboration
4	C's left subtree's root	D	9	F's left subtree	HGK
3'	B's right subtree	DC	8'	E's right subtree	HGKF
1'	A's left subtree	BDC	6'	A's right subtree	EHGKF
5	Tree's root	A			

Fig. 1.53: Description of the in-order traversal process.

– Program implementation

```
/*=====================================================================
Functionality: Recursive algorithm on pre-order traversal of trees
Function input: Root node of the tree
Function output: None
Output on the screen: The sequence of nodes obtained via pre-order traversal
on the tree
=====================================================================*/
void PreOrder(BinTreeNode *t)
{
   if(t)
   {
      putchar(t->data);
      PreOrder (t->lchild);
      PreOrder (t->rchild);
   }
}   // Pre-order traversal
```

The execution process of the DLR recursive traversal program is shown in Fig. 1.56. In the figure, for brevity's sake, the pre-order traversal function's name is expressed as pre, the root as T and the left/right subtrees as L/R.

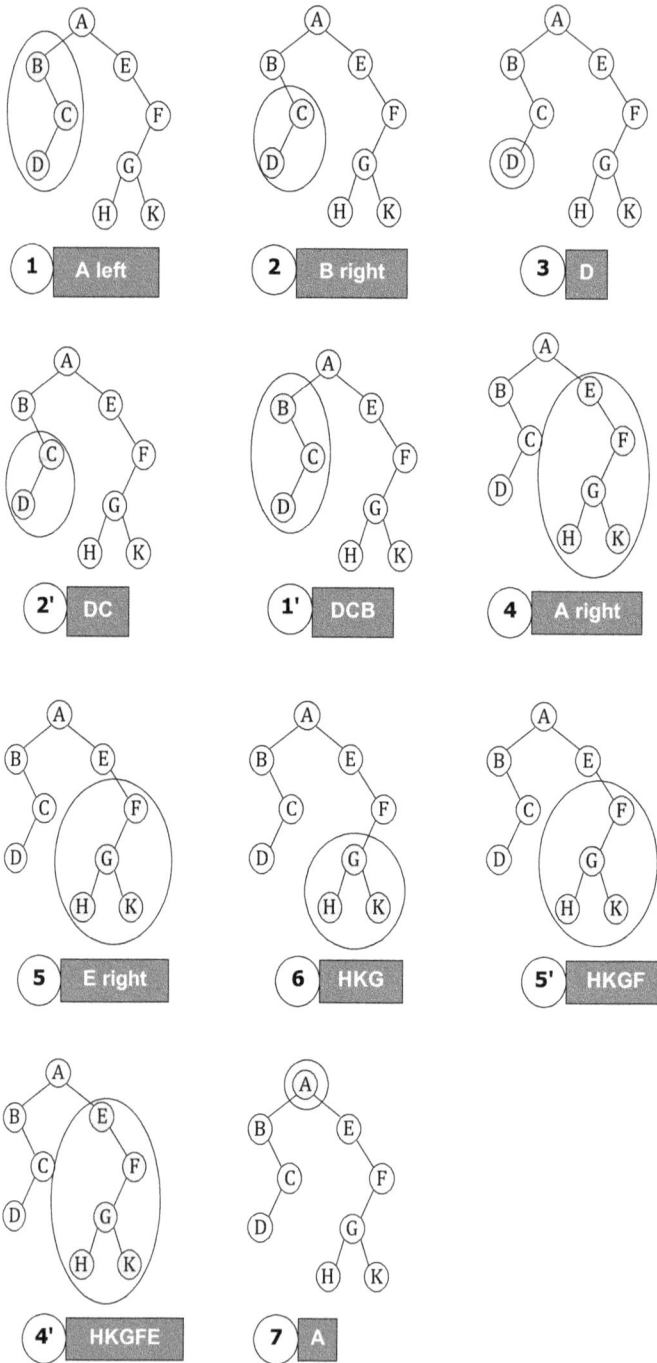

Fig. 1.54: The process of post-order traversal.

Scenario	In the circle	LDR Sequence	Scenario	In the circle	LDR Sequence
1	A's left subtree	continue elaboration	5	E's right subtree	continue elaboration
2	B's right subtree	continue elaboration	6	F's left subtree	HKG
3	C's left subtree 's root	D	5'	E's right subtree	HKGF
2'	B's right subtree	DC	4'	A's right subtree	HKGFE
1'	A's left subtree	DCB	7	Tree's root	A
4	A's right subtree	continue elaboration			

Fig. 1.55: Description of post-order traversal.

Table 1.5: Function framework design of pre-order traversal algorithm.

Functionality description	Input	Output
Output DLR traversal sequence: PreOrder	Root address	None
Function name	Parameters	Function type

Table 1.6: Pseudocode for pre-order traversal algorithm.

Description of the pseudocode
Base case for recursion: t is an empty node. Return
Condition for continuing recursion: Visit node t Pre-order traverse the left subtree of t Pre-order the right subtree of t

According to the structure of the tree given, the recursive process is carried out according to the direction pointed to by the arrows. During the DLR recursion process, the root node is visited first each time, then the left branch of the tree is visited until the left subtree is empty. It returns and then visits the right subtree.

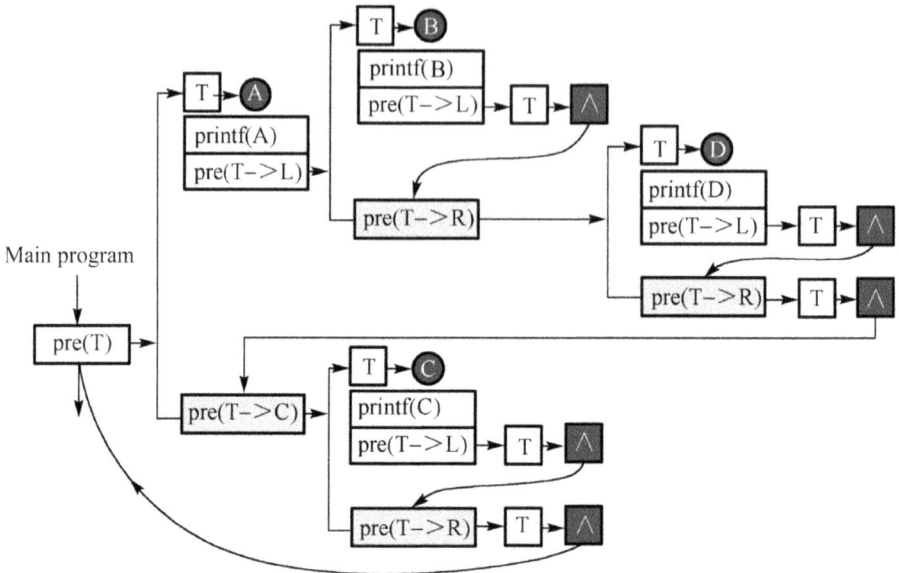

Fig. 1.56: Illustration of DLR recursion.

2. Recursive algorithms for in-order and post-order traversals

```
/*============================================================
Functionality: Recursive algorithm on in-order traversal of tree
Function input: Root node of the tree
Function output: None
Output on the screen: The sequence of nodes obtained via in-order traversal on
trees
==========================================================*/
void inorder(BinTreeNode *t)
{
  if (t)
  {
    inorder(t->lchild);
    putchar(t->data);
    inorder(t->rchild);
  }
}
/*============================================================
Functionality: Recursive algorithm on post-order traversal of tree
Function input: Root node of the tree
Function output: None
```

Output on the screen: The sequence of nodes obtained via post-order traversal on trees
===*/

```
void postorder (BinTreeNode *t)
{
  if ( t )
  {
    postorder(t->lchild);
    postorder(t->rchild);
    putchar(t->data);
  }
}
```

3. Analysis on traversal

Observing the three recursive traversal algorithms listed above, if we remove the print statement, then we can see that the general routes of visit of these three algorithms are the same. It is just that the moment of visiting the root node is different. The traversal of the nodes in the figure is shown in Fig. 1.57: In the route from the root node to the end, each node will be visited three times, expressed via numbered nodes, for example, the A1A2A3 corresponding to node A. "(1)" means that the subtree is empty and is only visited once.

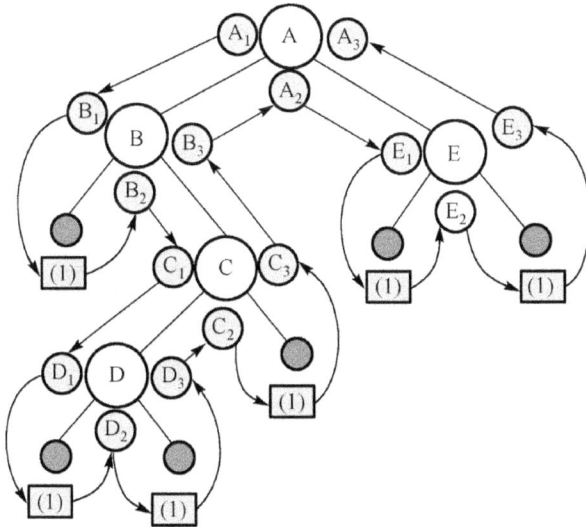

Fig. 1.57: Path of traversal on binary tree.

The visit of the first time is pre-order traversal; the visit of the second time is in-order traversal; the visit of the third time is post-order traversal. The sequence of pre-order traversal is A1B1C1D1E1, that is, ABCDE; the sequence of the in-order traversal is B2D2C2A2E2, that is, BDCAE; the sequence of the post-order traversal is D3C3B3E3A3, that is, DCBEA.

The time efficiency of traversal on binary tree is $O(n)$, since each node is only visited once; the space efficiency is the maximum auxiliary space possible occupied by the stack, which is $O(n)$.

!

Think and Discuss Can we establish all the nodes of a binary linked list and finish the linking of the corresponding nodes using the idea of traversal?
Discussion: See the following example.
Example 1.3 Construct binary linked lists with traversal

Solution: The basic idea of this algorithm is to construct all the nodes of a binary linked list and finish the linking of the corresponding nodes, according to the ordering of pre-order traversal.
   Using Fig. 1.58 as a sample, the pre-order sequence of the tree will be ABD@F@@@CE@@@.

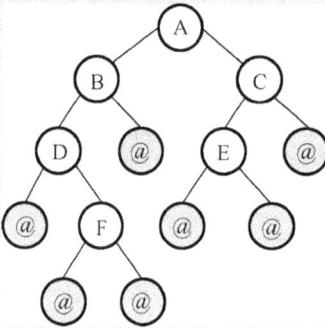

Fig. 1.58: Tree structure 2.

The program implementation is as follows:

```
/*====================================================================
Functionality: Construct binary linked list with pre-order traversal
Function input: (Root node of the binary linked list)
Function output: Root node of the binary linked list
Keyboard input: The sequence of nodes when the tree is pre-order traversed.
When the subtree is empty, input @
=====================================================================*/
BinTreeNode *CreatBTree_DLR(BinTreeNode *root )
{
    char ch;
    scanf("%c",&ch);
```

```
if (ch=='@') root=NULL;
// ch=='@' The subtree is empty, then set root=NULL and finish with this node
else
{
    root=( BinTreeNode *)malloc(sizeof(BinTreeNode));  // Construct a node
    root->data = ch;
    // Construct the linked list for the left subtree, and assign the root
    // pointer of the left subtree to the left child field of the root node
    root->lchild=CreatBTree_DLR(root->lchild);
    // Construct the linked list for the right subtree, and assign the root
    // pointer of the right subtree to the right child field of the root node
    root->rchild=CreatBTree_DLR(root->rchild);
}
    return (root);
}
```

1.6.5.3 Nonrecursive algorithm for depth-first-search traversal

Stack is the most common auxiliary structure for recursion. We can use a stack to record nodes that are still to be traversed, so that we can visit them later. We can change the recursive depth-first-search algorithm to a nonrecursive one.

1. Nonrecursive pre-order traversal

The traversal order on the nodes of the binary tree is to follow along the left link, and push the node onto the stack when visiting it, until the left link is empty. Afterward, the nodes are popped out of the stack. Popping a node indicates that this node and its left subtree have already been visited, and we should visit the right subtree of this node. The concrete steps are as follows:

1. The current pointer points to the root node.
2. Print the current node, assign the current pointer to the left child and push it onto the stack, repeat (2), until left child is NULL.
3. Pop the stack one by one, each time pointing the current pointer to the right child.
4. If the stack is nonempty or the current pointer is non-NULL, go to (2); otherwise, finish.

When we meet a node, we visit this node and push this node onto the stack, and then we traverse its left subtree.

After traversing its left subtree, we pop this node from the stack and traverse the right subtree structure according to the address indicated by its right link.

Program implementation

```
*===============================================================================
Functionality: Non-recursive function for pre-order traversal of trees
Function input: Root node of the tree
Function output: None
Screen output: The sequence of nodes obtained via pre-order traversal on the tree
==============================================================================*/
#define MAX 20
void PreOrder_NR(BinTreeNode *root)
{
  BinTreeNode *Ptr;
  BinTreeNode *Stack[MAX];   // Definition of stack
  int top=0;                 // Stack-top pointer

  Ptr=root;
  do
  {
    while( Ptr!=NULL)
    // The tree node is non-empty, traverse its left subtree
    {
      printf("%c", Ptr->data) ;      // Print the value of the node
      Stack[top]=Ptr;                // Push the tree node onto the stack
      top++;
      Ptr=Ptr->lchild;              // Inspect the left subtree
    }
    if (top>0)        // The stack is non-empty, pop
    {
      top--;
      Ptr=Stack[top];
      Ptr=Ptr->rchild;
      // Inspect the right subtree of the node at the top of the stack
    }
  }   while( top>0 || Ptr!=NULL);
}
```

2.   Nonrecursive in-order traversal

Whenever we meet a node, we push it onto the stack and traverse its left subtree. After we have traversed the left subtree, we pop the node and visit it and then we traverse its right subtree according to the address indicated by its right link.

3.   Nonrecursive post-order traversal

Whenever we meet a node, we push it into the stack and traverse its left subtree. After the traversal is finished, we cannot immediately visit this node which is stored at the top of the stack, but instead should traverse the right subtree of this node according to the address indicated by its right link. Only after traversing its right subtree, can we pop this node and visit it. Besides, we need to add a feature indicator on each element in the stack, so that when the node is popped, we can decide whether this node belongs to a left subtree (then we need to continue traversing the right subtree), or to a right subtree (then we have traversed both the left and right subtrees of this node). When the feature is marked as left, then we have entered the left subtree of this node and will return from the left subtree; when the feature is marked as right, then we have entered the right subtree of this node and will return from the right subtree.

Testing the traversal application on trees

```
/*========================================================================
Functionality: Various traversal algorithms on tree Test function:
1. Construct a binary linked list with pre-order traversal: CreatBTree_DLR
2  Sequence from non-recursive pre-order traversal: PreOrder_NR
3. Sequence from recursive pre-order traversal: PreOrder
4  Sequence from recursive in-order traversal: inorder
5. Sequence from recursive post-order traversal: postorder
=====================================================================*/
Test program
#include "stdio.h"
#include <stdio.h>
#include <stdlib.h>
typedef struct node
{ char data;
   struct node *lchild,*rchild;
} BinTreeNode;

int main()
{
   BinTreeNode *RPtr;
   printf("Constructing the tree, please input the sequence of nodes according
   to pre-order traversal order\n");
```

```
    RPtr=CreatBTree_DLR(RPtr);
    printf("\n Result of non-recursive pre-order traversal");
    PreOrder_NR(RPtr);
    printf("\n Result of recursive pre-order traversal");
    PreOrder(RPtr);
    printf("\n Result of recursive in-order traversal");
    inorder(RPtr);
    printf("\n Result of recursive post-order traversal");
    postorder(RPtr);
    printf("\n");
    return;
}
```

Test data (the structure of the corresponding tree is shown in Fig. 1.58)
Input: ABD@F@@@CE@@@ (the sequence obtained by pre-order traversal on tree)
   Results:
– Construct the tree and input the sequence obtained via pre-order traversal
– ABD@F@@@CE@@@
– Result of nonrecursive pre-order traversal: ABDFCE
– Result of recursive pre-order traversal: ABDFCE
– Result of recursive in-order traversal: DFBAEC
– Result of recursive post-order traversal: FDBECA

1.6.6 Applications of tree traversal

1.6.6.1 Get the depth of the binary tree
1. Get the depth of the binary tree with pre-order traversal
The depth of the binary tree is the maximum value among the node levels of the binary tree. We can calculate the level of each node in the tree by pre-order traversal, the maximum value from which would be the depth of the binary tree. The recursion process is shown in Fig. 1.59. Whenever we visit a node, the count level is incremented by 1. h will be taken from the bigger level value from left and right subtrees. We set the initial value of level as 0.

The result of pre-order traversal on the tree shown in Fig. 1.59 is ABDEGHCF. The correspondence between the level of each node level and the height h is given in Table 1.7.

The pseudocode description of the algorithm is shown in Table 1.8.

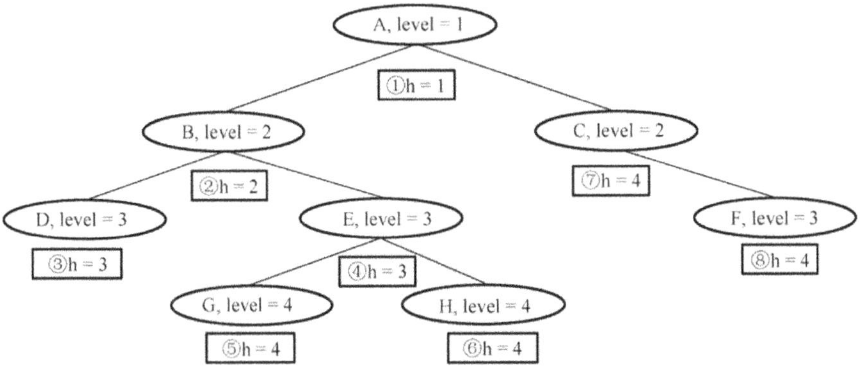

Fig. 1.59: Obtaining the depth of a binary tree via pre-order traversal.

Table 1.7: The correspondence between the level number and height of a binary tree.

DLR sequence	A	B	D	E	G	H	C	F
Level	1	2	3	3	4	4	2	3
Maximum height	1	2	3	3	4	4	4	4

Table 1.8: Pseudocode for obtaining the depth of a binary tree via pre-order traversal.

Obtaining the depth of a binary tree via pre-order traversal

Input: The current root node of the tree; the level of the current root node level
Let the node pointer of the tree = BinTreeNode *p, The height of the tree = h, The level of the tree = level

---

The base case of recursion: P is an empty node, return h;

Condition for the recursion to continue: level is incremented by 1, h takes on the value of the bigger h between the left and the right subtrees
The left child of p is nonempty, pre-order traverses the left child of p;
The right child of p is nonempty, pre-order traverses the right child of p;

Program implementation

```
/*================================================================
Functionality: Obtain the depth of the binary tree via pre-order traversal
Function input: Root node
Function output: Depth of the tree
Screen output: (Node value of the leaf, number of levels, current height of the
tree) - Useful for debugging
================================================================*/
```

```
int h=0;
// Global variable. The accumulated depth of the tree
int TreeDepth_DLR(BinTreeNode *p, int level )
{
  if ( p!= NULL)
  {
    level++;
    if( level>h ) h=level;
    putchar(p->data);
    printf(" level=%d, h=%d\n",level,h);
    h=TreeDepth_DLR( p->lchild, level );
    // Calculate the depth of the left child tree
    h=TreeDepth_DLR( p->rchild, level );
    // Calculate the depth of the right child tree
  }
  return h;
}
```

2. Obtaining the depth of the binary tree via post-order traversal

We can calculate the height of the left and right children of each node, and then take the bigger value between the two as the height of the tree. The recursion process is shown in Fig. 1.60.

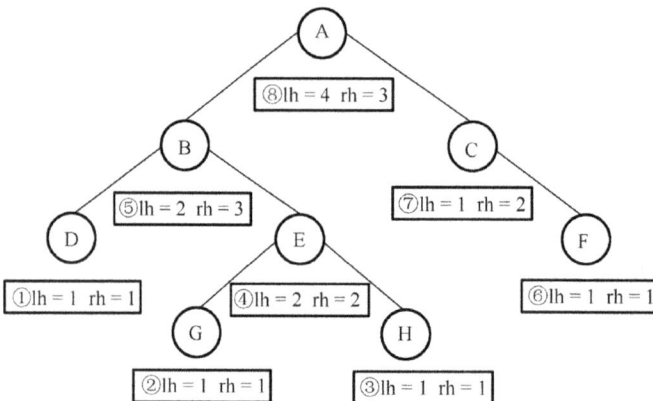

Fig. 1.60: Obtaining the depth of the binary tree via post-order traversal.

The result obtained via post-order traversal on the tree is DGHEBFCA. We start calculating the heights of the left and right subtrees from the bottommost node. Let the height of the left and right subtrees be lh and rh, respectively, then the height of the left and right subtrees at each node is given in Table 1.9.

Table 1.9: Heights of the left and right subtrees of the binary tree.

DLR sequence	D	G	H	E	B	F	C	A
Height of the left subtree lh	1	1	1	2	2	1	1	4
Height of the right subtree rh	1	1	1	2	3	1	2	3

The pseudocode description of the algorithm is given in Table 1.10.

Table 1.10: Pseudocode for obtaining the depth of the binary tree via post-order traversal.

Obtaining the depth of the binary tree via post-order traversal
Input: The current root node of the tree Let node pointer = BinTreeNode *p, height of the left subtree = lh, height of the right subtree = rh
The base case of recursion: p is an empty node, return 0;
Conditions for the recursion to continue: The left child of p is not empty, we obtain lh via post-order traversal; The right child of p is not empty, we obtain rh via post-order traversal; Return the bigger value between the heights of left and right subtrees

We record the heights of the left and right subtrees in two variables lh and rh. Then, although they are both local variables, they can be compared at each level.

Program implementation

```
/*=============================================================
Functionality: Obtain the depth of the binary tree via post-order traversal
Function input: Root node
Function output: Depth of the tree
Screen output: (Value of the leaf node, heights of the left and right subtrees) -
Useful for debugging
=============================================================*/
int TreeDepth_LRD(BinTreeNode *p )
{
  if (p!=NULL)
  {
    int lh = TreeDepth_LRD( p->lchild );
    int rh = TreeDepth_LRD( p->rchild );
    putchar(p->data);
    printf(":lh=%d   rh=%d\n",lh+1,rh+1);
```

```
        return lh<rh? rh+1: lh+1;
    }
    return 0;
}
```

1.6.6.2 Obtain the numbers of leaves

We visit each node of the tree via traversal. Because of the special nature of leaf node, we can obtain the total number of leaves in a tree.

Example 1.4 Obtain the total number of leaf nodes in the binary tree and print out the values of leaf nodes.

Solution: If both the left and right pointers of the node root are empty, then it is also a leaf. We can use any kind of traversal algorithm to look for the leaves, count and print them.
The condition for deciding whether it is a leaf node: root -> lchild = = NULL && root -> rchild = = NULL
or we may also write in C: !root-> lchild && !root-> rchild
Method 1 Obtain the number of leaf nodes in the tree via recursive pre-order traversal.

```
/*========================================================================
Functionality: Obtain the number of leaf nodes in the tree via recursive pre-
order traversal
Function input: Root node of the binary tree
Function output: None (We use a global variable to pass on the total number of
leaf nodes)
Screen output: The value of the leaf node
========================================================================*/
int sum=0;
// Pass on the number of leaf nodes via a global variable
void LeafNum_DLR(BinTreeNode *root)
{   if ( root!=NULL )
    // The condition for the binary tree to be non-empty, same as if(root)
    {   if(!root->lchild && !root->rchild)
        // If it's a root node, then count and print
        {
            sum++;
            printf("%c ",root->data);
        }
        LeafNum_DLR(root->lchild);
        // Recursively traverse the left subtree until we reach its leaves
```

```
    LeafNum_DLR(root->rchild);
    // Recursively traverse the right subtree until we reach its leaves
  }
}
```

Method 2 Obtain the number of leaf nodes in the tree via recursion.

```
/*================================================================
Functionality: Obtain the number of leaf nodes in the tree via recursion
Function input: Address of the root node
Function output: Number of leaf nodes
==============================================================*/
int LeafNum(BinTreeNode *root )
{ if (root ==NULL) return(0);
  else if (root ->lchild==NULL && root ->rchild==NULL)
  return(1);
  else return(LeafNum(root->lchild)+LeafNum(root->rchild));
}
```

Test for the function to obtain the number of leaf nodes.

```
/*================================================================
Functionality: Test for obtaining the number of leaf nodes
Test function:
1. Recursively obtain the number of leaf nodes in the tree: LeafNum
2. Obtain the number of leaf nodes in the tree with pre-order traversal:
LeafNum_DLR
==============================================================*/
int main()
{
  BinTreeNode *RPtr;
  int i;
  RPtr=CreatBtree_DLR(RPtr);
  LeafNum_DLR(RPtr);
  printf("LeafNum_DLR:%d\n ",sum);
  i=LeafNum(RPtr);
  printf("LeafNum:%d \n",i);
  return 0;
}
```

Test cases. The tree structure is shown in Fig. 1.58.

```
Input: ABD@F@@@CE@@@
Output:
F E
LeafNum_DLR: 2
LeafNum: 2
```

1.7 Application of trees

The application of trees can be viewed as special ways of usage via the addition of various extensional or restrictive conditions on the basis of the structure and basic operations of the tree. This section introduces some classical applications with a particular focus on Huffman trees. The contents related to search operations on trees will be introduced in the chapter on search.

1.7.1 Expression tree

Expression tree refers to the representation of a mathematical expression using a binary tree. The operands are the leaf nodes of the tree, while the operators are the nonleaf nodes.

The earliest usage of traversal is to obtain the values of the expressions stored in the computer. For example: $(a + b \times (c - d)) - e/f$. The expression is represented with a tree shape, as shown in Fig. 1.61. The operators in the tree are put on the nonleaf nodes, while the numbers to be operated upon are put on the leaf nodes. If we traverse the tree in pre-order, in-order or post-order manners, respectively, then the sequences obtained are called prefix expression (also called Polish expression), infix expression, postfix expression (also called reverse Polish expression).

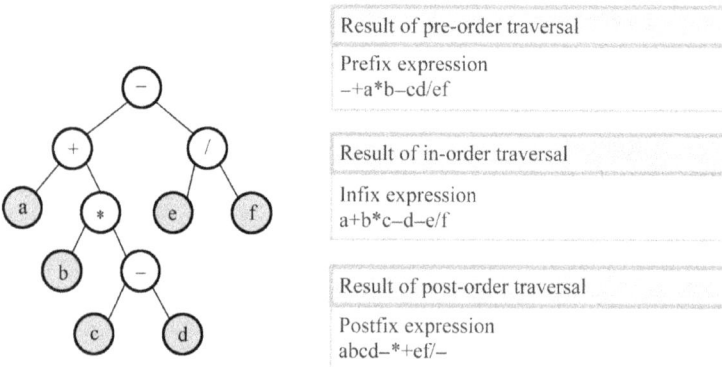

Result of pre-order traversal

Prefix expression
−+a*b−cd/ef

Result of in-order traversal

Infix expression
a+b*c−d−e/f

Result of post-order traversal

Postfix expression
abcd−*+ef/−

Fig. 1.61: Tree-form representation of arithmetic expressions.

Infix expression is the most common form of arithmetic expressions, without the brackets. For a computer, it is easier to obtain the value of the expression with post-fix expressions.

Think and Discuss How to obtain the corresponding expression tree from an expression?
Discussion: Expressing a mathematical expression as an expression tree, the key problem is to put the operator of the last operation as the root of the binary tree. Using this operator as the boundary, the part before it (here called expression 1) will be the expression represented by the left subtree of the binary tree, the part after it (here called expression 2) will be the expression represented by the right subtree. We can then recursively apply the above rules to the (sub) expression 1 and (sub) expression 2 to construct further left/right subtrees.
Example 1.5 Input an arithmetic expression, output the expression tree corresponding to this expression.
(We skip over the check over the soundness of this expression.)
(There are three occasions where the expression might be unsound:
1. Left and right brackets do not match;
2. Variable name is illegal;
3. There are no variables or numbers that participate in the calculation at both sides of an operator.)

Solution: Analyzing Fig. 1.62, we can see that the root node of the expression and the root node of its subtree are the operators. Their order in the tree is from back to front according to the order of the operations. The leaves of the expression tree are the variables or numbers participating in the calculation.

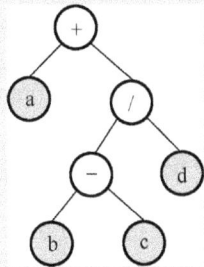

Fig. 1.62: Tree of expressions.

For example, the calculation order of the expression $a + (b - c)/d$ is: first find the "+" which is at the lowest precedence as the root node, then ascertain the ranges of the left/right subtrees of the expression string to be a and $(b - c)/d$, then search for the operator with the lowest prece-dence as the root nodes of the subtrees, until there are no operators in range. Then, the remain-ing variables are numbers or the leaves of the expression tree.

Algorithm Description

1. Suppose an array stores the various characters of the expression string, lt, rt are the left and right pointers of the node, variables left, right are used to store the left and right boundaries of the range of characters each time.
2. Set the initial value of the left boundary as 1, and the initial value of the right boundary as the length of the string.
3. Check whether left and right brackets match. If they do not match, then we suppose there is some error in the input.
4. Search for the operators with the lowest level of precedence within the range of the left and right boundaries. At the same time, check whether there are variables or numbers that participate in the calculation. If there is none, then the expression entered is unsound; if there are, set this operator as the root node of the current subtree, set the left subtree pointer and its left/right boundaries, set the right subtree pointer and its left/right boundaries.
5. If the expression has no operators in the range denoted by left and right boundaries, then this is a root node. We check whether the variable name or the number is sound.
6. Jump to step 4, until all the characters in the expression string are processed.

1.7.2 Threaded binary tree

The pre-order, in-order and post-order traversal methods on binary trees were realized either via recursion or nonrecursive implementations based on stack. These two methods are classical processing methods based on the recursive features of the tree itself. No matter it is the recursion itself which takes up space on the stack or the stack defined by the user, $O(n)$ space complexity will be needed. The various operations on tree introduced above were mostly based on recursion or stack.

In actual applications, there are systems with limited time/space capabilities, for example, embedded system or system on chip. They usually require algorithms to "have few jumps, use no stack, be nonrecursive." Compared with the operations on binary trees based on recursion, nonrecursive pattern will have a wider application [2]. In the following, we analyze whether we can realize the operations on binary trees in a "nonrecursive, use no stack" manner.

1.7.2.1 Start from the problem of searching the preceding/succeeding nodes in binary linked lists

We know that the traversal of binary trees is to arrange the nodes of the binary tree into a linear sequence according to certain rules. The obtainment of the pre-order, in-order or post-order sequences of nodes in a binary tree is actually a linearization operation on a nonlinear structure. For the normal nodes in the tree, there are two possible relations between each node and another node: parent or child. The predecessor and successor relations of the nodes are also based on linear list. The discussion on the predecessor or successor of a particular node in the tree structure is carried out on the basis of the sequence obtained by the traversal on the binary tree. For example, in Fig. 1.63, in the pre-order traversal sequence ABCDEF of the binary tree, the predecessor of C is B, the successor is D.

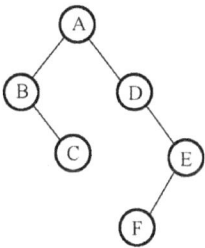

- •Pre-order traversal sequence ABCDEF
- •In-order traversal sequence: BCADFE
- •Post-order traversal sequence: CBFEDA

Fig. 1.63: The binary tree and its various sequences obtained via traversal.

Think and Discuss
1. Is it convenient to look up the predecessor and successor of a node?
2. What are the potential problems with using the linear storage method of tree to solve the lookup of the predecessor/successor of a node?

Discussion 1: In the storage structure of binary tree, we can only find the information about the left and right children of a node, instead of obtaining the predecessor and successor information about the node in any kind of sequence. We can only obtain this information during the traversal process. When the operations on the binary tree frequently involve the lookup of a predecessor or a successor of a node, using the binary linked list as the storage structure would lead to much inconvenience during searching, and the speed will be slow.

Discussion 2: The lookup on a sequential list is relatively inefficient, with a time complexity of $O(n)$. The lookup on tree belongs to "ranged lookup," which should have a higher efficiency than generic sequential lookup. For detailed information, please refer to the relevant contents in Chapter 4.

Think and Discuss How do we simplify the operations of looking up the predecessors and the successors of a binary linked list?

Discussion: When we scan the binary tree with a certain traversal order, during the scan, we can record the clues (threads) about the predecessor and successor by adding data items in each node, so that we can facilitate the lookup of predecessors and successors in subsequent operations. The concrete implementation method can be seen below.

1.7.2.2 The implementation of threaded binary linked list

Plan 1: Add predecessor/successor pointer fields.

We can add pointer fields in nodes of a binary linked list in order to store information about predecessor and successor nodes, as shown in Fig. 1.64, where Lthread indicates the address of the predecessor node, and Rthread indicates the address of the successor node.

Referencing the tree structure in Fig. 1.65, according to the node design plan in Fig. 1.64, we can give out the actual storage form of the result of pre-order traversal on this tree structure as shown in Fig. 1.66.

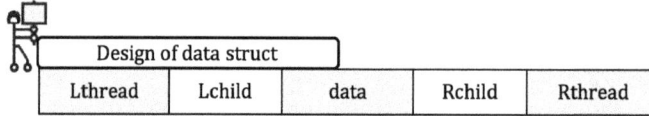

Fig. 1.64: Node design in plan 1 of threaded binary tree.

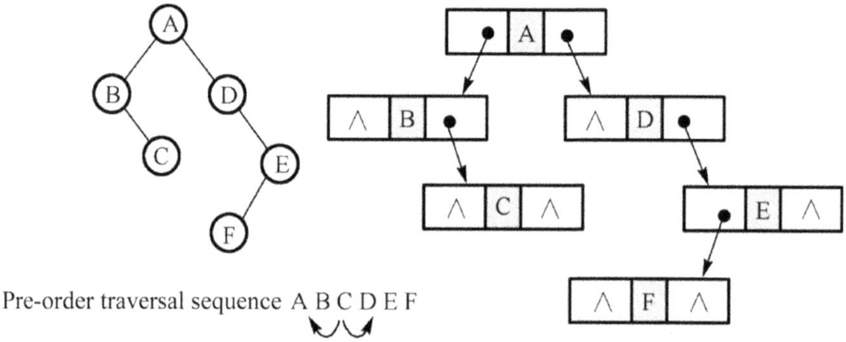

Pre-order traversal sequence A B C D E F

Fig. 1.65: Storage of the tree with binary linked list.

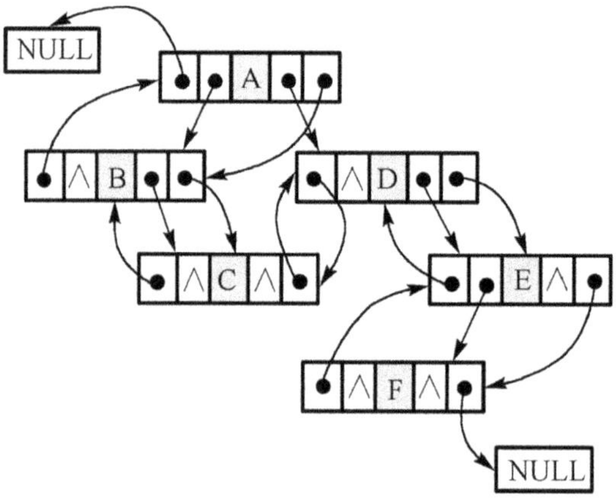

Fig. 1.66: Storage method 1 of the predecessors and successors of the nodes in a tree structure.

Think and Discuss

Question 1: Would there be any overlap between the predecessor/successor threads and the information stored by child pointers?

Discussion: According to Fig. 1.66, we can list the predecessors, successors and children of all nodes in one table, and then analyze them.

From Table 1.11, we can observe that there is no overlap between predecessor nodes and the left/right children information. However, there might be some overlap between the children of the node and the successor. Concerning the pre-order traversal of the tree, the relation between the children of the node and the successors is as such: the successor of any node is its left child; if the left child is empty, then it will be its right child; if the right child is also empty, then it is only the successor itself. (What is this?)

Table 1.11: The relationship between the current node and the predecessor, the successor and the children during pre-order traversal.

Node	Predecessor	Left child	Right child	Successor
A	Empty	B	D	B
B	A	Empty	C	C
C	B	Empty	Empty	D
D	C	Empty	E	E
E	D	F	Empty	F
F	E	Empty	Empty	Empty

Since the children of a node overlap with the successor, we encounter the next question.

Question 2: Can we make any improvements to the methods of adding pointer fields for the predecessors and successors?

Discussion: Since the children of the node overlap with the successor, we only need to use the pointer field of one of the children. We can consider utilizing the pointer fields of the original node to store threads for the predecessor and the successor. See the following Plan 2:

Plan 2: Utilize the pointer fields of the original node.

Plan 1 stores the predecessor/successor node information by adding pointer fields. This lowers the utilization rate of the storage space.

In Fig. 1.66, the pre-order traversal sequence of the corresponding tree is ABCDEF. Take the example of the node C: it has neither left subtree nor right subtree, and the corresponding pointer fields are both empty. The predecessor of C is B, and its successor is D. To record the information about the addresses of the predecessor and successor nodes of node C, we can utilize this empty link field to store the address of B and D, that is, add the threads for the predecessor and the successor. For node B, lchild field is empty; thus, we can use it to record the address of its predecessor node A. Since C is its right child, that is, its successor node, there is no need for us to record the address of its successor. Following this pattern, we can conveniently find the corresponding successor for any node of the tree, as shown in Fig. 1.67.

Pre-order traversal sequenle A B C D E F

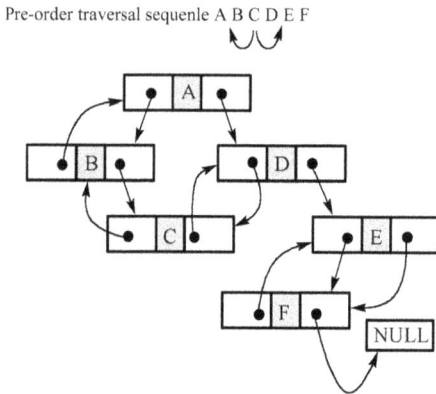

Fig. 1.67: The predecessor and successor of the nodes of a tree structure 2.

Think and Discuss

Question 1: Can we differentiate whether we have stored a child or a thread in the plan above which utilizes the pointer field of the original node?

Discussion: If we use the method in Fig. 1.67 to record the predecessor and successor of a node, we cannot differentiate whether we have stored the address of left/right child or the predecessor/successor. To solve the problem, we can set up a corresponding marker field to indicate the property of the data in the pointer field, for example, 0 to represent child and 1 to represent thread, as shown in Fig. 1.68.

Question 2: Does the plan using the pointer fields of the original node guarantee that the threads for all predecessors will be recorded?

Discussion: There exist situations where a node has a left child and thus its predecessor cannot be recorded, for example, node E in Fig. 1.68, which does not have a pointer field to record the address of its predecessor D. However, note that for pre-order traversal, it suffices to have the thread for the successor of a node. Therefore, even when there is some "gap" in the threads for the predecessor, it does not impact the operation of pre-order traversal.

The process of scanning the binary tree according to a certain traversal sequence and adding the predecessor/successor threads to each node is called the threading of binary tree. A binary tree with threads added is called a threaded binary tree. The purpose of threading is to simplify and speed up the lookup of the predecessors and successors of a node in a binary tree. Also, it is more convenient to traverse a threaded binary tree, since recursion will not be needed, and the program runs more efficiently.

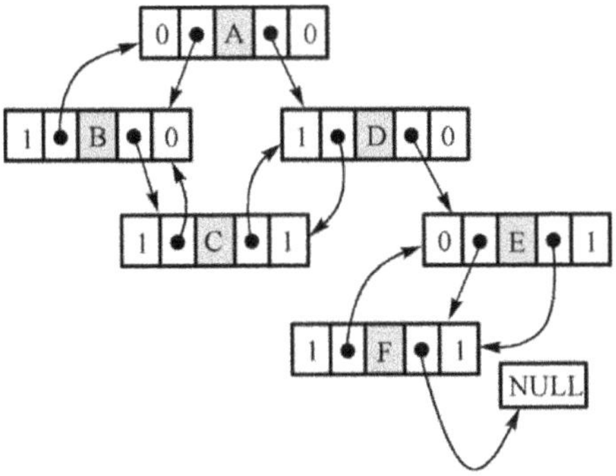

Fig. 1.68: The predecessor and successor of the nodes of a tree structure 3.

Term Explanation Threaded binary tree
A binary tree with additional pointers that directly point to the predecessor and successor of the node is called a threaded binary tree.

According to the differences in the contents of the threads, threaded binary trees can be further divided into pre-order threaded binary tree, in-order threaded binary tree and post-order threaded binary tree. The structural design of a node of a threaded binary tree is shown in Fig. 1.69. The markers Ltag and Rtag are used to indicate whether the corresponding pointer field stores a thread or a child.

Structure design

Ltag	Lchild	data	Rchild	Rtag

Ltag	0: Lchild field points to the left child of the node
	1: Lchild field points to the predecessor of the node
Rtag	0: Rchild field points to the right child of the node
	1: Rchild field points to the successor of the node

Fig. 1.69: Node struct design for type 2 of threaded binary tree.

Example 1.6 Draw the in-order binary linked list and in-order threaded binary tree of the binary tree in Fig. 1.65.

Solution: The in-order traversal sequence of the binary tree is BCADFE.
1. Draw the binary linked list according to the shape of the binary tree.
2. Draw the arrow direction of the thread field according to the definition of node of threaded binary tree.

The result is shown in Fig. 1.70.

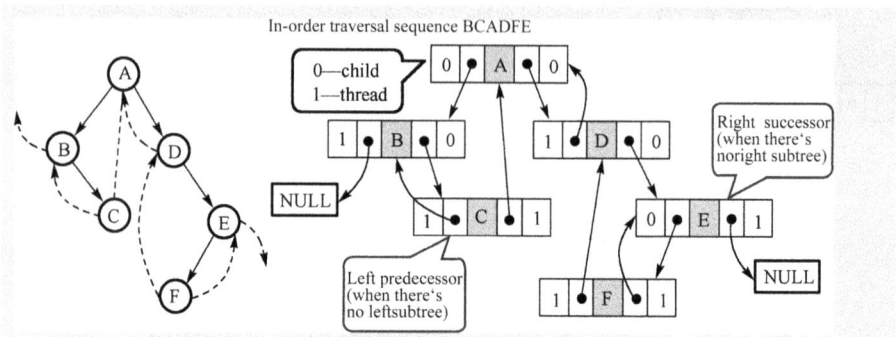

Fig. 1.70: In-order threaded binary tree and in-order binary linked list.

Think and Discuss How to find the predecessor and successor of the node in the traversal sequence of a threaded binary tree?

Discussion: Using the in-order threaded binary tree in Fig. 1.70 as an example, the right links of all the leaf nodes in the tree serve as the thread, and therefore the RightChild field of the leaf node points to the successor node of the node. For example, the successor of node C in Fig. 1.70 is node A, and the successor of node F is node E.

When an internal node has its right thread tag as 0, we know its RightChild pointer points to its right child; therefore, we would not be able to obtain the successor node from RightChild. An example is node D.

However, we know from the definition of in-order traversal that the successor of node D, node F, should be the first node to be visited when traversing the right subtree of D, that is, the most bottom-left node of the right subtree of D.

Similarly, the pattern when finding the predecessor node in an in-order threaded tree is: If the node's left tag is marked as 1, then LeftChild stores the thread which directly points to its predecessor node. Otherwise, the node last visited when traversing the left subtree, that is, the most bottom-right node of the left subtree, is its predecessor node. The predecessor and successor of each node in Fig. 1.70 are given in Table 1.12.

The marker for right thread is 1: the right child is the successor.

The marker for right thread is 0: the leftmost node in the right subtree is the successor.

The leftmost child of the right subtree: the first node to be visited when we in-order traverse the right subtree.

The marker for left thread is 1: the left child is the predecessor.

The marker for left thread is 0: the rightmost node in the left subtree is the predecessor.

The rightmost child of the left subtree: the last node to be visited when we in-order traverse the left subtree.

From this we can see that if the height of the threaded binary tree is h, then in the worst-case scenario, we can find the predecessor or successor node of a node in $O(h)$ time. When we traverse an in-order threaded binary tree, there is no need to store the information of the subtree to be visited with recursion/stacks, like when we traverse nonthreaded trees.

Table 1.12: The predecessor and successor in sequence obtained via in-order traversal.

	B	C	A	D	F	E
Predecessor marker	1	1	0	1	1	0
Predecessor	Empty	B	The rightmost node of the left subtree, C	A	D	The rightmost node of the left subtree, F
Successor marker	0	1	0	0	1	1
Successor	The leftmost node of the right tree, C	A	The leftmost node of the right tree, D	The leftmost node of the right tree, F	E	Empty

The method to obtain in-order traversal sequence from in-order traversal threaded binary tree:
- Visit the root node
- Find the predecessor sequence of the root
- Find the successor sequence of the root

The method to obtain pre-order traversal sequence from pre-order traversal threaded binary tree:
- Visit the root node
- Find the successor sequence of the root

The method to obtain post-order traversal sequence from post-order traversal threaded binary tree:
- Visit the root node
- Find the predecessor sequence of the root

1.7.2.3 Summary of threading

Since the essence of threading is modifying the empty pointers in binary linked list into threads that point to the predecessor or successor of a node, and we can only obtain the information about the predecessor/successor nodes during traversal, the threading process is the process of modifying empty pointers during the traversal process. In order to record the order of node visits during the traversal, we can add a pointer pre which always points to the just-visited node. When pointer p points to the currently visited node, pre points to its predecessor. Therefore, we can also deduce that the successor of the node pointed to by pre is the current node pointed to by p. In this way, we can carry out the threading during traversal process. For the lookup of predecessor and successor, threaded tree is better than the nonthreaded tree. But threaded tree also has its shortcomings. When performing insertion and deletion operations, threaded trees cost more than nonthreaded trees. The reason is that when we perform insertion and deletion in a threaded tree, we need to modify the corresponding threads in addition to modifying the corresponding pointers.

1.7.3 Huffman tree and Huffman encoding

Huffman tree is the tree with the shortest weighted paths. It is also called optimal tree. One of its usage is to construct the compression codes used in telecommunication. Before we introduce the concepts on Huffman tree and Huffman encoding, let us first introduce the relation between telecommunications and encodings.

Knowledge ABC Data compression
Data compression is a technique that, by reducing the redundancy in the data stored in the computer or the data in communication transmission, increases data density, and eventually decreases storage space needed for the data. Data compression has very widespread application in file storage and distributed systems.

Knowledge ABC Encoding problem in communication
Transmission refers to widely existent process of information relay. The process of information transmission can be briefly described as: Information source -> Information tunnel -> Information host, as shown in Fig. 1.71. Information source refers to where the information comes from, the recipient of information is called information host. "Information source" is the publisher of information, while "information recipient" is the receiver of information. Information tunnel is the pathway through which information is transmitted; it constitutes the transmission media through which the signal passes by from the origin to receiver.

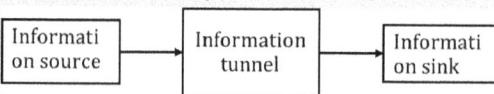

Fig. 1.71: Simplified model of information transmission process.

The information symbols sent from the information source are usually not suited to the transmission by the signal tunnel. In this case, we need encoding at information source, to convert the information symbols sent from the information source into symbols suited to transmission by signal tunnel. After the recipient has received them through the transmission via the signal tunnel, according to the reverse process of encoding, it can revert the information into its original form. This process is shown in Fig. 1.72. (This is a simplification of the digital communication model proposed by Shannon and Weaver in 1949.)

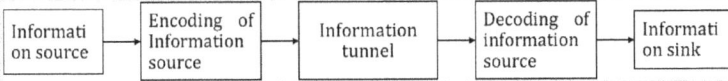

Fig. 1.72: Model of information transmission process.

The most common information tunnel is binary information tunnel, that is, an information tunnel that can transmit two basic symbols. The basic symbols of a binary information tunnel are usually represented as 0, 1.

We will look at the application of data structure in the design of encoding as follows:

Example 1.7 Design of fixed-length binary encoding
Suppose a set of source symbols has only four characters A, B, C, D. In order to differentiate between different characters, suppose the digits of binary encoding needed by each character is n:

$$\because 4 = 2^n \therefore n = 2$$

Suppose the encoding of A, B, C, D are, respectively, 00, 01, 10 and 11. If there is a text "ABACCD" to be transmitted, as shown in Fig. 1.73, then the sender should transmit the encodings according to the agreed-upon encoding plan. The receiver, after receiving the binary string, can translate it by each two digits, and obtain the original text, as shown in Fig. 1.73. If there are 26 different characters in the text, then the length of encoding for each character will be 5 (24 < 26 < 25). We call this as encoding method, where each character has the same code length as fixed-length encoding.

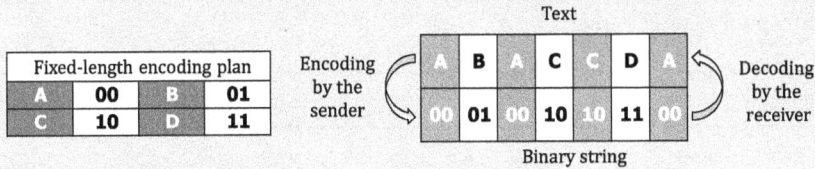

Fig. 1.73: Fixed-length encoding.

Think and Discuss How is the efficiency of fixed-length encoding?
Discussion: In actual applications, some characters (e.g., e, s, t) appear with relatively high frequency, while some characters are relatively few. As shown in Fig. 1.74, if we re-encode all the characters so that we use fewer bits to represent characters with a higher number of occurrence, and use relatively more bits to represent characters with low occurrences, then we should be able to store the same amount of information with relatively small storage space, in order to store the same amount of information with relatively small storage space, so that the total length of the encoding used for the text is shortened and the communication efficiency improved.

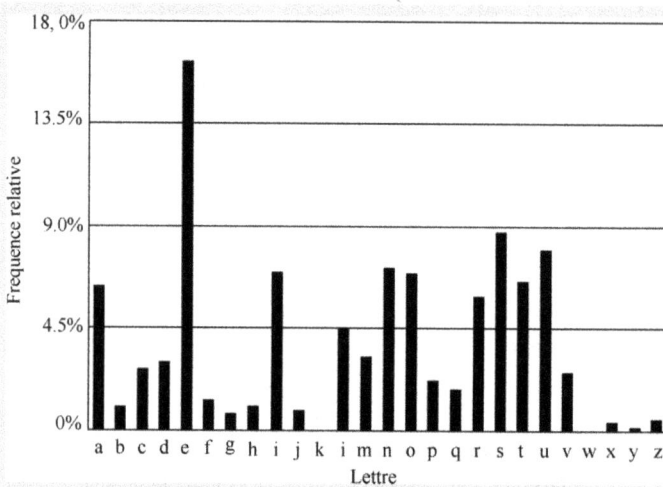

Fig. 1.74: Table of character efficiency.

For example, Table 1.13 lists the frequency of appearance of different characters in a document (suppose there are 100,000 characters in total in the document and only five characters, a, b, c, d, e are present, then there is a significant difference between the lengths obtained via fixed-length encoding and variable-length encoding, that is, fixed-length 300,000 bits and variable length 224,000 bits.

Fixed-length encoding: $3*(17 + 12 + 12 + 27 + 32) = 300$Kbits.

Variable-length encoding: $2*(17 + 27 + 32) + 3*(12 + 12) = 224$Kbits.

Table 1.13: Example of file encoding.

	a	b	c	d	e
Frequency (thousands)	17	12	12	27	32
Fixed-length encoding	000	001	010	011	100
Variable-length encoding	00	010	011	10	11

We might summarize the above question as such: we already know the set of characters and their relative frequencies for a text. Try to encode this set of characters, so that the total length of the text is the shortest.

We can discuss variable-length encoding via the following example:

Example 1.8 Design of variable-length binary encoding
Use the previous example, in which the symbol set of information source contains only four characters: a, b, c, d.

Based on the fact that the characters appear in the descending order of a, c, b, d, if we design the encodings of a, b, c, d to be 0, 00, 1, 01, then the above text ABACCDA becomes 000011010. Such a text can be translated in multiple ways (see Fig. 1.75).

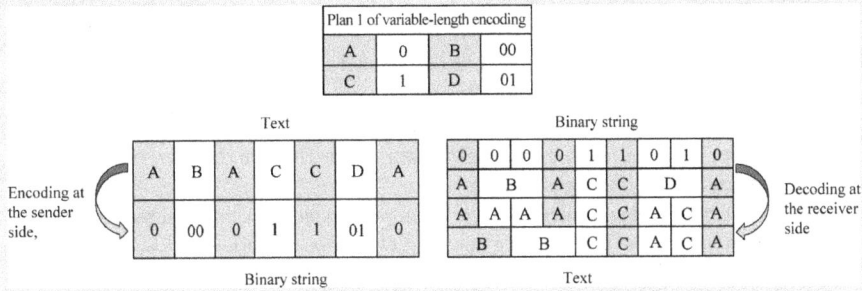

Fig. 1.75: Variable-length encoding.

If there are multiple possible translations, then such an encoding plan is not really usable.

Think and Discuss
1   What is the reason for the existence of multiple possible translations in the example above?

Discussion: Before we analyze this question, we can first analyze the way translation in fixed-length encoding works. We can describe the path of the translation with a tree, as shown in Fig. 1.76. If the receiving end receives 00011011, then it can start from the root, and retrieve encodings one by one from the text to be translated. If the encoding is "0," then it goes left; if the encoding is "1," it goes right. Whenever it reaches a leaf, it translates to a character. It repeats the above process until the translation ends. It thus obtains ABCD in the end.

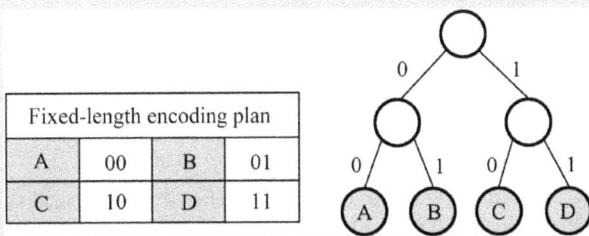

Fig. 1.76: Translation tree for fixed-length encoding.

The problem of code translation is a search problem on string, that is, multipattern matching on string.

Let us see the translation scenario for variable-length encoding plan 1 in Fig. 1.77. The encoding of A is 0 and the encoding of B is 00. The beginning parts of encodings for A and B are identical. Therefore, when B is encountered, the receiver cannot distinguish whether its first bit represents A or is a prefix for B.

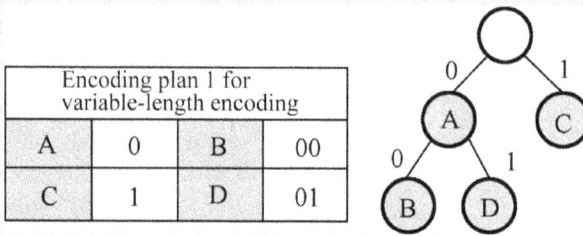

Fig. 1.77: Translation tree for variable-length encoding plan 1.

After comparing Figs. 1.76 and 1.77, we can see that, in order for the translation to be unique, then there cannot be total overlap between each path that arrives at a character node, that is, a character node cannot appear at a branching node.

Therefore, when designing a variable-length encoding, we must do it so that the encoding for any character cannot be the prefix for the encoding of another character, that is, prefixes must differ. Only in this way can we ensure the uniqueness of translations. Such encoding is called different-prefix encoding, and also called different beginning code.

2. What kind of design would ensure prefix encoding that has unique translations?

Discussion: Let us continue to use the data in the previous example for the discussion.

According to the rule that "character node cannot appear in a branching node," we should let the characters with the most appearances (i.e., frequencies) be arranged as close to the root node as possible. Then we obtain the translation tree listed in Fig. 1.78.

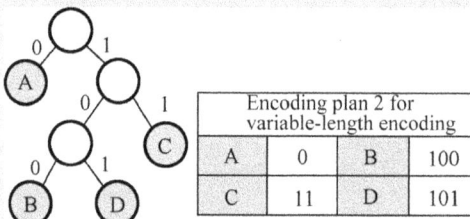

Fig. 1.78: The translation tree of variable-length encoding plan 2.

In the encoding plan 2, although the encoding length of character A is shortened, the encoding length of characters B and D is increased. Will it reduce the overall encoding length? Still using Example 5.7 as an example, we would obtain a translation as shown in Fig. 1.79 if we follow encoding plan 2. The conclusion is that it is indeed shorter.

Huffman encoding is an encoding method proposed by Huffman in 1952. This method constructs the encoding scheme so that the average length of different characters is the shortest, completely according to the appearance frequencies of characters. It is also called optimal encoding.

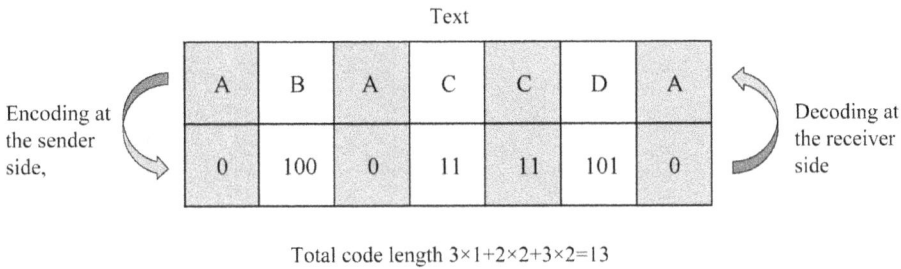

Fig. 1.79: Testing for variable-length encoding plan 2.

1.7.3.1 The concept of Huffman tree

Before we give the definition of Huffman tree, let us first explain several related concepts:

- Path: several edges from one node to another
- Path length: the number of edges on the path is called the length of the path
- Path length of a node: the length of the path from the root to this node
- The weight of a node: a ratio coefficient with a certain meaning
- The path length with weight of a node: the product of the path length from root to this node and the weight of this node

Note: The "meaning" of the weight would be related to the actual corresponding problem.

Huffman tree is also called the optimal binary tree. It is a binary tree with the smallest weighted path length (WPL).

The WPL of a tree is the sum of the products of all leaf nodes with its length to the root node, denoted as:

$$WPL = \sum_{i=1}^{n} W_i \times L_i$$

where n is the number of leaf nodes; W_i the weight of the ith leaf node, $i = 1, 2, . . ., n$; L_i the path length of the ith leaf node, $i = 1, 2, . . ., n$.

We can prove that the WPL of the Huffman tree is the smallest.

Figure 1.80 gives different binary tree forms constructed by four leaf nodes with different weights ($W = 7, 5, 4, 2$), and the corresponding WPL values.

Normally speaking, when we use n ($n > 0$) leaf nodes with weight to construct a binary tree, we add the restriction that except for these n leaf nodes, there can only be nodes with degree 2 in the tree. We can construct many binary trees that satisfy this condition, among which the binary tree with the smallest WPL will be the Huffman tree or the optimal binary tree.

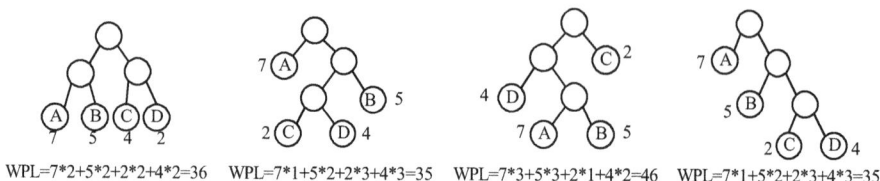

WPL=7*2+5*2+2*2+4*2=36   WPL=7*1+5*2+2*3+4*3=35   WPL=7*3+5*3+2*1+4*2=46   WPL=7*1+5*2+2*3+4*3=35

Fig. 1.80: Several binary trees constructed by four leaf nodes.

! **Think and Discuss** How is a Huffman tree constructed?
Discussion: Please see the construction method of Huffman tree.

1.7.3.2 Algorithm for the construction of Huffman tree

According to the definition of Huffman tree, in order to minimize the WPL value of a binary tree, we must make the leaf nodes with bigger weights closer to the root node, and leaf nodes with smaller weights farther from the root node. According to this characteristic, Huffman proposed a method to construct the optimal binary tree. The basic idea is as follows:

1. gAccording to the weights of the given n leaf nodes, we can view them as n binary trees with only one root node. Suppose F is the set composed of these n binary trees.
2. Select two trees with the smallest root node values as the left, right subtrees, and construct a new binary tree. Set the weight of the root of the new binary tree as the sum of the weight values of the roots of left and right subtrees.
3. Remove these two trees from F, and add the new constructed tree into F.
4. Repeat 2 and 3, until there is only one tree left in F.

Huffman algorithm constructs the binary tree that represents the optimal prefix encodings in a bottom-up manner.

Example 1.9 Constructing a Huffman tree
Construct Huffman trees for the set $W = (6, 5, 3, 4)$ and the set $W = (7, 2, 4, 5)$

Solution: According to the construction algorithm for Huffman trees, the construction process can be seen in Fig. 1.81.

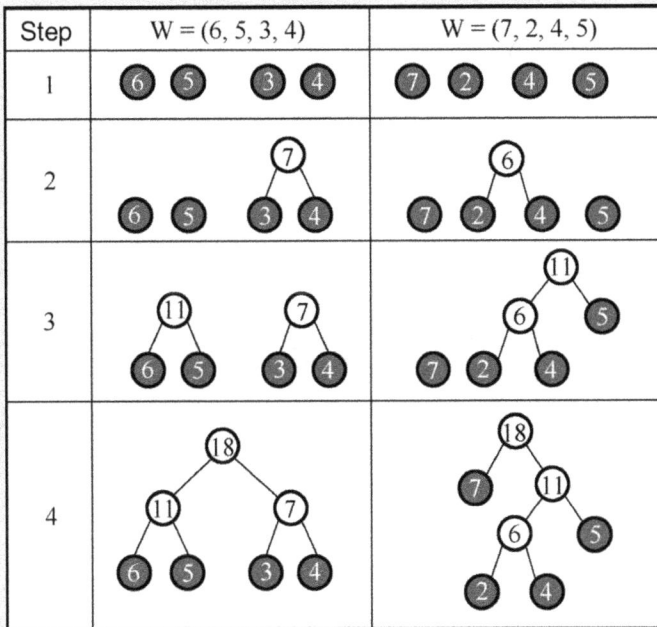

Fig. 1.81: The construction of Huffman tree.

Think and Discuss Is the form of Huffman tree unique?

Discussion: In Fig. 1.81, we can see that during the construction process, there is no particular restriction on the left–right order of the leaf nodes. Therefore, due to the differences in this order, there will be multiple forms of Huffman trees constructed. Generally, nodes with bigger weights will be closer to the root.

1.7.3.3 The algorithmic implementation of Huffman tree

Method 1 Constructing Huffman tree with priority queue

Use priority queue to finish the construction process of Huffman tree. Suppose the weight of the node is its priority. The steps are as follows:

1. Add n leaf nodes into the priority queue, then all n nodes will have a priority P_i, $1 \le i \le n$.
2. If the number of nodes in the queue >1, then:
   i. Remove the two nodes with smallest P_i in the queue.
   ii. Generate a new node, which would be the parent node of the two removed nodes in i. Therefore, the weight of this new node will be the sum of the weights of two nodes in i.
   iii. Add the node generated in ii. into the priority queue.
3. The one node that is left in the priority queue in the end will be the root node of the tree.
4. If we save the nodes into a vector, in the order in which they were removed in the above steps, we would have a Huffman tree.

Method 2 Constructing Huffman tree with struct

The way to realize Huffman encoding is to create a binary tree. The nodes of such trees can be stored in a struct, as shown in Fig. 1.82.

List structure of Huffman tree

data	weight	parent	lchild	rchild

Data structure design

```
typedef struct          //Struct for node
{
      char data;          // Node value
      int weight;         // Weight value
      int parent;         // Parent node
      int lchild;         // Left child node
      int rchild;         // Right child node
} HTree_Node;
```

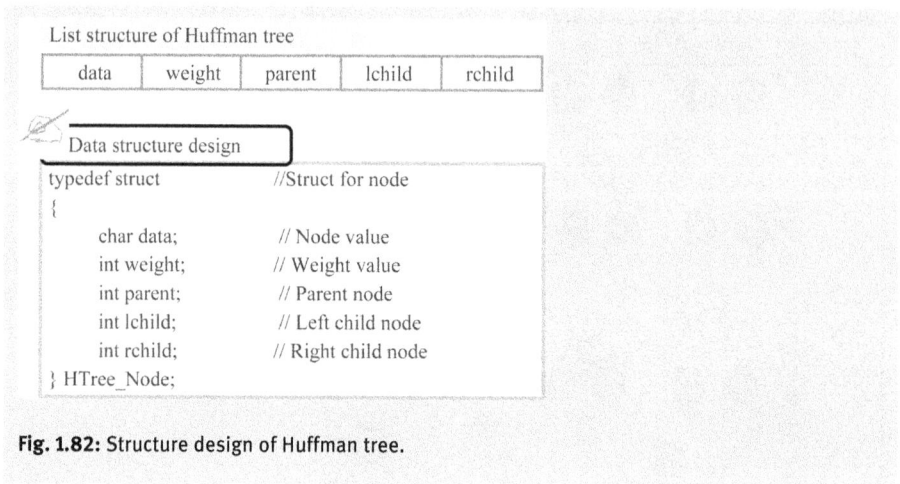

Fig. 1.82: Structure design of Huffman tree.

1. Data structure description

```
#define N 4              // Number of leaf nodes. It should also be the maximum
                         // number of digits in this Huffman encoding
#define M 2*N-1          // The total number of nodes in the tree
#define MAXSIZE 128      // The maximum length of the string to be encoded or the
                         // string of Huffman encoding
```

The struct that stores Huffman encoding is defined as the following:

```
typedef struct   // Encoding struct
{
    char bit[N]; // Store the Huffman encoding
    int start;   // Read the Huffman encoding in bit from the start_th digit
} HCode_Node;
```

2. Function framework design: see Table 1.14.

Table 1.14: Function framework design of Huffman encoding.

Functionality description	Input	Output
Constructing Huffman tree	n weight values	Huffman tree
Function name	Parameters	Function type

3. Illustration of the steps of the algorithm

Using the example of four leaf nodes with weights 7, 5, 2, 4, the steps of constructing a Huffman tree are shown in Fig. 1.83.

(1) Initial state

index	weight	parent	lchild	rchild
0	7	-1	-1	-1
1	5	-1	-1	-1
2	2	-1	-1	-1
3	4	-1	-1	-1
4		-1	-1	-1
5		-1	-1	-1
6		-1	-1	-1

(2) Process

index	weight	parent	lchild	rchild
0	7	-1	-1	-1
1	5	-1	-1	-1
2	2	4	-1	-1
3	4	4	-1	-1
4	6	-1	2	3
5		-1	-1	-1
6		-1	-1	-1

$p_1 \rightarrow$ 2, $p_2 \rightarrow$ 3, $i \rightarrow$ 4

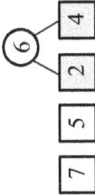

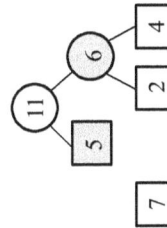

(3) Process

index	weight	parent	lchild	rchild
0	7	-1	-1	-1
1	5	5	-1	-1
2	2	4	-1	-1
3	4	4	-1	-1
4	6	5	2	3
5	11	-1	1	4
6		-1	-1	-1

$p_1 \rightarrow$ 1, $p_2 \rightarrow$ 5, $i \rightarrow$ 5

(4) Final state

index	weight	parent	lchild	rchild
0	7	6	-1	-1
1	5	5	-1	-1
2	2	4	-1	-1
3	4	4	-1	-1
4	6	5	2	3
5	11	6	1	4
6	18	-1	0	5

$p_1 \rightarrow$ 0, $p_2 \rightarrow$ 5, $i \rightarrow$ 6

Fig. 1.83: Steps of constructing a Huffman tree.

– Initialization

Initialization: set the three pointers for parent and left and right children to –1.

Input: read the weights of n = 4 leaf nodes and store them into the first four components of the vector. They are the weight values of the four isolated root nodes in the initial forest.

– Process of merging

We perform $n - 1$ merges in total on the trees in the forest. The new nodes produced are put into the ith component of the vector tree, respectively ($n \le i < 2n - 1$). Each merge is carried out in two steps.

1. In all the nodes of the current forest tree[n], select two root nodes with the smallest and second smallest weights tree[p1] and tree[p2] as the candidates to be merged, where $0 \le p_1, p_2 \le i - 1$.
2. Treat the two trees with roots tree[p1] and tree[p2] as left and right subtrees and merge them into a new tree.

Parent: The root of the new tree is the new node tree[i]. Therefore, we should set the parent of both tree[p1] and tree[p2] as i. Child: We should set the lchild and rchild fields of tree[i] as p_1 and p_2, respectively. Weights: The weight of the new node tree [i] should be set to the sum of the weights of tree[p1] and tree[p2].

Note that after the merging tree[p1] and tree[p2] will no longer be roots in the current forest, since their parent pointers both point to tree[i]. Therefore, they would not be selected as merge candidates in the next merge.

– Program implementation

```
/*====================================
Functionality: Creation of Huffman tree
Function input: (Huffman tree)
Function output: (Huffman tree)
===================================*/
void create_HuffmanTree(HTree_Node hTree[])
{
   int i,k,lnode,rnode;
   int min1,min2;
   printf("data weight parent lchild  rchild\n");
   for (i=0;i<M;i++)
     hTree[i].parent=hTree[i].lchild=hTree[i].rchild=-1;
   // Set the initial value
   for (i=N;i<M;i++)
   // Construct the Huffman tree
   {
     min1=min2=32767;
```

```
// The range of int is -32768~32767
lnode=rnode=-1;
// lnode and rnode record the locations of
// two nodes with the smallest weights
for (k=0;k<=i-1;k++)
{
   if (hTree[k].parent==-1)
   // Only search among the nodes that don't have a parent node
   {
      if (hTree[k].weight<min1)
      // If the weight is smaller than the weight of
      // the smallest left node
      {
         min2=min1;  rnode=lnode;
         min1=hTree[k].weight;lnode=k;
      }
      else if (hTree[k].weight<min2)
      {
         min2=hTree[k].weight;rnode=k;
      }
   }
}
// The parent node of the two smallest nodes would be i
// hTree[lnode]. parent=i;  hTree[rnode].parent=i;
// The weight of the parent node of the two smallest node
// would be the sum of the weights of the two smallest nodes
// hTree[i].weight=hTree [lnode].weight+hTree[rnode].weight;
// Assign the lchild and rchild pointers of the parent node
// hTree[i]. lchild=lnode;  hTree[i].rchild=rnode;
}
for (i=0;i<M;i++)
{
   printf("%4c%6d%6d%6d%6d\n",hTree[i].data,hTree[i].weight,
   hTree[i].parent,hTree[i].lchild,hTree[i].rchild);
}
}
```

1.7.3.4 Huffman encoding

In the previous discussion about whether the result of the encoding is unique, we can see that, in order to avoid prefix encodings, the design of the character encoding should start from the leaf nodes. The essence of Huffman encoding is to retrace via the parents the leaf nodes back to the root node. At each retreat, we pass an edge in the Huffman tree, obtaining one digit of Huffman encoding value. Since the Huffman encoding of a character is the 01 sequence composed of all the edges on the path from the root node to the corresponding leaf node, the edge code we obtained first would be the lower digit code of the encoding we are getting; the edge code we obtained later would be the higher digit code of the encoding we are getting.

If we specify that a left branch in the Huffman tree represents 0, right branch represents 1, then the sequence of 0 and 1 constituting edges from the root node to each leaf node would be the encoding of the character corresponding to this node, as shown in Fig. 1.84.

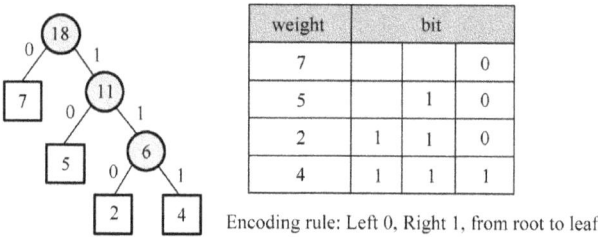

weight			bit	
7				0
5			1	0
2		1	1	0
4		1	1	1

Encoding rule: Left 0, Right 1, from root to leaf

Fig. 1.84: Huffman encoding.

Algorithm description
1. In the Huffman tree structure already constructed, starting from the leaf node L, find its parent F. Based on F, we check whether L is the left child or the right child of F: in the case of left child, the encoding would be 0; in the case of right child, the encoding would be 1.
2. Let L = F, repeat the above process, until L is the root node. We get an encoding from lower digits to higher digits.

Taking the leaf node with weight 5 in Fig. 1.84 as an example, we illustrate the encoding process, as shown in Fig. 1.85.

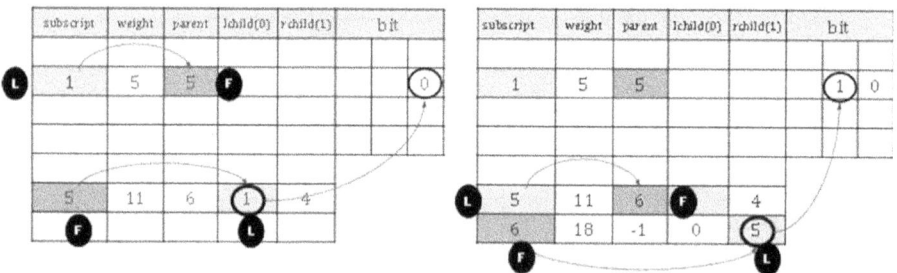

Fig. 1.85: Huffman encoding process.

- The weight of leaf L is 5, its corresponding parent F = 5 (5 refers to its index).
- Finding the row with index 5, we discovered that L is its left child after looking at the left and right columns. Therefore, we fill in 0 at the corresponding position in the bit table.
- L = F
- The weight of the leaf L is 11, the corresponding parent F = 6 (index).
- Finding the row with index 6, we discovered that L is its right child after looking at the left and right columns. Therefore, we fill in 1 at the corresponding position in the bit table.
- L = F
- The weight of the leaf L is 18, the corresponding parent F = −1 (index), which means L is already a root node. This encoding process ends.

The result of the completed encoding is shown in Fig. 1.86, where start is the starting index position of the corresponding character encoding in bit array, to facilitate decoding.

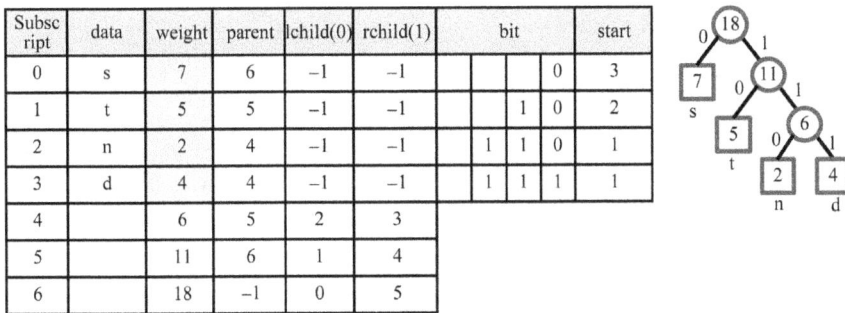

Subscript	data	weight	parent	lchild(0)	rchild(1)	bit			start
0	s	7	6	−1	−1			0	3
1	t	5	5	−1	−1		1	0	2
2	n	2	4	−1	−1	1	1	0	1
3	d	4	4	−1	−1	1	1	1	1
4		6	5	2	3				
5		11	6	1	4				
6		18	−1	0	5				

Fig. 1.86: Result of Huffman encoding.

```
/*===============================================================
Functionality: Encoding of the Huffman character set
(Calculating the Huffman encoding according to the Huffman tree)
Function input: Huffman tree, (Code table for Huffman encoding)
Function output: (Code table for Huffman encoding)
==============================================================*/
void create_HuffmanCode(HTree_Node hTree[], HCode_Node hCode[])
{
    int i,F,L;
    //F: Index for the parent, L: Index for the leaf node
    HCode_Node hc;
```

```
for (i=0;i<N;i++)      //N: Number of leaf nodes
{
    hc.start=N;  L=i;
    F=hTree[i].parent;
    while (F != -1)     // Loop until the root node is reached
    {
        if (hTree[f].lchild==L)      // Process the left child node
            hc.bit[--hc.start]='0';
        else        // Process the right child node
            hc.bit[--hc.start]='1';
        L=F;
        F=hTree[f].parent;
    }
    hCode[i]=hc;
}
}
```

1.7.3.5 Decoding of Huffman tree
There are different approaches to the decoding of Huffman encoding.

Decoding Method 1
Starting from the root of the Huffman tree, retrieve the encoding from the text to be decoded one by one. If the encoding is "0," then walk toward the left; if the encoding is "1," then walk toward the right. Once the leaf node is reached, then one character is translated. We would start again from the root and repeat this process until the whole encoded text ends.

Using Fig. 1.86 as an example, if the encoded text received is 110100, then according to the decoding process described above, the steps are as follows.
- Step 1: Start from the root node $F = 6$ (which corresponds to the row with index 6 of Huffman encoding in Fig. 1.86; here, we represent the node with an index). The first digit of the encoded text is "1," which corresponds to the right child, which is a nonleaf node. We change F to 5.
- Step 2: $F = 5$. The second digit of the encoded text is "1," which corresponds to the right child, which is a nonleaf node. We change F to 4.
- Step 3: $F = 4$. The third digit of the encoded text is "0," which corresponds to the left child, which is a leaf node. We recover the corresponding character as "*n*."

With these three steps, the decoding process for one character ends. If there is remaining encoded text, then we set $F = 6$, and repeat the above steps. The complete decoding process is shown in Fig. 1.87. We obtain a decoded text of "nts."

Step	Text	Left(0)	Right(1)	Subscript	Leaf node	Decoding result
1	1		√	5	no	
2	1		√	4	no	
3	0	√		2	yes	n
4	1		√	5	no	
5	0	√		1	yes	t
6	0	√		0	yes	s

Fig. 1.87: Decoding steps for Huffman encoding.

The implementation steps for Huffman encoding in static storage structure are the following:
- Starts from the root node F, find its child Ch according to the encoded text
- If the encoding is 0, Ch is the left child
- If the encoding is 1, Ch is its right child
- Let F = Ch and repeat the above process, until Ch is a leaf node

The program implementation of this decoding process can be realized by the reader on its own.

Decoding Method 2
In Fig. 1.86, the bit array in the Huffman encoding table compares the encoded text bit by bit, starting from the start position. If the strings of codes(bits) are the same, then the corresponding character is found. The program implementation is as follows:

```
/*===========================================================
Functionality: Translate the Huffman encoding into characters
Function input: Huffman tree, Huffman encoding
Function output: None
==========================================================*/
void decode_HuffmanCode(HTree_Node hTree[ ], HCode_Node hCode[ ])
{
  char code[MAXSIZE];
  int i,j,k,m,x;
  scanf("%s",code);
  //Save the 01 string to be decoded into code array
  while(code[0]!='#')   // '#' is the termination indicator
  {
```

```
for (i=0;i<N;i++)
{
    m=0;    //m is the counter for the number of codes
    j=0;    //j records the number of codes for the corresponding character
    for (k=hCode[i].start; k<N; k++, j++)
    // Compare the codes for the characters from start
    {
        if(code[j]==hCode[i].bit[k]) m++;
        // When it's identical to the input string, m is added by 1
    }
    if(m==j )
    // When the input string has the same length
    // as the stored encoding string
    {
        printf("%c",hTree[i].data);
        for(x=0;code[x-1]!='#';x++)
            // Delete the string used by code array
        {
            code[x]=code[x+j];
        }
    }
}
}
}
```

Some functions added for the ease of testing:

```
/*============================================================
Functionality: Output Huffman encoding table
Function input: Huffman tree、Code table for Huffman encoding
Function output:  (Code table for Huffman encoding)
Screen output: Huffman tree、Code table for Huffman encoding
=============================================================*/
void display_HuffmanCode(HTree_Node hTree[ ],HCode_Node hCode[ ])
{
    int i,k;
    printf(" Output Huffman encoding:\n");
```

```
    for (i=0;i<N;i++)
    // Output all the data in data
    {
        printf("    %c:\t",hTree[i].data);
        for (k=hCode[i].start; k<N; k++)
        // Output encodings for all the data in data
        {
            printf("%c",hCode[i].bit[k]);
        }
        printf("\n");
    }
}
/*==============================================================
Functionality: Encode the given string
Function input: Huffman tree、Code table for Huffman encoding
Function output: None
Keyboard input: The string to be encoded
Screen output: The Huffman encoding string for the input string
==============================================================*/
void edit_HuffmanCode(HTree_Node hTree[],HCode_Node hCode[])
{
    char string[MAXSIZE];
    int i,j,k;
    scanf("%s",string);
    // Save the string to be encoded into string array
    printf("Output encoding result:\n");
    for (i=0;string[i]!='#';i++)
    //# is the termination indicator
    {
        for (j=0;j<N;j++)
        {
            if(string[i]==hTree[j].data)
            // Find the same index as the input character
            {
                for (k=hCode[j].start; k<N; k++)
                {
                    printf("%c",hCode[j].bit[k]);
                }
            }
        }
    }
}
```

Test the main function

```
int main()
{
    int i;
    char str[]={'s','t','n','d'};
    int fnum[]={7,5,2,4};
    HTree_Node hTree[M];
    // Establish the struct for Huffman tree
    HCode_Node hCode[N];
    // Establish the struct for encoding table
    for (i=0;i<N;i++)
    // Save the initialized data into the HTree_Node struct
    {
        hTree[i].data=str[i];
        hTree[i].weight=fnum[i];
    }
    create_HuffmanTree(hTree);
    //Create Huffman tree
    create_HuffmanCode(hTree,hCode);
    // Create Huffman encoding table
    display_HuffmanCode(hTree,hCode);
    // Display Huffman encoding table
    printf("Please input the string to be encoded (use # for termination):\n");
    edit_HuffmanCode(hTree,hCode);
    // Encode the input string
    printf("\nPlease input the encoding (use # for termination):\n");
    decode_HuffmanCode(hTree,hCode);
    // Decode the input Huffman encoding string
}
```

1.7.3.6 Applications of Huffman tree

During the processing of many problems, a huge amount of condition checks is needed. The structural design of these checking structures directly impacts the execution efficiency of the program. Using Huffman tree, we can optimize the algorithm efficiency.

Example 1.10 Application of Huffman tree – algorithm for checking the best
Write a program to convert points on a 100-point scale into five ranks (Excellent, Good, OK, Pass, Fail) and output the result.

Solution: To convert points on a 100-point scale into five ranks, the most straightforward program implementation is to use if clauses. We can describe the checking process with binary trees, as shown in Fig. 1.88.

In reality, the distribution of grades of students is not uniform. Its discrete distribution can be approximated mathematically with normal distribution. The distribution of grades on five ranks is shown in Fig. 1.89.

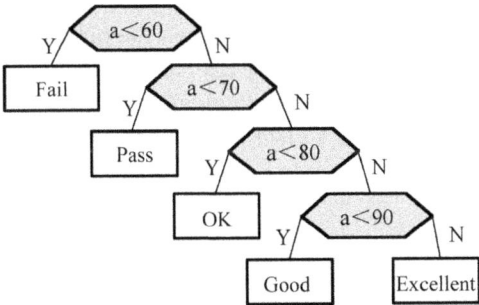

Fig. 1.88: The process of converting from 100-point scale to five ranks.

Points	0~59	60~69	70~79	80~89	90~100
Percentage	0.05	0.15	0.40	0.30	0.10

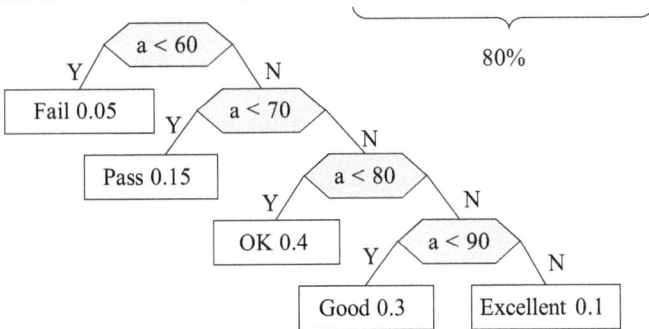

$$\text{WPL} = \underline{0.05 \times 1 + 0.15 \times 2 + 0.4 \times 3 + 0.3 \times 4 + 0.1 \times 4} = 3.15$$

Fig. 1.89: The distribution of grades and the process.

Think and Discuss Is the conversion process that processes the data from small to large efficient? **Discussion:** 80% of the data need three or more comparisons to reach the final output. **!**

According to the decision process shown in Fig. 1.89, we can obtain the below relation.

The number of comparisons needed for a conversion = The path length from root to the corresponding node

Considering the probability of each grade appearing in the five ranks, then:

The average number of comparisons needed by a grade = The WPL of the binary tree

that is WPL = $0.05 \times 1 + 0.15 \times 2 + 0.4 \times 3 + 0.3 \times 4 + 0.1 \times 4 = 3.15$

If we view the probability distribution of grades as the weights, then the process of grade conversion can be represented with a weighted binary tree. According to the construction method of Huffman trees, we should be able to obtain the most efficient decision algorithm. The improved flowchart is shown in Fig. 1.90. This time, the WPL value is 2.05.

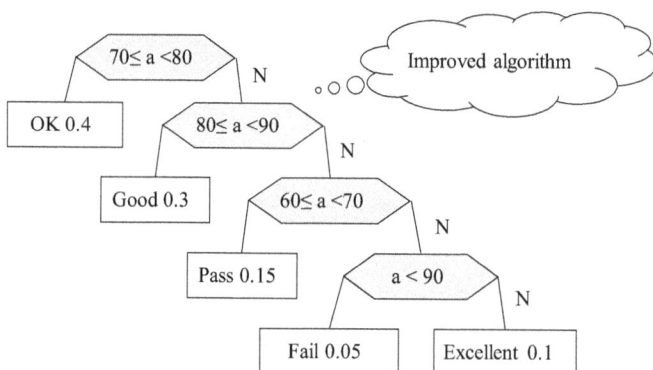

WPL=$0.4 \times 1 + 0.3 \times 2 + 0.15 \times 3 + 0.1 \times 4 + 0.05 \times 4 = 2.05$

Fig. 1.90: Improved processing flow.

Think and Discuss Does the processing efficiency really improve, when we change it according to the probability distribution?
Discussion: Although the WPL value of this tree is 2.05, which is smaller than the value 3.15 in Fig. 1.89, the number of comparisons at each node is increased. Therefore, we cannot ascertain an improvement in efficiency.

We can make an improvement by keeping the number of comparisons at each node to 1, as shown in Fig. 1.91. At this time, the WPL value is 2.2, but it is completely comparable with Fig. 1.89. We can see that the algorithm efficiency indeed increased with optimal binary tree method.

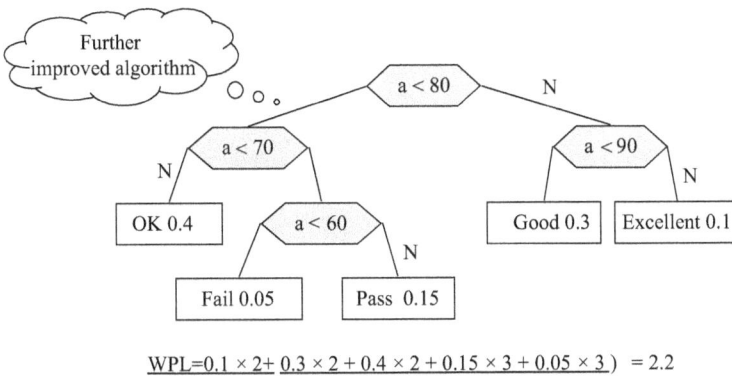

WPL=$0.1 \times 2 + 0.3 \times 2 + 0.4 \times 2 + 0.15 \times 3 + 0.05 \times 3$) = 2.2

Fig. 1.91: Improved-again processing flow.

1.8 Generalized lists

First, let us look at a concrete example. Somebody has traveled to certain places both within and outside of China. The destinations within China can be categorized by provinces, for example, he has been to Beijing; Shanghai; Nanjing and Suzhou of Jiangsu; Hangzhou and Ningbo of Zhejiang; Xi'an, Xianyang, Baoji of Shanxi. Outside of China, in Europe he has been to Paris of France and London of Great Britain. We can concisely list the information and classify it in the method listed below by adding brackets around the same group of information:

China [Beijing, Shanghai, Jiangsu (Nanjing, Suzhou), Zhejiang (Hangzhou, Ningbo), Shanxi (Xi'an, Xianyang, Baoji)]

Europe [Netherlands (Amsterdam, Eindhoven), Belgium (Brussels), France (Paris), Great Britain (London)]

The general forms are

Form 1: Country [direct-controlled municipality, province (city 1, city 2, ..., city n)]
Form 2: Continent [country (city 1, city 2, ..., city n)]

To store the above information into a computer, we need to first analyze their features. The different types of information above have different types, which are stored in the forms of combinations or inclusions. For example, we may view province as a type of node, and city another type of node. The "levels" of different nodes are different, for example, province contains city. This is different from the linear list we have learned before, where nodes are all of the same type of data. The difference lies in that it relaxes the restriction on the elements in the list, for example, it allows the elements to be other structures such as linear list, and it does not require the data elements to be of the same type. Such a linear structure that contains substructures is an extension on linear list, and it is called generalized list.

Previously, we have studied contents such as stack, queue, array and string. They are all various forms and structures derived from corresponding changes on linear lists, as shown in Fig. 1.92.

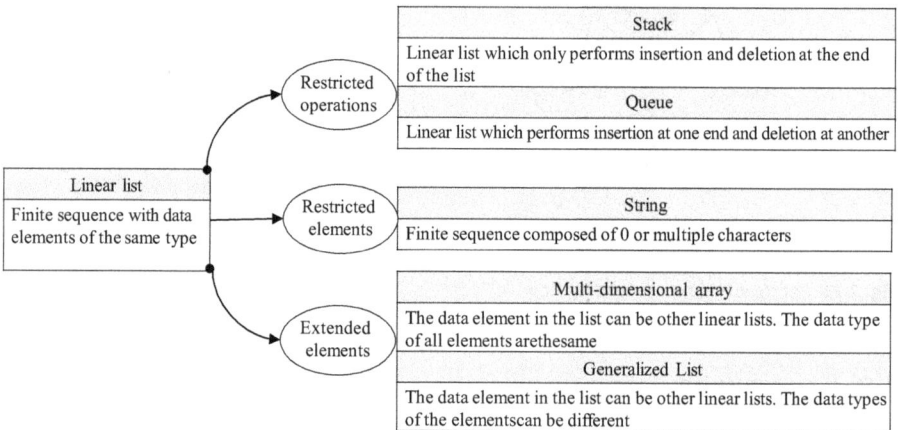

Linear list Finite sequence with data elements of the same type	Restricted operations	Stack
		Linear list which only performs insertion and deletion at the end of the list
		Queue
		Linear list which performs insertion at one end and deletion at another
	Restricted elements	String
		Finite sequence composed of 0 or multiple characters
	Extended elements	Multi-dimensional array
		The data element in the list can be other linear lists. The data type of all elements arethesame
		Generalized List
		The data element in the list can be other linear lists. The data types of the elementscan be different

Fig. 1.92: Linear list and the related forms.

A linear list is a finite sequence of n elements $a_1, a_2,\ldots, a_n$. All the elements in this sequence have the same data type and can only be atom items. A so-called atom item can be a number or a structure, and is structurally indivisible. If we relax this restriction on elements, and allow them to have complex structures on their own, then we get the concept of generalized list.

In the following, we discuss the recursive definition, storage structure, basic operations and classical applications of generalized list.

1.8.1 Definition of generalized List

Generalized list is an extension of linear list. It is also called lists. (The plural form is used in order to signify the difference from the generic name, list.)

1.8.1.1 Definition of generalized list
A generalized list is a finite sequence of n $(n \geq 0)$ data elements, denoted as:

$$LS = (a_1, a_2, \ldots, a_n)$$

Note:
1. Name of the generalized list – LS
2. Head of the list – when the generalized list LS is not empty, the first element is called the head of LS

3. Tail of the list – the generalized list consisting of all the elements in the original generalized list, except for the head
4. Length – the number of direct elements in the generalized list LS
5. Depth – the maximum number of embedded layers in the brackets of the generalized list LS. An atom has a depth of 0. An empty list has a depth of 1.
6. Singular element – a_i can be a singular data element. A singular element is also called an atom; it refers to a structurally indivisible number or structure.
7. Sublist – a_i can also be another generalized list.

Convention: In order to differentiate between atom and sublist, uppercase characters are used to represent sublist, lowercase characters are used to represent atoms.

1.8.1.2 Features of generalized list
1. Generalized list is a multilayered structure. This is because an element of a generalized list can be a sublist, and a sublist can also have other sublists as its elements.
2. A generalized list can be shared by other generalized lists. Such shared generalized list is called reentrant list.
3. A generalized list can be a recursive list. A generalized list can be its own sublist. Such generalized list is called recursive list. The depth of a recursive list is infinite, while its length is finite.
4. When a generalized list corresponds to a tree structure, this generalized list is a pure list.

Generalized list is an extension of linear list and tree. The relation between various lists is as follows:

$$\text{Linear list} \in \text{Pure List (Tree)} \in \text{Reentrant list} \in \text{Recursive list}$$

1.8.1.3 Representational methods of generalized list
1. Common representations of generalized list
There are two types of common representations of generalized lists: without name and with name. The general rules for representation can be described with syntax diagram.
    Method 1: the representation of generalized list without name is shown in Fig. 1.93.
Method 2: The representation of generalized list with name
    If we specify that any list must have a name, then, in order to indicate the name of each list as well as illustrate its composition, we can add the name of the list before each list (see Fig. 1.94). The corresponding example is given in Table 1.15.

2. The graphical representation of generalized list
If we use circle and square to represent list and singular element, respectively, and use line segments to connect a list and its elements (the element node should be

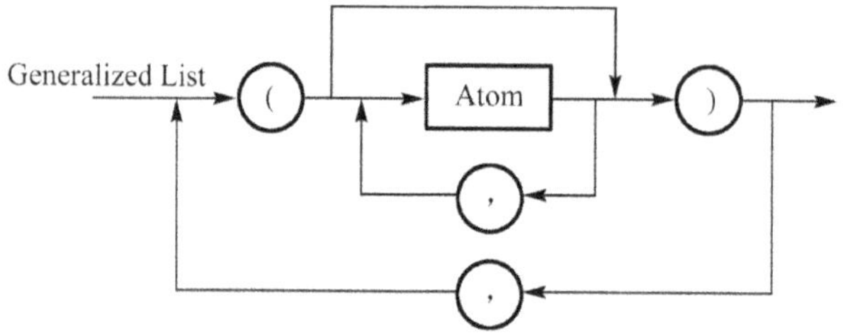

Fig. 1.93: Grammar of list l.

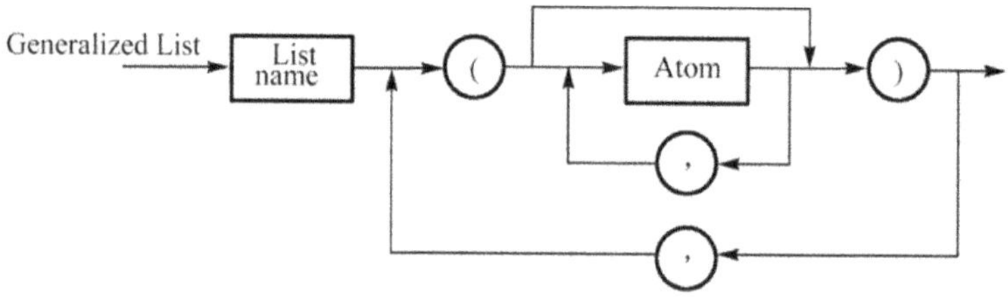

Fig. 1.94: Syntax diagram of generalized list 2.

Table 1.15: Examples of representations of generalized list.

Common representation	Equivalent representation	Representation with name	Length	Atoms	Sublists
E = ()		E()	0	None	Non
L = (a, b)		L(a, b)	2	a, b	None
A = (x, L)	A = (x, (a, b))	A(x, L(a, b))	2	x	L
B = (A, y)	B = ((x, (a, b)), y)	B(A(x, L(a, b)), y)	2	y	A
C = (A, B)	C = ((x,(a,b)),((x,(a,b)),y))	C(A(x,L(a,b)), B(A(x,L(a,b)),y))	2	None	A,B
D = (a, D)	D = (a, (a, (a, (…))))	D(a, D(a, D(…)))	2	a	D

Note: E is an empty list – the number of elements in the list is 0. L is a linear list – all the elements in the list are atoms. C is a shared list – generalized lists A and B are sublists of C, then there is no need to list out the values of sublists in C; instead we can refer to the sublists directly via their names. In actual application, we can utilize the sharing feature of generalized list to reduce data redundancy in storage structures and thus save storage space. D is a recursive list – its own sublist. Its length is 2 and its depth is ∞, making it an infinite generalized list.

below the node of its list), then, drawing figures in this way, we can see that the graphical representations of generalized list is usually a tree structure, where the root node represents the whole generalized list, the various branch nodes represent the corresponding sublist and the leaf nodes represent singular elements or empty lists. The examples of generalized lists represented by figures are shown in Fig. 1.95. This section does not discuss the two situations (d) and (e).

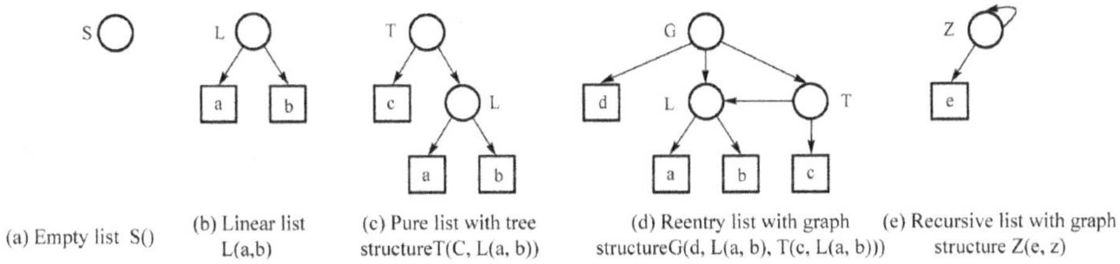

(a) Empty list S()   (b) Linear list L(a,b)   (c) Pure list with tree structure T(C, L(a, b))   (d) Reentry list with graph structure G(d, L(a, b), T(c, L(a, b)))   (e) Recursive list with graph structure Z(e, z)

Fig. 1.95: Graphical representation of generalized list.

Generalized list is a recursively defined linear structure, as it uses list again when describing a list. This is significantly different from linear list. Generalized list is also a multilayered linear structure, and we can also say it is a nonlinear structure; thus, it is appropriate for describing and solving problems about linear structures with layered features. Such a recursive definition can concisely describe huge and complex structures. In general, the structure of generalized list is very flexible. Under certain premises, it is compatible with common data structures such as linear list, array, tree and graph.

1.8.1.4 The relation between generalized list and other data structures
1.   Relation with linear list
When we restrict each item of a generalized list to be only basic element and not sublist, a generalized list regresses into a linear list: $(a_1,a_2,\ldots,a_n)$

2.   Relation with two-dimensional array
When we view each row (or each column) of the array as a sublist, the array is a generalized list:

$$((a_{11},a_{12},\ldots,a_{1n}), (a_{21},a_{22},\ldots,a_{2n}),\ldots,(a_{n1},a_{n2},\ldots,a_{nn}))$$

3.   Relation with tree
Tree can also be represented with generalized list.

1.8.1.5 Definitions of operations on generalized list
The basic operations on generalized list include creation of generalized list, output of generalized list, obtaining the length and depth of generalized list, lookup and deletion of elements from the generalized list.

1.8.2 The storage of generalized list

The data elements in a generalized list can have different structures, and a neighboring relation can be either linear or nonlinear. Under such complex circumstances, how should we design its storage structure? According to our already familiar general storage structures, there are sequential storage and linked storage. Which method would be appropriate?

> **!** **Think and Discuss**
>
> 1. How is the feasibility of employing sequential storage structure for generalized list?
>
> **Discussion:** Previously when we discussed storage of trees, we used sequential storage. This is because it is only a regular and singular nonlinear structure, that is, each node does not contain any substructure. On the contrary, elements in a generalized list can still contain substructures. This enables the neighboring relations to be either linear or nonlinear. Also, in a generalized list, it is usually the case that linear and nonlinear structures exist at the same time. We may use sequential structure to represent its exceptional cases (e.g., pure list and linear list), but it is very difficult to use it as a general storage mechanism for generalized list.
>
> 2. How is the feasibility of employing linked storage structure for generalized list?
>
> **Discussion:** The feature of linked storage method, first, is the flexibility in node assignment, second is that whenever nonlinear structure is stored, it is only needed to increase node pointer fields to extend linear structure to nonlinear structure. Such a feature fits exactly to the description requirements of such a complex structure as generalized list. It makes it easy to solve the sharing and recursion in generalized list, and therefore we usually use linked storage structure to store generalized list.
>
> 3. There are two types of nodes in the generalized list. How do we deal with them in storage?
>
> **Discussion:** The first type of node is atom and the second type of node is sublist. According to the storage principle "store the values and store the relations," we need to first clarify what is the information inherent to the node and what are the relations. The key element of an atom is the value; a sublist should include multiple pointer fields; see plan 1 of Fig. 1.96.
>
> Based on the requirements of C language on pointer types, to realize the connection of these two types of nodes in a linked list, we need to package atom nodes and sublist nodes with a unified "specification," that is, we must use one same kind of struct, so that we can fulfill the required storage structure. (The actual implementation of the storage of logical structures is closely related to the features of the language. The features of the data types of a language can determine the detailed structure used for storage.)
>
> From plan 1 in Fig. 1.96 we can see that both the type and number of the data elements of these two types of nodes are different. How can we make such different data into one unified structure? In C language, there is a "union" type, which is able to achieve what we want. We can add a field tag to distinguish the meaning of the data in the union. Then, node structure can be designed as plan 2 in Fig. 1.96.

From the discussion above we can see that, in order to employ linked storage method on the storage of generalized list, we need to construct sublist node and atom node into a unified form of node. The problem that follows would be to construct the linked list according to the connections between nodes.

List pointer	Pointer field 1	Pointer field 2	...		List node	Tag	Pointer field 1	Pointer field 2	...
	Ptr1	Ptr2				tag=1	Ptr1	Ptr2	

Atom node	Element value		Atom node	Tag	Element value
	data			tag=0	data

Structure design	Integer type	Union type

Fig. 1.96: Design plan for the node storage structure of linked storage of generalized list.

Think and Discuss How are the nodes in a generalized list connected with each other?

1. Analysis of node relations in the head–tail representation method of generalized list
   i. Analysis of node relations in the head–tail decomposition method of generalized list
   According to the definition of generalized list, if the generalized list is not empty, then we can decompose it into list head and list tail; in the opposite direction, a pair of defined head and tail can uniquely define a generalized list: Generalized list = head + tail.
   We can recursively use the head–tail representation method of generalized list, in order to decompose generalized lists into atoms eventually.
   For example: E = (a, (b, c, d)). Recursively decomposing it with head–tail representation method, we obtain:
   - Head(E) = a   Tail(E) = E1(b,c,d)
   - Head(E1) = b   Tail(E1) = E2(c,d)
   - Head(E2) = c   Tail(E2) = E3(d)

The corresponding diagram of decomposition is shown in Fig. 1.97.

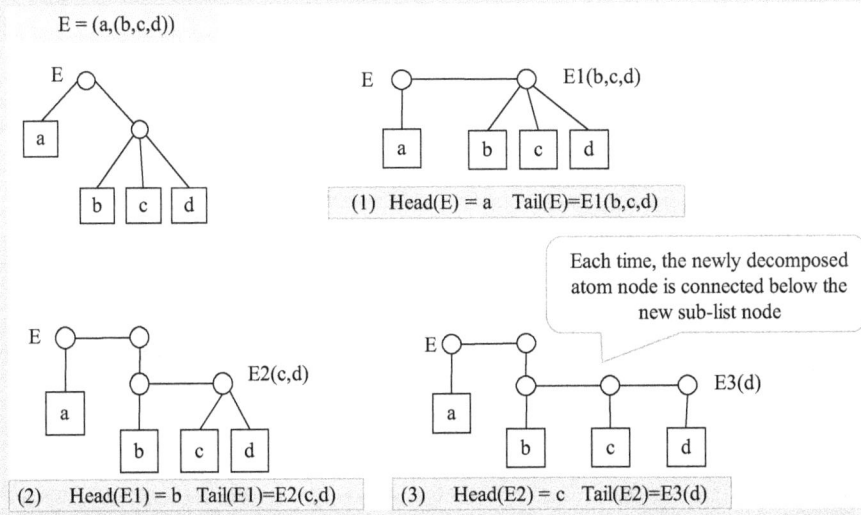

Fig. 1.97: Decomposition diagram for the head–tail representation method of generalized list.

   ii. Analysis on the relations between nodes in the sublist
   The analysis on sublist is done by decomposing a nonempty generalized list into n juxtaposed sublists, until they are decomposed into atoms:
   Generalized list = Sublist 1 + Sublist 2 + ... + Sublist n

Figure 1.98 gives the decomposition diagram of generalized list E = (a, (b, c, d)) and L = (a, (x, y), ((z))).

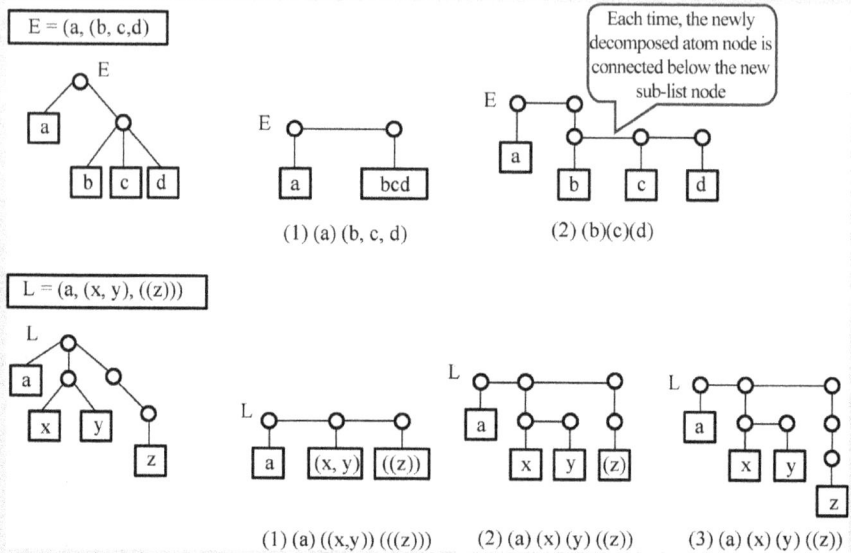

Fig. 1.98: The decomposition diagram of sublist analysis method of generalized list.

From the decomposition diagram, we can see that the two decomposition methods for head–tail representation method arrive at the same result eventually.

2. Analysis on the connections between nodes in graphical representation method of generalized list

We can first analyze the two simple forms: linear list and pure list, since the corresponding graphs for these two forms are the general forms of a tree. We can directly use the storage plan for tree, and then test whether the storage of reentrant list and recursive list can be realized using the same plan.

To store a generic tree with homogeneous structure, we need to deal with the degree of the tree, which determines the number of pointer fields in the list node. Therefore, it is not a commonly applicable method. If we convert the generic tree into a binary tree for storage, then the storage of its linked list will be done with binary linked list, which is a common solution.

From the description of the storage relations of trees and binary trees in Section 5.2.4, we know that the storage structures of a tree constructed with child–sibling representational method and that constructed as the binary linked list for the corresponding binary tree are completely the same. In the linked storage of generalized list, we usually call this storage method "child–sibling representation method," as shown in Fig. 1.99 for a concrete example.

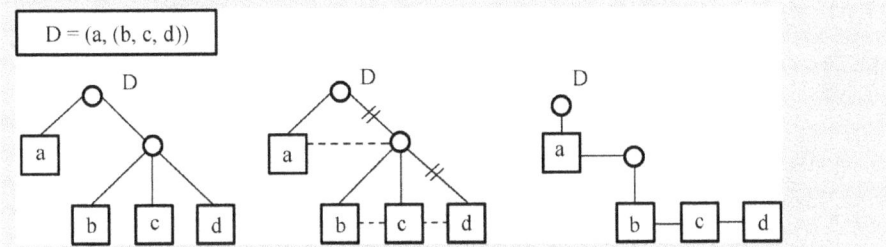

Fig. 1.99: Child–sibling representational method of generalized list.

When we compare two storage structures, they have their respective features. For the first storage structure, except for the head pointer of the empty list, which is empty, for any nonempty list, its head pointer points to one list node. From this kind of storage structure, it is easy to distinguish the level of the atom and sublist; the number of list nodes at the highest level is the length of the generalized list. Therefore, certain operations on the generalized list become relatively convenient, for example, obtaining the length and depth of the generalized list and obtaining the head and tail of the list. However, its shortcomings are also obvious: there are a lot of list nodes in the storage structures; they occupy a lot of storage space, and they do not correspond to the number of pairs of brackets in the generalized list.

The features of the second storage structure are exactly the reverse: the number of sublist nodes is low and the same as the number of pairs of brackets in the generalized list. However, it is inconvenient to write recursive algorithms on this structure.

Through the above discussion, we can see that the linked storage structure of generalized list can be divided into two different storage methods. One is called head–tail representational method, which can be further divided into "head–tail decomposition method" and "sublist decomposition method"; the other is called child–sibling representational method.

1.8.2.1 The head–tail representational method of generalized list

Through the analysis on the structure of generalized list above, we can see that, no matter "head–tail decomposition method" or "sublist decomposition method," the node structure will be the same, and needs at most two pointer fields to point to the head pointer and the node of the succeeding list, as shown in Fig. 1.100. Take the example of list node 1: its head node is atom "a," its succeeding node is list node 2; the head node of list node 2 is node 4, and its succeeding node is node 3.

Therefore, the number of pointer fields in plan 2 of Fig. 1.98 can be ascertained as 2, and the description of the definition of its form is shown in Fig. 1.101, where:

- Head pointer hPtr —— address of the first node of the sublist
- Tail pointer tPtr —— saves the address of the succeeding node on the same level (the last node on one level has the successor pointer as NULL)

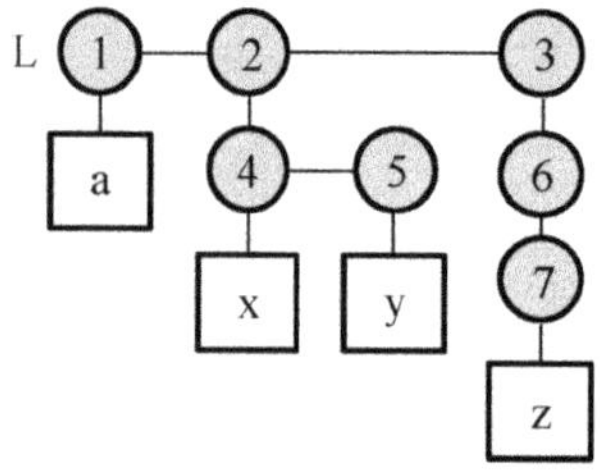

Fig. 1.100: Analysis on the pointer field of list node represented with head–tail representational method.

List node	Tag	List head pointer	List tail pointer
	tag=1	hPtr	tPtr

Atom node	Tag	Element value
	tag=0	data

Structure design	Integer type	Union type

Fig. 1.101: Node structure under head–tail representational method.

- Tag tag – tag to distinguish list node from atom node
- Value of element data – stores the value of the atom

Description of node type

```
typedef enum {ATOM,LIST} ElemTag; //ATOM==0: atom, LIST==1: sub-list
Typedef struct GLNode
{
    ElemTag tag;      // tag field
    union
    {
        AtomType data;
        struct {struct GLNode *hPtr, *tPtr;} ptr;
    };
} Glist;
```

1. Head–tail decomposition method

Example 1.11 Give the illustration of linked storage of generalized list E = (a, (b, c, d)) using head–tail decomposition method.

Solution: The decomposition process is shown in Fig. 1.97, and the final linked form is shown in Fig. 1.102.

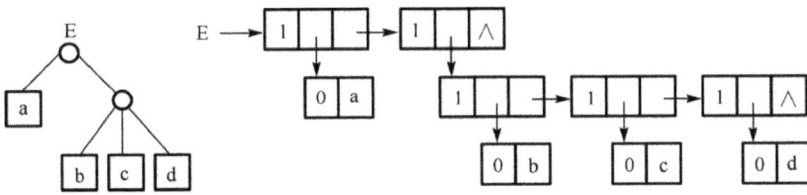

Fig. 1.102: Example of head–tail decomposition method.

2. Sublist decomposition method

Example 1.12 Sublist decomposition method: give the linked storage form illustration of generalized list L = (a, (x, y), ((z))).

Solution: The decomposition process is shown in Fig. 1.98. The final linked form is shown in Fig. 1.103.

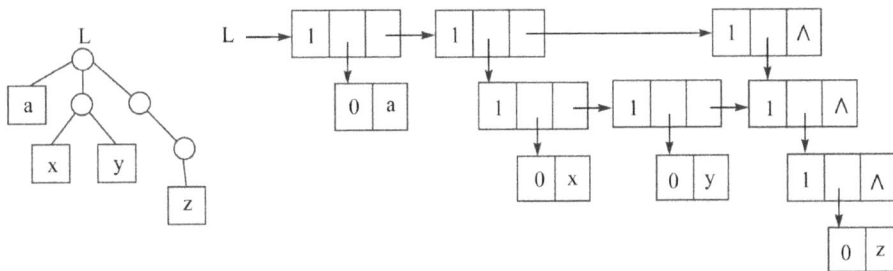

Fig. 1.103: Example of sublist decomposition method.

1.8.2.2 The child–sibling representational method of generalized list

Another representational method of generalized list is called child–sibling representational method. In child–sibling representational method, there are also two forms of node:

1. List node – used to express lists; it has one pointer to point to the child and another pointer to point to its sibling
2. Atom node – a pointer to point to the sibling and the value of this element

To distinguish these two types of nodes, we need to set a tag field in the node. If the tag is 1, then this node has child node; if the tag is 0, then this node does not have any child node. The structure form is shown in Fig. 1.104.

List node	Tag	Left child pointer	Sibling pointer
	tag=1	firstchild	sibling

Atom node	Tag	Element value	Sibling value
	tag=0	data	sibling

Structure design	Integer type	Union type	Node type pointer

Fig. 1.104: Node structure of child–sibling representational method.

```
typedef enum {ATOM, LIST} ElemTag; //ATOM=0: Singular element; LIST=1: Sub-list
typedef struct GLENode
{
    ElemTag tag;    // Tag field, used to distinguish atom node and list node
    union
    {
        // The union of atom node and list node
        AtomType data;        // The value field for atom node
        struct GLENode *firstchildPtr;    // The head pointer of list node
    } val;
        struct GLENode *siblingPtr;       // Points to the next node
} GList    // Type of generalized list
```

Example 1.13 Express the storage of the following different generalized lists with child–sibling representational method.
   Empty list: A = ()
   Linear list: L = (a, b, c, d)
   Pure list: D = (a, (b, c))
   Pure list: D = (A, B, C) = ((), (e), (a, (b, c)))
   Reentrant list: G(d, L(a, b), T(c, L(a, b)))
   Recursive list: E = (a, E)

Solution: The corresponding storage structure for each list is shown in Fig. 1.105.
From the illustration of storage structure in Fig. 1.105, we can see that, when we use child–sibling representational method, the left bracket in the expression "(" corresponds to the node with tag = 1 in the storage, and the highest level node's top field is always NULL.

The generalized list which corresponds to tree is a pure list. It restricts the sharing and recursion of components of the list; a list that allows for node sharing is called reentrant list; a list that allows recursion is called recursive list.

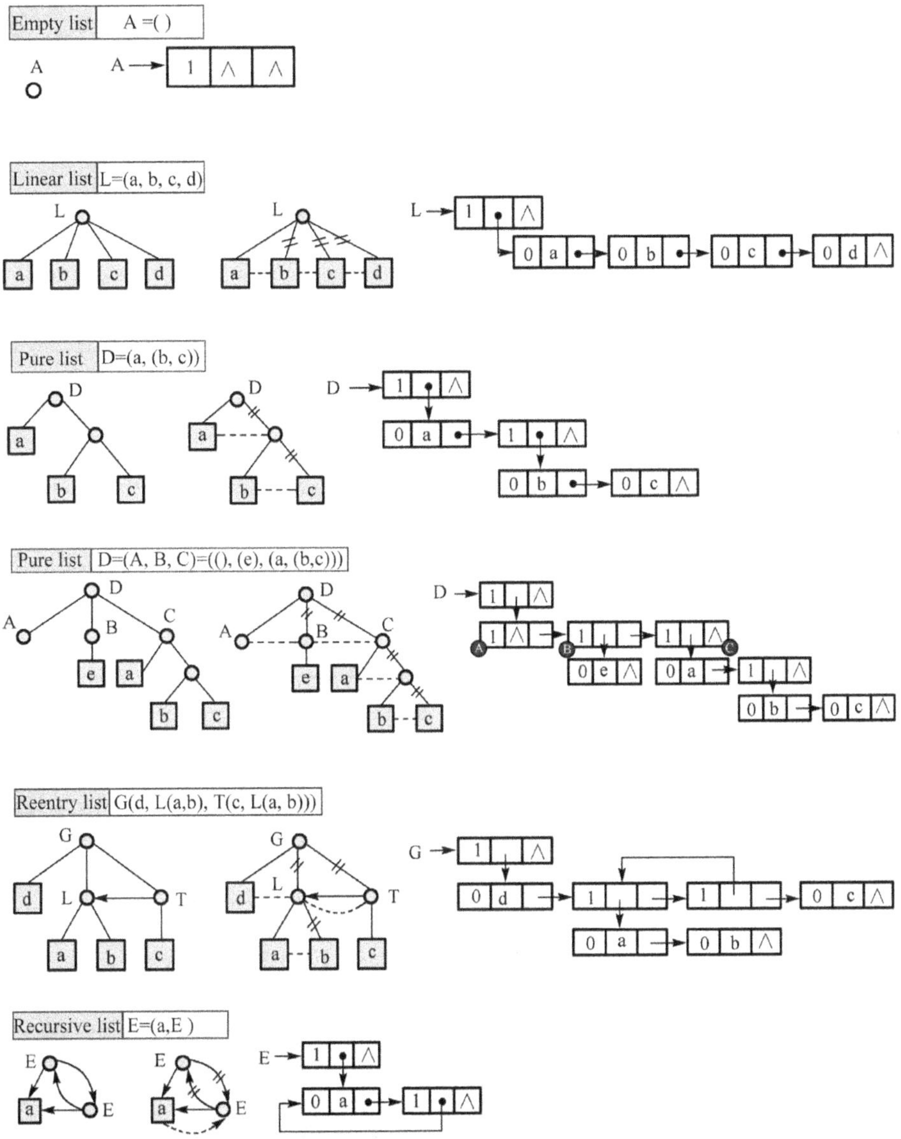

Fig. 1.105: Example of child–sibling storage of generalized list.

The relations between different lists: Linear list $\in$ Pure List (Tree) $\in$ Reentrant list $\in$ Recursive list.

1.9 Chapter summary

The connections between the main contents of this chapter are shown in Figs. 1.106 and 1.107.

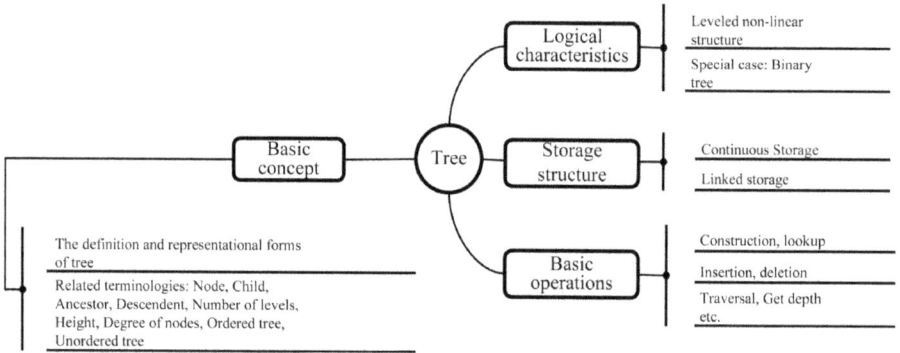

Fig. 1.106: The connections between various concepts of tree.

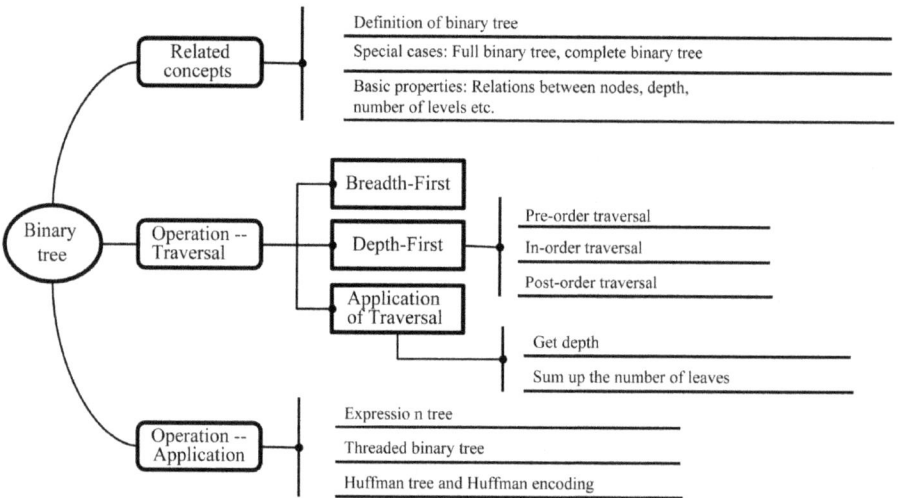

Fig. 1.107: The connections between various concepts of binary tree.

The sequential storage structure of binary tree is to store all the nodes of the binary tree according to the level order into continuous storage units (before storage it needs to be expressed as a complete binary tree). The storage structure of a tree is mostly linked storage.

In nonlinear data structure, to solve the majority of problems, one needs to start from traversal. Traversal algorithm is the core idea and foundation of all algorithms in this case. The three traversal methods "pre-order, in-order, post-order" and their variations can solve most of the problems related to binary trees.

There are three variants to depth-first traversal according to the different orders of visiting the nodes: pre-order traversal (or pre-root traversal), in-order traversal (or in-root traversal) and post-order traversal (or post-root traversal). The time complexity is $O(n)$.

Classic tree traversal algorithms are implemented with recursion. The nonrecursive depth-first traversal of binary tree is normally implemented with stack. The nonrecursive breadth-first traversal is commonly implemented with queue.

There are subtrees in a tree. They are aligned according to levels. The ordering between ancestors and descendants are strict and must not be mixed up.

The form of binary tree is the simplest. Please study its features carefully.

For the storage of nodes with sequential list, the connections are hidden in the index locations.

The binary linked list is very vivid, and the connections between branching structures are recorded in the left and right child fields.

Construct linked lists with queues. This linear structure can store nonlinear relations within.

Dequeue tree nodes one by one and add pointer fields, we obtain the binary linked list.

The structure of the tree is recursive. There are two types of plans for operations on them.

The horizontal processing based on level is based on the structure of sequential list and is bottom-up.

The vertical processing runs from the root and follows the paths; it is based on linked list and recursion.

Pre-, in-, post-order traversals run through the left and right children and the root recursively. They go through each node only once.

For the application of trees, Huffman encoding is the focus.

For a Huffman tree, the WPL will be the smallest.

The design for variable-prefix encoding method is skillful and classic.

To construct a Huffman tree, find the nodes with smallest and second smallest weights and combine them together into a new root; refine step by step.

The left branch is 0, and the right branch is 1. Please mark them well.

The 01s from root to leaf, connected together, form the encoding string for the character on the leaf.

Before transmitting the message to be sent, it needs to be written according to the encoding scheme into a sequence of 01s, and sent via wired or wireless transmission to the receiver side.

The receiver cannot understand the 10 01 strings. He/she needs to translate them via the Huffman tree, distinguish them from the root and along the paths, turn left on seeing 0 and right on seeing 1, until the character decoded on the leaf is revealed.

Trees that contain other trees are recursive trees; lists that contain other lists are generalized lists.

Generalized list has a wide range of meanings, which include various relations, both linear and nonlinear.

It is quite a feat to use head–tail method and child–sibling method to describe the complicated node connections.

There are two types of elements: atom and sublist, which are very different. It is quite a hassle to store them; we use a tag to distinguish them and the node structure design is unified via union.

(Note: In Huffman encoding, it is principally OK to either use 0 for left branch and 1 for right branch, or use 1 for left branch and 0 for right branch.)

1.10 Exercises

1.10.1 Multiple-choice questions

1. Tree is most suitable for representing ( ).
    (A) ordered data elements
    (B) unordered data elements
    (C) data with branching leveled relations between elements
    (D) data with no connection between elements

2. Binary tree is a nonlinear data structure; therefore, ( ).
    (A) it cannot be stored with sequential storage structure
    (B) it cannot be stored with linked storage structure
    (C) both sequential and linked storage structures can be used
    (D) neither sequential nor linked storage can be used

3. In the following scenarios, the ones that can be called binary trees are ( ).
    (A) tree whose each node has at most two subtrees
    (B) Huffman tree
    (C) ordered tree whose each node has two subtrees
    (D) tree whose each node has only one subtree

4. An empty tree without any node ( ).
    (A) is a tree
    (B) is a binary tree
    (C) is a tree and a binary tree
    (D) is neither a tree nor a binary tree

5. After converting a tree to a binary tree, the form of this binary tree is ( ).
    (A) unique
    (B) the one that has multiple possibilities

(C) the one that has multiple possibilities, but in all of them, the root node has no left child

(D) the one that has multiple possibilities, but in all of them, the root node has no right child

6. The depth of the binary tree is k, then the binary tree has at most ( ) nodes.
   (A)$2k$ (B)2^{k-1} (C)$2^k - 1$ (D)$2k - 1$

7. In a full binary tree with five levels, the total number of nodes is ( ).
   (A) 31   (B) 32   (C) 33   (D) 16

8. To store all the nodes in a complete binary tree left to right, level by level into an one-dimensional array R[1...N], if the node R[i] has a right child, then its right child is ( ).
   (A) R[2i–1]   (B) R[2i + 1]   (C) R[2i]   (D) R[2/i]

9. Suppose a, b are two nodes in a binary tree. During in-order traversal, the condition for a to be before b is ( ).
   (A) a is to the right of b
   (B) a is to the left of b
   (C) a is the ancestor of b
   (D) a is the descendant of b

10. From the pre-order traversal sequence and post-order traversal sequence of a binary tree, we ( ) uniquely determine this binary tree.
    (A) can   (B) cannot

11. The in-order traversal sequence of a certain binary tree is ABCDEFG, its post-order traversal sequence is BDCAFGE, then the number of nodes in its left sub-tree is ( ).
    (A) 3   (B) 2   (C) 4   (D) 5

12. If we construct a Huffman tree with {4, 5, 6, 7, 8} as the weights. Then this tree's WPL is ( ).
    (A) 67   (B) 68   (C) 69   (D) 70

13. According to the definition of binary tree, there are ( ) types of binary trees with three nodes.
    (A) 3   (B) 4   (C) 5   (D) 6

14. Starting from root, number a complete binary tree with 100 nodes left to right on each level. The number of the root node is 1, then the number of the left child of the node with number 49 is (Ê).
    (A) 98   (B) 99   (C) 50   (D) 48

15. The result of pre-order traversal on a certain binary tree is ABDEFC, the result of in-order traversal on it is DBFEAC, then the post-order traversal would yield ( ).
    (A) DBFEAC (B) DFEBCA (C) BDFECA (D) BDEFAC

1.10.2 Practical problems

1. A full k-nary tree with depth L has the below properties: The nodes on level L are all leaves, on the remaining levels, each node has k nonempty subtrees. If we number all the nodes from 1, according to level order, then:
   (1) What is the number of nodes on each level?
   (2) What is the number of the parent node (if exists) of a node numbered n?
   (3) What is the number of the ith child node (if exists) of a node numbered n?
   (4) What is the condition for the node numbered n to have a right sibling? What is the number of the right sibling?
2. We already know that a certain tree with degree m has n_1 nodes with degree 1, n_2 nodes with degree 2, ..., n_m nodes with degree m. How many leaf nodes does this tree have?
3. Write out the pre-order, in-order and post-order traversal sequences for the tree shown in Fig. 1.108.
4. Try to find all the binary trees satisfying each of the following conditions, respectively:
   (1) The pre-order traversal sequence is the same as the in-order traversal sequence
   (2) The in-order traversal sequence is the same as the post-order traversal sequence

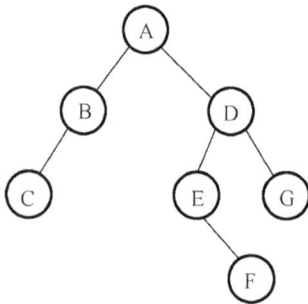

Fig. 1.108: Binary tree.

(3) The pre-order traversal sequence is the same as the post-order traversal sequence

5. Answer the following questions for the tree shown in Fig. 1.109:
   (1) Write down its binary group representation.
   (2) What are its root node, leaf nodes and branching nodes, respectively?
   (3) What are the degree and depth of node F?
   (4) What are the siblings and children of node B?
   (5) Is there any path from node B to node N? If yes, please write down the path.
   (6) Is there any path from node C to node L? If yes, please write down the path.

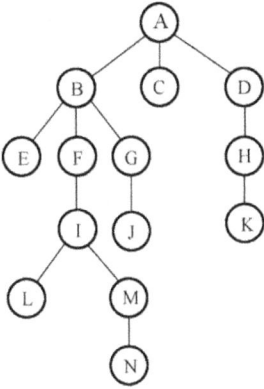

Fig. 1.109: A tree.

6. Suppose the storage structure for a certain binary tree is as shown in Fig. 1.110.

	1	2	3	4	5	6	7	8	9	10
left	0	0	2	3	7	5	8	0	10	1
data	j	h	f	d	b	a	c	e	g	i
right	0	0	0	9	4	0	0	0	0	0

Fig. 1.110: Storage structure for a binary tree.

Where *t* is the root node pointer; left and right are the left and right pointer fields of a node, respectively; data is the data field of a node. Please answer the following questions:
(1) Draw the logical structure of this binary tree.
(2) Draw the post-order threaded tree of this binary tree.

7. We already know that the in-order sequence and the post-order sequence of a certain binary tree are D, G, B, A, E, C, H, F and G, D, B, E, H, F, C, A, respectively. Draw this binary tree.

8. Suppose a text transmitted is composed of 10 different characters, the frequencies of which are 8, 21, 37, 24, 6, 18, 23, 41, 56, 14, please design the corresponding Huffman encoding for these 10 characters.
9. In Table 1.16, M and N are two nodes in a binary tree, respectively. The number of rows i = 1, 2, 3, 4 represents four kinds of correspondence between M and N, the number of columns j = 1, 2, 3 represents the ordering between M and N in pre-order, in-order and post-order traversals, respectively. Please tick the box whenever i and j have the specified relation. For example, if you think N is the ancestor of M, and N can be visited before M in in-order traversal, then tick the box (3, 2).

Table 1.16: Order of visit in different types of traversals.

	N is visited first in pre-order traversal	N is visited first in in-order traversal	N is visited first in post-order traversal
N to the right of M			
N is an ancestor of M			
N is a descendent of M			

10. Draw the sequential storage structure and the binary linked list storage structure of the binary tree in Fig. 1.111, respectively. Write down the pre-order, in-order and post-order traversal sequences of this binary tree.

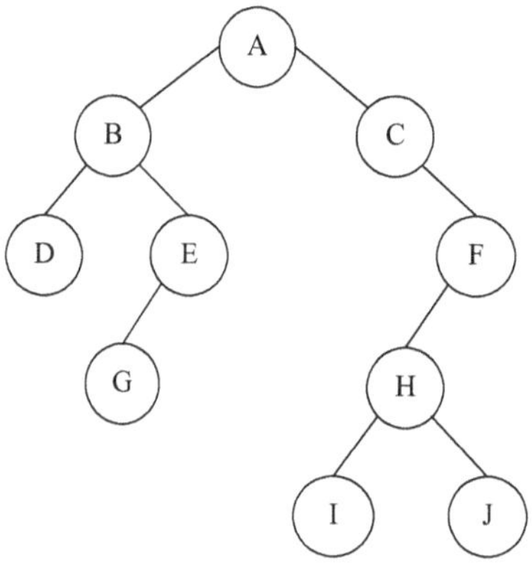

Fig. 1.111: A binary tree.

11. Suppose the corresponding binary tree to forest F is B, which has m nodes. The root of B is p. The number of nodes in the right subtree of p is n. What is the number of nodes on the first tree in the forest F?

12. Suppose a text transmitted consists of characters {a, b, c, d, e, f, g, h}. These eight characters have the possibilities of appearance in the text {0.07, 0.19, 0.02, 0.06, 0.32, 0.03, 0.21, 0.10}.
    (1) Design Huffman encoding for these eight characters.
    (2) Get the WPL of the Huffman tree.
    (3) If we use three-digit binary numbers to perform fixed-length encoding on these eight characters, then what is the ratio of the average encoding length of Huffman encoding to fixed-length encoding?

1.10.3 Algorithm design questions

1. We know a binary tree which employs binary linked list storage structure. The pointer to the storage address of the root is t. Try to write an algorithm that checks whether this binary tree is a complete binary tree.
2. We know a binary tree with binary linked list storage structure. Try to write an algorithm that interchanges all the positions of the left and right subtrees, that is, the left subtree of a node becomes the right subtree of a node and the right subtree of a node becomes the left subtree of a node.
3. Suppose the in-order and post-order traversal sequences of a binary tree are, respectively, G L D H B E I A C J F K and L G H D I E B J K F C A, complete the following operations:
    (1) Draw the illustration of the logical structure of the binary tree.
    (2) Draw the in-order threaded binary tree of the question above.
    (3) Draw the illustration of the linked storage structure of in-order threaded binary tree, and use C language to represent it.
    (4) Give the algorithm which uses in-order thread to obtain the in-order successor of a node.
4. Try to come up with an algorithm which inserts a new node x as the left child of node s into an in-order threaded binary tree, such that the original left child of s will become the left child of x.
5. Mentors of undergraduates. In the education reform of higher education institutions, many schools implemented mentor mechanism for undergraduates. The students in a class are given to several teachers, with each teacher guiding n students. If this teacher also guides postgraduates, then the postgraduates can also directly guide the undergraduates. The data elements in this problem have the following forms generally:
    (1) A mentor with postgraduates
    (Teacher, ((Postgraduate 1, (Undergraduate 1, ..., Undergraduate m1)), (Postgraduate 2, (Undergraduate 1, ..., Undergraduate m2)) ...))

(2)   A mentor without postgraduates

(Teacher, (Undergraduate 1, ..., Undergraduate m))

The natural data of a mentor only include his name and title; the natural data of a postgraduate student only include his name and class; the natural data of an undergraduate only include her name and class.

Check how many postgraduates and undergraduates a mentor guides.

2 Nonlinear structure with arbitrary logical relations between nodes – graph

Main Content
- Definition of the logical structure of graph
- The storage structure of graph
- The traversal of graph
- The applications of graph – minimal generated tree, shortest path, topological ordering and key path

Learning Aims
- Grasp the features of the logical structure of graph
- Grasp the storage structure of graph
- Grasp the traversal of graph
- Grasp the related applications of graph

!

In the data structure graph, the correspondence relations between nodes is multiple to multiple. This is also a classic structure. Broadly speaking, all kinds of nodes and their connections can be viewed as a graph.

2.1 Graph in actual problems and its abstraction

In many problems, the connection between information objects can be represented in graphs after abstraction, for example, the problem about traffic lights in the introductory chapter. One thing to notice is that the graph in data structure is not the same as the graph in an Euclidean plane, but one in topological sense. For example, a metro line map utilizes principles from typology. When we take a bus and refer to the line plan, we only need to find the start and end stations, which line to take and how many stations are in between. We do not need to pay attention to the physical route on an actual map. In the problem of querying the line map, we only care about whether there are some connections between points and points, and completely ignore their size, shape and spatial distance as concrete entities.

In topology, no matter how the size and shape of a graph change, as long as the number of points and line segments in it does not change, they are equivalent graphs, as shown in Fig. 2.1. Although the shape and size of the circle, square and triangle are different, under topological transformation, they are all equivalent shapes. There is no unbendable element in topology. The size and shape of each figure can be changed.

Let us first see some concrete examples of graph.

https://doi.org/10.1515/9783110676075-002

Fig. 2.1: Topologically equivalent graphs.

2.1.1 Introductory example of graph 1 – coloring problem of map

In 1852, Francis Griffiths who graduated from the University of London joined a research institution to study the coloring of maps. He discovered an interesting phenomenon: "It seems that each map needs to be colored in four colors, such that countries bordering each other can be painted with different colors," as shown in Fig. 2.2.

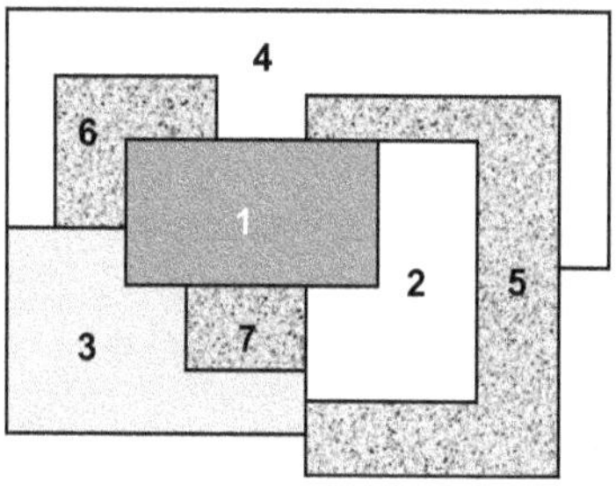

Fig. 2.2: Map coloring problem.

Can we rigorously prove this conclusion mathematically? The famous mathematicians at that time, such as De Morgan and Hamilton, could not solve this problem. In the more than 100 years that followed, a lot of mathematicians tried to prove this problem, but without success. After the advent of computers, due to the rapid increase in calculation speed, the brute-force proving of four-color conjecture was rapidly accelerated. In 1976, American mathematicians Kenneth Appel and Wolfgang Haken used 1,200 h and carried out more than 10 billion checks on two computers at UIUC, and finally proved the problem. The four-color conjecture became the four-color theorem. This result shook the world.

How should one let a computer prove the four-color conjecture? We can first try to verify that the four-color conjecture is indeed plausible. Figure 2.3 is the the region adjacency map of a certain area. The result of coloring is that we indeed only need four colors to differentiate between different adjacent regions.

Solving the problem with computer, we need to first abstract the known conditions, that is, the connection between information, and store the abstraction into the computer. In terms of map coloring, how should we carry out the abstraction and establishment of model? We can contract each region in the map into a vertex, and connect two adjacent regions with a line segment. In this way, we can abstract a region map into a topological map. The coloring is represented as giving different colors to two vertices at two ends of a line segment in the graph. Figure 2.3 can be abstracted into the region adjacency relation map shown in Fig. 2.4.

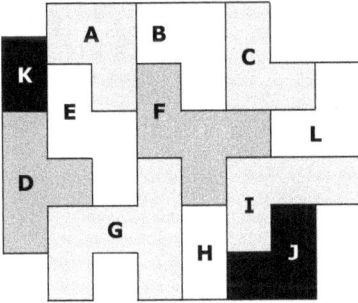

Fig. 2.3: The adjacency map.

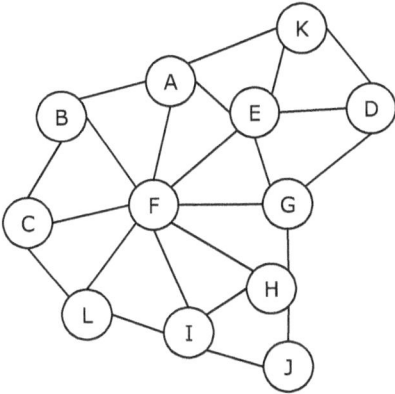

Fig. 2.4: The abstracted adjacency graph.

2.1.2 Introductory example of graph 2 – search engine and the structure of website

Nowadays, the Internet develops very rapidly. We use search engines such as Yahoo, Google and Baidu to conveniently query all kinds of desired information. However, for the developers of search engines, it is a huge challenge to effectively retrieve and organize the massive amount of information from the Internet.

The principle of search engine is to search the Internet and collect information according to a certain search strategy. It stores information in a huge database after organization and processing. After the user enters keyword to query at the search page, the search engine will find the indexes of all web pages that match this keyword from its database, and show them according to certain ranking rules. Different search engines will have different indexing databases and different ranking rules.

The software that performs search of web information is normally called "web spider" or "web crawler." A web spider traverses the content of a website according to the links within the site. The link structure inside a website refers to the topological structure of interconnections between pages, such as the one shown in Fig. 2.5. It is built upon the foundation of directory structure. We can view each page as a fixed point. A link is a line connecting two fixed points. A point can be connected with another point, and can also be connected with multiple points.

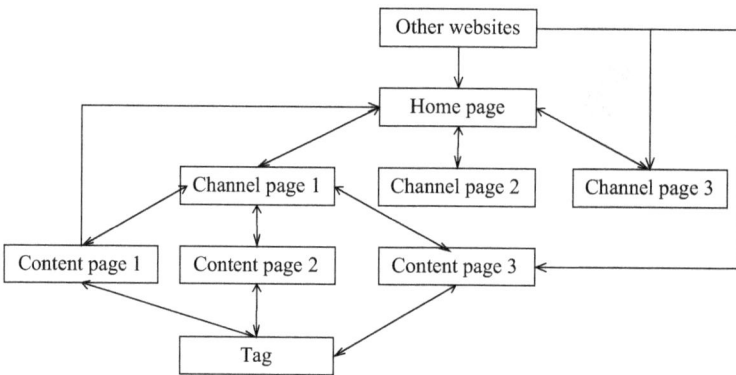

Fig. 2.5: Link structure within a website.

Links between websites

Normally, websites will give the link addresses of other related websites, as shown in Fig. 2.6(a). The connection relations between different sites can be abstracted in Fig. 2.6(b).

> **Knowledge ABC** Topological structure of computer network
>
> The topological structure of computer network is to use topological relations between points and lines that are not related to particular size/shape, to abstract the computers and communication devices in the network into points and to abstract the transmission medium into a line. The geometrical shape constituting the points and lines is the topological structure of the computer network. The topological structure of network reflects the structural relations between various entities in the network. It is the first step in constructing the computer network, and it is the foundation for implementing all kinds of network protocols (network protocols are the set of rules, standards or conventions for data exchanges in computer networks). It greatly impacts the performance, reliability and transmission cost of the network.

Links
> Ministry of Education, P. R. China
> Ministry of Foreign Affairs, P. R. China
> State Administration of Foreign
  Experts Affairs, P. R. China
> Shaanxi Administration of Foreign
  Experts Affairs
> Foreign Affairs Office, Shaanxi
  Provincial People's Government
> Xi'an Tourism China

(a) Website links

(b) Abstraction of links between websites

Fig. 2.6: Connections between websites.

2.1.3 Introductory example of graph 3 – shortest flight connection

Whenever you want to take a plane to reach a certain city but there is no direct flight. You can "detour" via other cities to reach the destination. The problem then is – there are flights between certain cities. We want to get the shortest flight connection between specific two cities. First, we need to abstract the problem model: we can use point to represent cities, and use line segment to connect two points whenever there is a direct flight between two cities, and denote the distance value besides the line. For example, in Fig. 2.7, we can treat each city as a vertex, and the edge between vertices as the direct flights between cities. Then we have:

City: Vertices = {a, b, c, d, e}

Flights: Edges between vertices = {ab, ad, bc, be, de}

Problem description: Get the shortest flight connection from d to c.

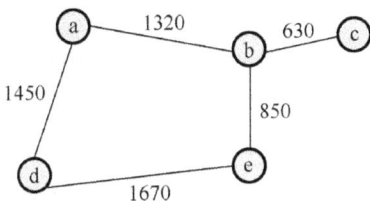

Fig. 2.7: Graph of flights.

> **!** **Think and Discuss** What features does the data structure graph, which is abstracted out of concrete problems, have?
> **Discussion:** Graph in data structures is a graph in a topological sense. Its features are:
> - Represent objects with points. The objects with direct connections are connected with lines.
> - The length and curve of the lines in the graph and the positions of the nodes are irrelevant. Each line has nodes at both ends.

2.2 The logical structure of graph

2.2.1 The definition and basic terminologies of graph

Graph is a nonlinear data structure more complex than the linear list and tree-like structure. Graph does not put any restriction on the number of predecessors and successors of a node. There can be arbitrary relations between each data element. It describes a "many-to-many" relation.

Theories related to graph are illustrated and proved in detail in graph theory section of "discrete mathematics" class. In data structures, we mainly discuss the storage and operations of graph on computers.

2.2.1.1 Definition of graph

In a graph structure, it is possible to relate any two arbitrary data elements. Therefore, to describe such a relation, we cannot use the immediate predecessor and immediate successor of linear structures. It is also inconvenient to describe it with the layer relation in tree structures. We can describe the "many-to-many" relation between the nodes in a graph with sets.

> **i** **Definition of Graph**
> A graph is a tuple < V, E >, noted as G = < V, E >, where:
>
> (1) $V = \{v_1, v_2, \ldots, v_n\}$ is a finite, nonempty collection. V_i is called a node, V is called the set of nodes.
> (2) E is a finite collection, called the set of edges. Each element in E can be regarded as a pair of vertices in V, called an edge.

> **i** **Knowledge ABC** Ordered pair
> Ordered pair is called a pair of numbers with a certain order. It is contrary to the concept of unordered pair.
> The condition for two ordered pairs to be equivalent is the equivalence of the elements that constitute the ordered pair as well as the equivalence of the order. For example, for ordered pair < a, b > and < b, a >, although they have the same elements, they have different orders; thus, they are two different ordered pairs.

Think and Discuss When the set of edges of a graph is empty, does graph G still exist? **Discussion:** It still exists, but in this case this graph has only vertices, that is, it is possible for a graph to have no edges and only vertices. This can be viewed as an exceptional case of graphs.

2.2.1.2 The representational formats of graph

A graph can be represented either with written symbols or figures. There are the two following forms:

1. Set representation: for a graph G, if we note it as G = <V, E>, and write out the set representations of V and E, then we call this a set representation of the graph.
2. Figure representation: we use circles to represent vertices in V, and use directed line segments from u to v to represent directed edge <u, v>; undirected line segments to represent undirected edge (u, v). This is called figure representation of graph.

2.2.1.3 Basic terminologies of graph

1. Undirected and directed graph

A graph can either be a directed graph or an undirected graph.

Undirected graph: If the edge from vertex V_I to V_j does not have a direction, then we call this edge an undirected edge, represented with unordered pair (V_i, V_j). If any edge between two nodes in the graph is an undirected edge, then we call this graph an undirected graph.

For example, for Fig. 2.8, $G_1 = (V_1, E_1)$, where

the set of vertices $V_1 = \{v_0, v_1, v_2, v_3, v_4\}$;

the set of edges $E_1 = \{(v_0, v_1), (v_0, v_3), (v_1, v_2), (v_1, v_4), (v_2, v_3)(v_2, v_4)\}$.

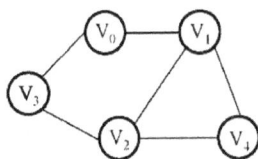

Fig. 2.8: Illustration of G_1.

Directed graph: if the edge from vertex V_i to V_j is directed, then we call this edge a directed edge, or an arc, represented with an ordered pair $<V_I, V_j>$. The first node in an ordered pair, V_i, is called the starting node (or arc tail); it is the end without the arrow in the figure. The second node in an ordered pair, V_j, is called the ending node (or arc head); it is the end with the arrow in the figure. If all the edges in a graph are directed edges, then it is a directed graph.

For example, for the directed graph G_2 in Fig. 2.9, $G_2 = (V_2, E_2)$, where

the set of vertices $V_2 = \{v_0\ v_1, v_2, v_3\}$;

the set of arcs $E_2 = \{<v_0, v_1>, <v_0, v_2>, <v_2, v_3>, <v_3, v_0>\}$.

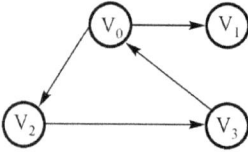

Fig. 2.9: Illustration of G_2.

Sparse graph and dense graph: a graph with very few edges/arcs (e.g., $e < n \log_2 n$, where n is the number of vertices and e is the number of arcs) is called a sparse graph. Otherwise, it is called a dense graph.

2. Vertex and edge

Adjacent vertices: the two vertices of an edge

Incident edge: If an edge $e = (v, u)$, then we say vertices v and u are incident to the edge e.

3. Weight and network

Weight of an edge: the data information relevant to an edge/arc is attached to this edge/arc of the graph.

Network: a graph with weights on its edges/arcs is called a network.

For example, the flight graph in Fig. 2.7 is a network. The number of kilometers at each edge of the graph is the weight value.

4. The degree, indegree and outdegree of a vertex

In an undirected graph:

The degree of vertex V = The number of incident edges

In a directed graph:

The outdegree of vertex V = The directed edges starting from V

The indegree of vertex V = The directed edges ending at V

The degree of vertex V = The indegree of V + The outdegree of V

5. Path and cycle

Path: in a graph, if we start from vertex v_1 and reach vertex v_n alongside certain edges (or arcs), through vertices $v_1, v_2, \ldots, v_{n-1}$, then we call this sequence of vertices $(v_1, v_2, v_3, \ldots, v_{n-1}, v_n)$ as a path from v_1 to v_n.

Cycle: If the first vertex on the path v_1 overlaps with the last vertex v_m, then we call such a path cycle or ring.

For example, in Fig. 2.8:

V_0, V_1, V_2, V_3 is a path from V_0 to V_3;

V_0, V_1, V_2, V_3, V_0 is a cycle.

In Fig. 2.9:

V_0, V_2, V_3 is a path from V_0 to V_3;

V_0, V_2, V_3, V_0 is a cycle.

6. Subgraph

Suppose that there are two graphs G = (V, E) and G_1 = (V_1, E_1). If $V_1 \subseteq V$, $E_1 \subseteq E$, and all the nodes incident to E_1 are in V_1, then we say G_1 is a subgraph of G.

For example, in Fig. 2.10, (b) and (c) are subgraphs of (a).

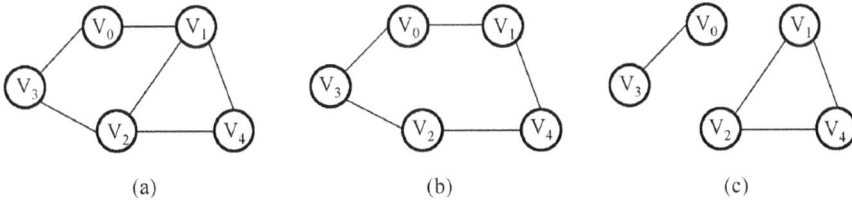

(a)   (b)   (c)

Fig. 2.10: Graph and subgraph.

7. Connected graph and strongly connected graph
    – Connected graph and disconnected graph

In a nondirected graph, if there is a path from vertex V_i to V_j, then we say V_i and V_j are connected. If any two vertices in the graph are connected, then it is called a connected graph. Otherwise, we call this graph a disconnected graph.

For example, in Fig. 2.11, (a) and (b) are undirected graphs, where (a) is a connected graph and (b) is a disconnected graph.

    – Connected component

A maximal connected subgraph of an undirected graph G is called a connected component of G. A connected graph has only one connected component, which is itself. A disconnected undirected graph can have multiple connected components.

    – Strongly connected graph

In a directed graph, if for each pair of nodes v_i and v_j, there is a path from v_i to v_j as well as a path from v_j to v_i, then we call this graph a strongly connected graph.

– Strongly connected component

A maximal strongly connected subgraph of a nonstrongly connected graph is a strongly connected graph that has only one strongly connected component, which is itself. A nonstrongly connected directly graph can have multiple strongly connected components.

For example, in Fig. 2.11, (c) and (d) are directed graphs, where (c) is a strongly connected graph, (d) is a nonstrongly connected graph,

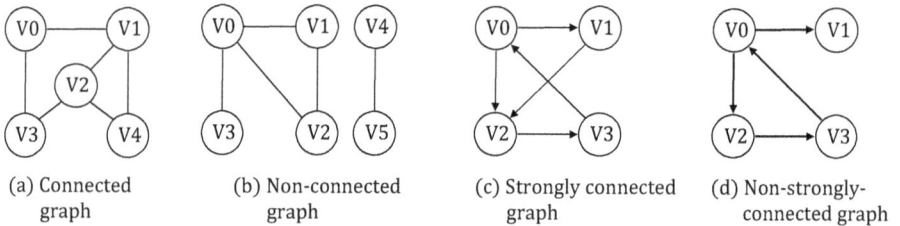

(a) Connected graph   (b) Non-connected graph   (c) Strongly connected graph   (d) Non-strongly-connected graph

Fig. 2.11: Connected graph and strongly connected graph.

The subgraph with nodes (V_0, V_2, V_3) is one of the strongly connected components.

8. Spanning tree

The minimal connected subgraph that includes all the nodes in the undirected graph G is called the spanning tree of G. The so-called minimal connected subgraph refers to the connected graph with all the vertices of the original graph and the least amount of edges. T is a spanning tree of G if and only if T satisfies the following conditions:
– T is a connected subgraph of G;
– T contains all the vertices of G;
– T does not contain any cycles.

For example, there can be multiple possible spanning trees for the graph in Fig. 2.12.

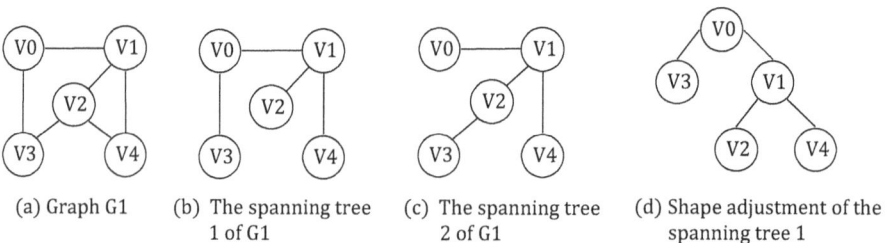

(a) Graph G1   (b) The spanning tree 1 of G1   (c) The spanning tree 2 of G1   (d) Shape adjustment of the spanning tree 1

Fig. 2.12: Examples of spanning trees.

2.2.2 Definitions of operations on graph

According to the structural characteristics of graph, the basic operations on graph include:
- creation of graph structure;
- traversal of graph;
- operations on vertices: lookup, value extraction, value assignment, insertion and deletion; and
- operations on edges: insertion, deletion, access or modification of weights.

2.3 The storage structure for graph and its implementations

When we discussed the storage structures for tree and linear lists, there were two types of storage: continuous storage and discrete storage, which were implemented with linear list and linked list concretely. What plan to use for the storage of graph? We can first test out the existing storage methods to see whether they are appropriate.

> **Think and Discuss**
>
> Question 1: Can we also use sequential storage structure on graph?
>
> **Discussion:** There is no ascertained relation between nodes in a graph (no root, no starting node, end node), and there could be a connection between any two vertices. Therefore, we cannot use sequential structure to store both node data and the relation between the nodes. However, if we store the information about vertices and edges separately into separate sequential structures, then this should be possible.
>
> Question 2: What information do we need to save to store a graph?
>
> **Discussion:** According to the storage principle of "store the data as well as the relations" and according to the definition of graph, a graph constitutes vertices and edges. Therefore, besides storing information of vertices, we also need to store the information of edges.

Due to the arbitrariness of the graph structure, there are multiple corresponding storage/representational methods. As long as the method corresponds to storage principles, it should be OK. Common storage structures for graph include adjacency matrix, adjacency list, cross-linked list and adjacency multilist.

There are some variables shared across various storage methods of graph in this book. Here we illustrate them together. The related types and common variables are given in Table 2.1.

Table 2.1: The related data types and common variables for graph.

	Type	Variable name	Number	Maximum number
Vertex	VexType	vertex	VERTEX_NUM	VERTEX_MAX_NUM
Adjacent Vertex	VexType	adjvex		
Edge	InfoType	edge		EDGE_MAX_NUM
Arc	InfoType	arc		

```
#define VERTEX_NUM 6        //The number of vertices of the graph
#define VERTEX_MAX_NUM 64 // The maximum number of vertices of the graph
#define EDGE_MAX_NUM 20    // The maximum number of edges of the graph
typedef char VexType;      // Data type of vertex
typedef int InfoType;      // Data type of edge, can contain information
                   // such as weight value, the existence/absence of edge etc.
```

2.3.1 Array representational method of graph 1 – adjacency matrix

From the definition of graph, we can see that the logical structure of a graph is divided into two parts: the set of vertices and the set of edges. A vertex is an element; therefore, we can use an one-dimensional array to store data for vertices. An edge is represented with two vertices incident to it, that is, an edge data element can have two data fields. Therefore, we can use a matrix to store the adjacency relations between vertices (i.e., edge or arc). This two-dimensional array is also called adjacency matrix. If two vertices are related, that is, there is an edge between them, then we can denote the value of the entry in the matrix as 1; otherwise, it would be 0.

Adjacency matrix is further divided into the type for directed graph and the type for undirected graph.

2.3.1.1 Definition of adjacency matrix

Suppose that the graph $G = (V, E)$ has n vertices, then the adjacency matrix AdjMatrix is an $n \times n$ square matrix, defined as

$$AdjMatrix[i, j] = \begin{cases} 1 & if (v_i, v_j) \in E \ (or < v_i, v_j > \in E) \\ 0 & Otherwise \end{cases}$$

When the graph has weight values, we can store the weights directly in the two-dimensional array:

$$AdjMatrix[i, j] = \begin{cases} w_{ij} & if (v_i, v_j) \in E \\ \infty & Otherwise \end{cases}$$

1. Adjacency matrix for undirected graph
In Fig. 2.13, the matrix element AdjMatrix[i, j] = 1 indicates that there is an edge (V_i, V_j) in the graph, and AdjMatrix[i, j] = 0 indicates that there is no such edge (V_i, V_j).

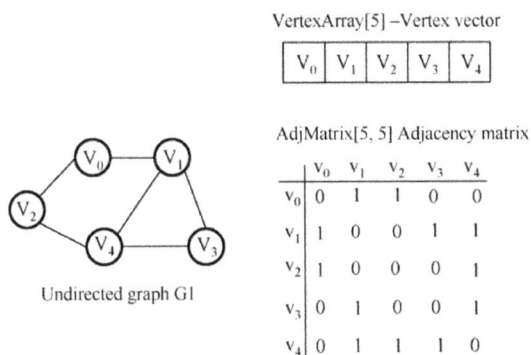

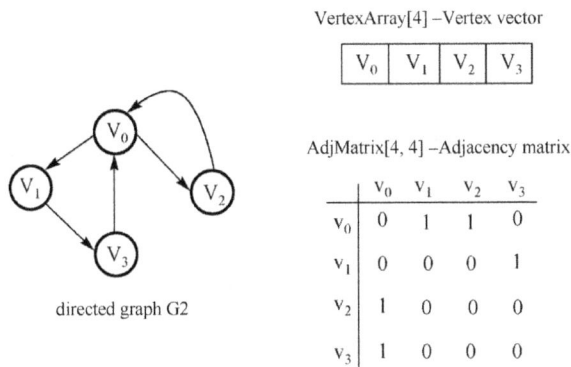

VertexArray[5] –Vertex vector

V_0	V_1	V_2	V_3	V_4

AdjMatrix[5, 5] Adjacency matrix

	V_0	V_1	V_2	V_3	V_4
V_0	0	1	1	0	0
V_1	1	0	0	1	1
V_2	1	0	0	0	1
V_3	0	1	0	0	1
V_4	0	1	1	1	0

Undirected graph G1

Fig. 2.13: Adjacency matrix for undirected graph.

Characteristics of adjacency matrix for undirected graph:
- The adjacency matrix for undirected graph is symmetric and can be stored with compression.
- The degree of vertex v_i in an undirected graph is the number of 1s in the ith row of the adjacency matrix.

2. Adjacency matrix for directed graph
In Fig. 2.14, the matrix element AdjMatrix[i, j] = 1 indicates that there is an arc $<V_i, V_j>$ in the graph, and AdjMatrix[i, j] = 0 indicates that there is no such arc $<V_i, V_j>$.

VertexArray[4] –Vertex vector

V_0	V_1	V_2	V_3

AdjMatrix[4, 4] –Adjacency matrix

	V_0	V_1	V_2	V_3
V_0	0	1	1	0
V_1	0	0	0	1
V_2	1	0	0	0
V_3	1	0	0	0

directed graph G2

Fig. 2.14: Adjacency matrix for directed graph.

Characteristics of adjacency matrix for directed graph:
- The adjacency matrix for directed graph is not necessarily symmetric
- In the directed graph:
    - The outdegree of the vertex v_i is the number of 1s in the ith row of the adjacency matrix;
    - The indegree of the vertex v_i is the number of 1s in the ith column of the adjacency matrix.

3. Adjacency matrix of network

In Fig. 2.15, the weights are stored at the entries of the adjacency matrix where there is an arc. The symbol "∞" indicates that there is no path.

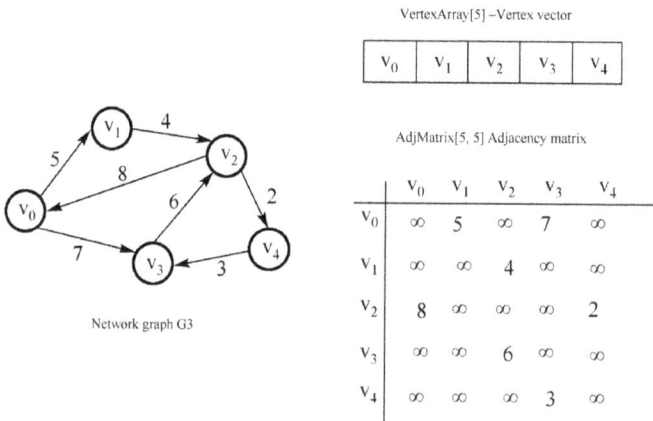

VertexArray[5] –Vertex vector

V_0	V_1	V_2	V_3	V_4

AdjMatrix[5, 5] Adjacency matrix

	V_0	V_1	V_2	V_3	V_4
V_0	∞	5	∞	7	∞
V_1	∞	∞	4	∞	∞
V_2	8	∞	∞	∞	2
V_3	∞	∞	6	∞	∞
V_4	∞	∞	∞	3	∞

Network graph G3

Fig. 2.15: Adjacency matrix for network.

2.3.1.2 The data structure description of adjacency matrix

```
// Adjacency matrix - AM
typedef struct
{   VexType VertexArray[VERTEX_NUM];        // Array for vertices
    InfoType AdjMatrix [VERTEX_NUM][ VERTEX_NUM];        // Adjacency matrix
} AM_Graph;
```

2.3.1.3 Complexity analysis of adjacency matrix

Using adjacency to represent graph is straightforward, convenient and facilitates computations.

1. Time complexity

Edge lookup: for the lookup of whether there exists an edge between any two vertices i and j in the graph, and the weight value on the edge, we can perform random access lookup according to the values of i and j. The time complexity is $O(1)$.

Computation of vertex degree: to compute the degree (or indegree/outdegree) of a vertex and its adjacent vertices, the time complexity is $O(n)$.

2. Space complexity: $O(n^2)$
If we use it to represent a sparse graph, then we might cause relatively huge space wastage.

2.3.2 Array representational method of graph 2 – array for set of edges

2.3.2.1 Design of edge set array
Edge set array only stores the information of all the edges (or arcs) of the graph. It stores the starting point, ending point (for undirected graph, we can choose any end of the edge as the starting point or the ending point) and related information on edges (e.g., weights). The order of edges in the array can be arbitrary, or it can be arranged according to the specific needs. An example of edge set array is shown in Fig. 2.16. Note that in the undirected graph G5, the contents stored in the edge set array are ordered ascendingly based on weights to facilitate computation. If we need to store information about vertices, we need a one-dimensional array with VERTEX_NUM elements.

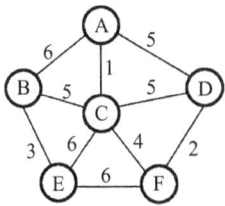

Vertex array: VertexArray[6]

A	B	C	D	E	F

Edge Set array: Edge Set[ ]

Start	A	D	B	C	A	B	C	A	C	E
End	C	F	E	F	D	C	D	B	E	F
Weight	1	2	3	4	5	5	5	6	6	6

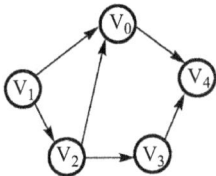

Undirected graph G5

Vertex array: VertexArray[5]

v0	v1	v2	v3	v4

Edge Set array: Edge Set[ ]

Start	v0	v1	v1	v2	v3	v2
End	v4	v0	v2	v3	v4	v0

directed graph G6

Fig. 2.16: Edge set array for graph.

2.3.2.2 Data structures description of edge set array

```
typedef struct        // Unit structure of edge set array
{  VexType start_vex;        // Starting point
   VexType end_vex;        // Ending point
   InfoType weight;        // We can set the weight item according to our needs
} EdgeStruct;
EdgeStruct Edge Set[EDGE_MAX_NUM];        // Edge set array
```

2.3.2.3 Complexity analysis for edge set array

The number of elements in the edge set array is bigger than the number of edges in the graph.

1. Time complexity

Calculation of vertex degree: if there are e edges and n vertices in a graph, we need to scan the whole array in order to lookup one edge or the degree of one vertex; thus, the time complexity is $O(e)$.

Lookup of vertex of edge: we need to scan the whole array to lookup one edge or the degree of one vertex, thus the time complexity is $O(e)$.

2. Space complexity

For the edge set array to represent a graph, it needs an edge array and a vertex array; thus, its space complexity is $O(n + e)$. In terms of space complexity, edge set array is more suitable in representing sparse graphs.

2.3.3 The linked list representational method of graph

2.3.3.1 The storage structure design of adjacency list

The arrangement of adjacency matrix is straightforward and easy to understand. However, if the graph structure needs to be dynamically generated in the process of problem solving, then whenever we add/delete a vertex, we need to change the size of the adjacency matrix. This is apparently very inefficient. Besides, the number of storage units occupied by adjacency matrix is only related to the number of vertices in the graph, and is unrelated to the number of edges (arcs). For sparse graphs with relatively fewer edges compared with vertices, this storage structure wastes relatively much space.

> **!** **Think and Discuss** What kind of storage method might save space when used on sparse graph?
> **Discussion:** The edge in the graph consists of each vertex V_i and its adjacent vertices. All adjacent vertices to each vertex V_i constitute a linear list. Considering the general case for graphs, for any vertex V_i, the number of adjacent vertices is uncertain. Therefore, we can choose singly

linked list for storage, that is, construct a singly linked list for each vertex (when there are n vertices, then we construct n singly linked lists). The nodes in the ith singly linked list contain all the adjacent nodes of the vertex V_i.

The remaining problem is how to organize these n linked lists together to facilitate the management.

In our experience, to manage n singly linked lists concurrently is to use "linked list representational method with row vector." This storage structure also applies to the storage of graph and is called adjacency list.

Besides adjacency list, there are cross-linked list and adjacency multilist. Those two representational methods are relatively complex, and we do not describe them in detail here.

Adjacency list is a linked storage structure for graph. An adjacency list consists of two parts: the head pointer list and the adjacency nodes linked list.

1. Adjacency list for undirected graph
As shown in Fig. 2.17, adjacency nodes store the index for vertices in the array.

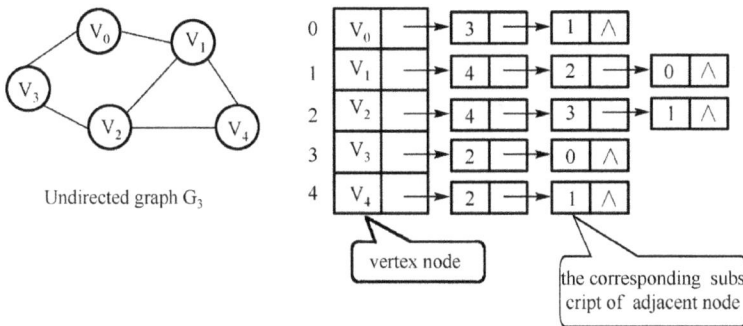

Fig. 2.17: Adjacency list for undirected graph.

Characteristics for adjacency list of undirected graph:
- If there are n vertices and e edges in the undirected graph, then its adjacency list needs n head nodes and $2e$ adjacency list nodes. It is suitable to store sparse graph.
- The degree of vertex V_i in the undirected graph is the number of nodes in the ith singly linked list.

For an undirected graph, using adjacency list for storage can still generate data redundancy. When the linked list pointed to a head node V_1 has an adjacent node pointing to V_4, the linked list pointed to a head node V_4 will also have an adjacent node pointing to V_1.

2. Adjacency list for directed graph (out-edge list)
Figure 2.18 is an adjacency list for directed graph.

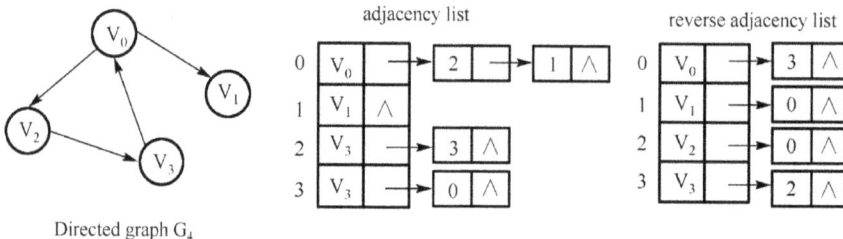

Fig. 2.18: Adjacency list and reverse adjacency list for directed graph.

Characteristics of adjacency list for directed graph:
- The outdegree of vertex V_i is the number of nodes in the ith singly linked list.
- The indegree of vertex V_i is the number of nodes in the whole singly linked list which have the value of their adjacency node field as i.
- It is easy to find the outdegree but hard to find the indegree.

3. Reverse adjacency list (in-edge list)
Sometimes, in order to ascertain the indegree of a vertex or the arcs that have the vertex as the arc head, we can construct a reverse adjacency list of the directed graph, as shown in Fig. 2.18. In this case, we can easily compute the indegree of a certain vertex, and it would be easy to check whether there is an arc between two vertices.

Characteristics of reverse adjacency list:
- The indegree of vertex V_i is the number of nodes in the ith singly linked list.
- The outdegree of vertex V_i is the number of nodes in the whole singly linked list which have the value of their adjacent node field as i.
- It is easy to find the indegree but hard to find the outdegree.

4. Weighted adjacency list
See Fig. 2.19. The data structure is basically the same as normal adjacency list. It just adds a data field in adjacent node in order to record the weights of the edges.

2.3.3.2 The description and construction of data structure of adjacency list
According to Fig. 2.20, we can write the definition of the data structure of adjacency list. The place with asterisk mark indicates that the corresponding variable is a pointer.

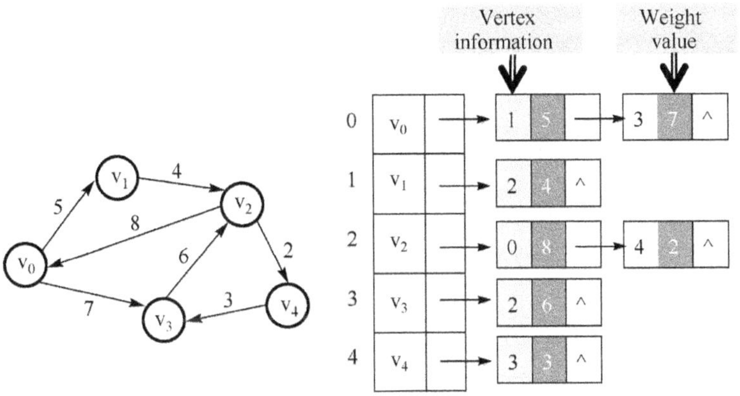

Fig. 2.19: Adjacency list for weighted graph.

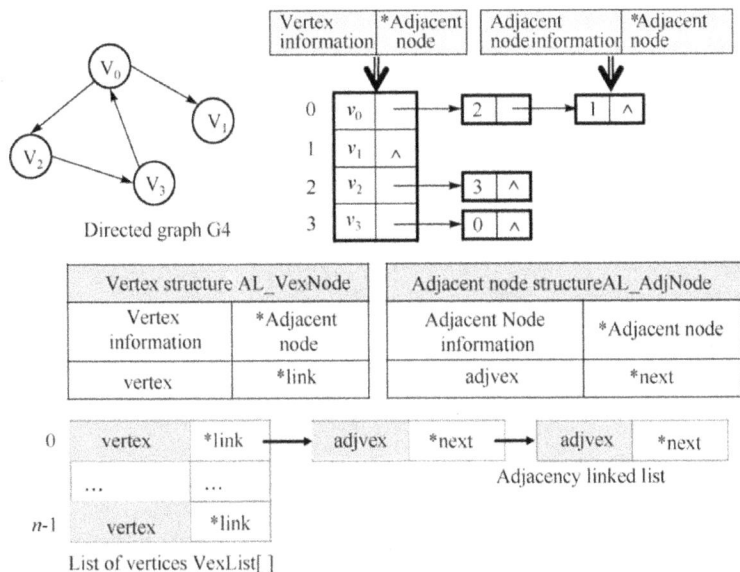

Fig. 2.20: The illustration of the storage structure of adjacency list.

```
// Adjacency List—AL
typedef struct AdjNode      // Structure of adjacent node
{
    int adjvex;             // Adjacent vertex
    AdjNode *next;          // Pointer for adjacent node
} AL_AdjNode;
typedef struct      // Structure for vertex node in the adjacency list
```

```
{
   VexType vertex;       // Vertex
   AdjNode *link;        // Head pointer for the adjacent node
} AL_VexNode;
typedef struct      // The structure for the adjacency list in general
{
   AL_VexNode VexList[VERTEX_MAX_NUM];      // Vertex list
   int VexNum, ArcNum;      // The number of vertices, the number of arcs (edges)
} AL_Graph;
#include <stdio.h>
#include <stdlib.h>

#define TRUE 1
#define FALSE 0
#define N 8      // The number of vertices
typedef int VexType;
int AdjMatrix[N][N];
// Adjacency matrix. The corresponding figure is shown in Fig. 2.21.
{{0,1,1,0,0,0,0,0},
 {1,0,0,1,1,0,0,0},
 {1,0,0,0,0,1,1,0},
 {0,1,0,0,0,0,0,1},
 {0,1,0,0,0,0,0,1},
 {0,0,1,0,0,0,1,0},
 {0,0,1,0,0,1,0,0},
 {0,0,0,1,1,0,0,0}};

typedef struct AdjNode      // Structure for adjacent node
{
   int adjvex;               // Adjacent vertex
   AdjNode *next;            // Pointer for adjacent node
} AL_AdjNode;

typedef struct      // Structure for vertex node in adjacency list
{
   VexType vertex;       // Vertex
   AdjNode *link;        // Head pointer for the adjacent node
} AL_VexNode;
/*=========================================
Functionality: Construct an adjacency list
Function input: None
```

```
Function output: None
Shared data: The adjacency matrix of the graph
==========================================*/
void Create_AdjList()
{
  AL_VexNode VexList[N]={0,NULL};      // Vertex list
  int j;
  AL_AdjNode *Ptr,*nextPtr;

  for(int i=0; i<N; i++)
  {
    VexList[i].vertex=i;
    VexList[i].link=NULL;
    j=0;
    while(j<N)
    {
      if (AdjMatrix[i][j]!=0)    // Has adjacent vertex
      {
        Ptr=(AL_AdjNode*)malloc(sizeof(AL_AdjNode));
        Ptr->adjvex=j;
        Ptr->next=NULL;
        if (VexList[i].link==NULL)
        // Add the adjacent vertex for the first time
        {
          VexList[i].link=Ptr;
          nextPtr=Ptr;
        }
        else
        {
          nextPtr->next=Ptr;
          nextPtr=Ptr;
        }
      }
      j++;
    }
  }
}
int main()
{
  Create_AdjList();
  return 0;
}
```

Test result: Using Fig. 2.21 as the test sample, the test result is shown in Fig. 2.22. Note that the number of the node in the figure has a difference of 1 with the subscript in the test result.

Fig. 2.21: Testing of adjacency list.

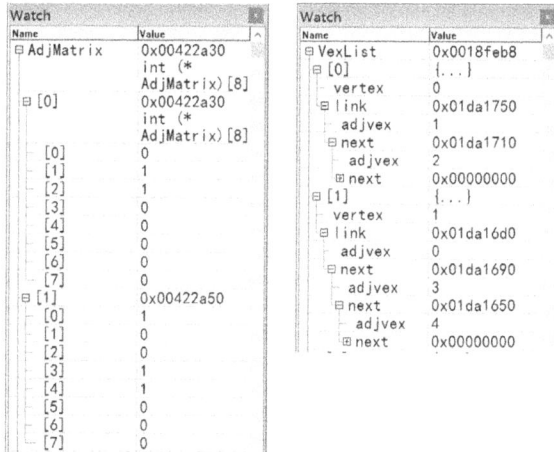

Fig. 2.22: Testing result of adjacency list.

2.3.3.3 Analysis of space complexity of adjacency list
In the adjacency list/reverse adjacency list representations of graph, the vector of head list needs to occupy the storage space for n head nodes. All edge nodes need to occupy $2e$ (undirected graph) or e (directed graph) edge node spaces. Therefore, its space complexity is $O(n + e)$.

The adjacency list relatively saves storage space for representing sparse graph, since it needs only relatively few edge nodes. If it is used to represent dense graph, then it will occupy relatively more storage space, and the lookup time for vertex will also increase.

2.3.4 Summary and comparison of various storage structures of graph

2.3.4.1 Storage structures focusing on vertices
There are certain correspondence relations between the adjacency list representation and the adjacency matrix representation of graph, although they are different methods:
1. The singly linked list for each vertex v_i in the adjacency list corresponds to the ith row of the adjacency matrix.
2. The whole adjacency list can be seen as the linked storage of adjacency matrix with row pointer vector.
3. The whole reverse adjacency list can be seen as the linked storage of adjacency matrix with column pointer vector.

Adjacency matrix and adjacency list are suitable to both undirected graph and directed graph.

2.3.4.2 Storage structures focusing on edges
Storage with cross-linked list: used for directed graph. The nodes for each arc only appear once. The outgoing arc and incoming arc of the vertices are stored with two separate linked lists.

Adjacency multilist: used for undirected graph. The incident edge (u, v) for the vertices is connected, respectively, to the two edge linked lists with u as the starting point and v as the end point. This is convenient for the processing of edge information of undirected graph.

2.3.4.3 Selection criteria for the storage structure of graph
The various storage and representational methods for graph, for example, adjacency matrix, edge set array, adjacency list, cross-linked list, adjacency multilist, have their pros and cons. In actual application, we need to select based on the density of the graph and the requirements of the calculation.

Adjacency matrix is a solid storage structure for graph. But for graphs whose number of sides is relatively fewer than the number of vertices, this storage method wastes a lot of space. In this case, we can use adjacency list for storage. Adjacency list is a storage method, which combines arrays and linked lists. Edge set array and adjacency list are suitable for representing sparse graph. Adjacency matrix is suitable for representing dense graph.

For directed graph, the adjacency list only facilitates obtaining outdegrees. If we want to get the indegrees, we must traverse the whole graph. On the contrary, reverse adjacency list makes it easy to obtain the indegrees, but we cannot easily obtain the outdegrees. Cross-linked list is a storage form which combines adjacency list and reverse adjacency list. The summary for storage methods of graph is given in Table 2.2, where n is the number of vertices and e is the number of edges.

Table 2.2: Storage methods of graph.

Storage method	Implementation method	Pros	Cons	Space complexity
Adjacency matrix	Two-dimensional array	Easy to check the relation between two vertices Easy to obtain the degrees of vertices	Occupies a lot of space	$O(n^2)$
Edge set array	Two-dimensional array	Relatively little space requirement	Not easy to check the relation between two vertices Not easy to obtain the degrees of vertices	$O(n + e)$
Adjacency list	Linked list	Saves space Easy to get the outdegrees of vertices	Not easy to check the relation between two vertices Not easy to obtain the indegrees of vertices	$O(n + e)$
Cross-linked list	Linked list	Saves space Easy to get the outdegrees and indegrees of vertices	Relatively complex structure	$O(n + e)$
Adjacency multilist	Linked list	Saves space Easy to check the relation between two vertices	Relatively complex structure	$O(n + e)$

2.4 Basic operations on graph

2.4.1 Operations on adjacency matrix

```
#include<stdio.h>
#include<stdlib.h>
#define VERTEX_NUM 64
typedef struct
{  char VertexArray[VERTEX_NUM];       //Vertex array
   int  AdjMatrix[VERTEX_NUM][ VERTEX_NUM];      // Adjacency matrix
   int  VexNum, EdgeNum;
} AM_Graph;
/*================================================================
Functionality: Construction of adjacency matrix for undirected graph
Function input: Address of the memory for adjacency matrix
Function output: None
Keyboard input: The value of vertex, the incident nodes for an edge
================================================================*/
void creat_AdjMatrix(AM_Graph *gPtr)
{
    int i,j,k;
    getchar();
    printf("Please input %d vertices:",gPtr->VexNum);
    for(i=0;i<gPtr->VexNum;i++)
        scanf("%c",&gPtr->VertexArray[i]);
    for(i=0;i<gPtr->VexNum;i++)
    for(j=0;j<gPtr->VexNum;j++)
        gPtr->AdjMatrix[i][j]=0;
    printf("Input the subscripts of groups of connected vertices:\n");
    for(k=0;k<gPtr->EdgeNum;k++)
    {
        scanf("%d%d",&i,&j);
        gPtr->AdjMatrix[i][j]=1;
        gPtr->AdjMatrix[j][i]=1;
    }
}
/*================================================================
Functionality: Output of adjacency matrix for undirected graph
Function input: Address of the memory for adjacency matrix
Function output: None
Screen output: The adjacency matrix
================================================================*/
```

```
void print_AdjMatrix(AM_Graph *gPtr)
{
    int i,j;
    printf("The finished adjacency matrix of the  undirected graph is:\n");
    for(i=0;i<gPtr->VexNum;i++)
    {
        printf("%c ",gPtr->VertexArray[i]);
        for(j=0;j<gPtr->VexNum;j++)
        printf("%d ",gPtr->AdjMatrix[i][j]);
        printf("\n");
    }
}

int main()
{
    AM_Graph Graph;
    printf("Please input the number of vertices and edges:");
    scanf("%d%d",&Graph.VexNum, &Graph.EdgeNum);
    creat_AdjMatrix(&Graph);
    print_AdjMatrix(&Graph);
    return 0;
}
```

Please input the number of vertices and edges: 5 6
Please input the names for five edges: abcde
Input the subscripts of groups of connected vertices:
0 1
0 2
0 4
1 2
2 3
3 4
The finished adjacency matrix of the undirected graph is
a 0 1 1 0 1
b 1 0 1 0 0
c 1 1 0 1 0
d 0 0 1 0 1
e 1 0 0 1 0

2.4.2 Operations on adjacency list

2.4.2.1 Construction of adjacency list

For the construction of the adjacency list of a graph, we need to know the number of vertices and number of edges of the graph, as well as the subscripts of the starting vertex and ending vertex of each edge.

The processing of directed graph is different from that of undirected graph: For an undirected graph, the two vertices of an edge v_1 and v_2 are adjacent vertices. During storage, we insert the vertex at the other side to the head of the singly linked list of the starting vertex v_1, that is, the ending vertex v_2; then we insert the vertex at the other side to the head of the singly linked list of the ending vertex v_2, that is, the starting vertex v_1. However, for directed graph, we only need to insert the ending vertex v_2 to the head of the singly linked list of the starting vertex v_1.

```c
#include<stdio.h>
#include<stdlib.h>
#define  VERTEX_NUM 6            // The number of vertices of the graph
#define  VERTEX_MAX_NUM 10       // The maximum number of vertices of the graph
#define  EDGE_MAX_NUM 20         // The maximum number of edges of the graph
typedef char VexType;      // Data type of the vertex
typedef int InfoType;          // Type for information attached to an arc (edge),
                               // e.g. weight value, the existence of the edge etc.

typedef struct AdjNode      // Struct for adjacent node
{
    int adjvex;      // Adjacent vertex
    AdjNode *next;       // Pointer for adjacent node
} AL_AdjNode;

typedef struct       // Struct for node of vertex of adjacency list
{
    VexType  vertex;       // Vertex
    AdjNode *link;        // Head pointer for the adjacent node
} AL_VexNode;

typedef struct       // The general struct for adjacency list
{
    AL_VexNode VexList[VERTEX_MAX_NUM];        // List of vertices
    int VexNum, ArcNum;       // Number of vertices, number of arcs (edges)
} AL_Graph;
```

```
/*=================================================================
Functionality: Construct the adjacency list structure for directed graph
Function input: None
Function output: Adjacency list
Keyboard input: The number of vertices, edges, information attached to edges
================================================================*/
AL_Graph *Create_AdjList()
{
    int n,e,i,v1,v2;
    AL_AdjNode * AdjPtr;     // Define adjacency node pointer
    AL_Graph *alPtr;         // Define adjacency list pointer
    alPtr=(AL_Graph *)malloc(sizeof(AL_Graph));
    // Allocate the memory space for the whole adjacency list
    printf("Please input the number of vertices of the graph: \n");
    scanf("%d",&n);          // Input the number of vertices
    for(i=0;i<n;i++)         // Initialize the memory space for adjacency list
    {
        alPtr->VexList[i].vertex=(char)i;      // Give value to the head node
        alPtr->VexList[i].link=NULL;           // Initialize the head node
    }
    printf("Please input the number of edges: \n");
    scanf("%d",&e);          // Input the number of edges
    printf("Please input information about the arc: \n");
    for(i=0;i<e;i++)
    {
        printf("Please input two ends of the arc, the first is the arc head,
        the second is the arc tail\n");
        scanf("%d%d",&v1,&v2);      // Input the two vertices for the edge
        AdjPtr =(AL_AdjNode *)malloc(sizeof(AL_AdjNode));
        // Allocate new adjacent node
        AdjPtr->adjvex=v2;
        //v2 will be the numbering for the new adjacent node
        AdjPtr->next=alPtr->VexList[v1].link;
        // The new adjacent node is linked into the list head of
        // the adjacency list with v1 as its vertex
        alPtr->VexList[v1].link=AdjPtr;
        // Point the head node to the new node
    }
```

```
    alPtr->VexNum=n;
    // Assign the number of vertices n to alPtr->VexNum
    alPtr->ArcNum=e;
    // Assign the number of edges e to alPtr->ArcNum
    return alPtr;    // Return this adjacency list
}
/*======================================================
Functionality: Output the adjacency list to the screen
Function input: Address of the adjacency list
Function output: None
=====================================================*/
void Print_AdjList(AL_Graph *algPtr)
{
    int i;
    AL_AdjNode *AdjPtr;
    printf("The adjacency list of the graph\n");
    for(i=1;i<=algPtr->VexNum;i++)
    // When i is smaller than the number of vertices
    {
        // Output the value of the ith vertex in the list of vertices
        printf("%d-",algPtr->VexList[i].vertex);
        // Get the head address of the ith adjacency linked list
        AdjPtr=algPtr->VexList[i].link;
        while(AdjPtr!=NULL)
        // The adjacency linked list is not empty
        {
            printf("%d-",AdjPtr->adjvex);       // Output the adjacent vertex
            AdjPtr=AdjPtr->next;    // Get the address of the next adjacent vertex
        }
        printf("--\n");
    }
}
int main()
{
    AL_Graph *gPtr;
    gPtr=Create_AdjList();
    Print_AdjList(gPtr);
    return 0;
}
```

Example 2.1 Arrangement of exam schedule.

There are nine courses in the communication engineering department that have midterm exams. The courses that are unable to have exams at the same time are listed in the set R, where the pair of numbers in the brackets indicates that there would be a conflict for the corresponding courses. How to arrange the schedule so that there is no conflict in the exams and the number of days used for the exam is kept minimum?

Numbering of the courses = {1, 2, 3, 4, 5, 6, 7, 8, 9}

Conflicting courses R = {(2, 8), (9, 4), (2, 9), (2, 1), (2, 5), (6, 2), (5, 9), (5, 6), (5, 4), (7, 5), (7, 6), (3, 7), (6, 3)}

Solution:

1. Analysis

From the set of courses and the set of conflicting courses, it is observed that we can correspond the courses to the set of vertices of a graph and the relations between courses to the edges of a graph. In this manner, we can express the information in the question and the connections with a graph, as shown in Fig. 2.23. When the vertices at two ends of an edge conflict with each other, we can mark them with different colors. Then, the scheduling problem is converted into a graph coloring problem – use fewest colors to color each vertex of the graph so that the adjacent vertices have different colors.

We can try to color the vertices in Fig. 2.23. If we start the coloring from an arbitrary vertex, we can realize that this would not guarantee that the number of colors is kept minimum. If we rank the vertices according to the degree (vertices with the same degree may be ranked arbitrarily), and color them one by one: we try to apply the first color to each vertex; if it conflicts with an already colored node, then we change to another color, until the colors do not conflict. This coloring method would achieve what we want. This kind of coloring algorithm for graph is the famous Welsh–Powell method. In Fig. 2.24, the sequence of vertices in the descending order of their degrees is {2, 5, 6, 7, 9, 3, 4, 1, 8}. The result of coloring is {2, 7, 4} {5, 3, 1, 8} {6, 9}.

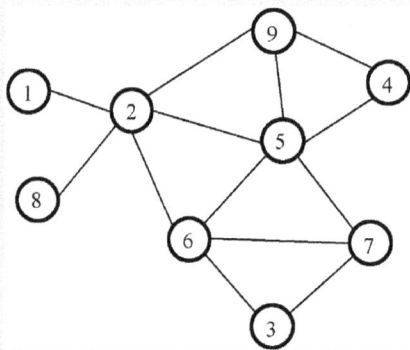

Fig. 2.23: Scheduling of exam.

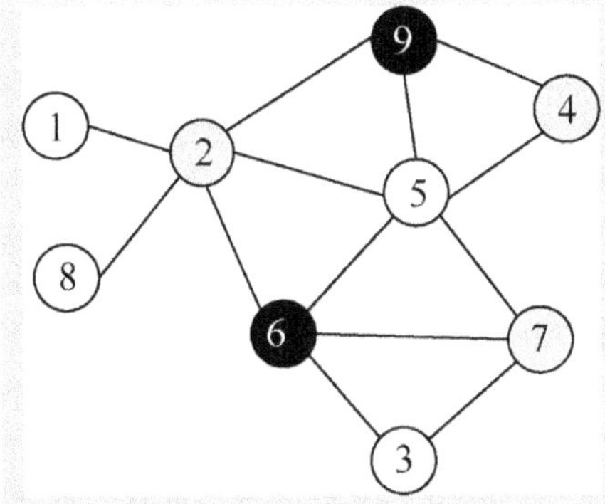

Fig. 2.24: Coloring of the graph.

2.4.2.2 Pseudocode description of the algorithm
For the coloring algorithm of the graph, see Table 2.3.

Table 2.3: Description of the coloring method for the graph.

Top-level pseudocode	Detailed description of pseudocode 1	Detailed description of pseudocode 2
Arrange the vertices of the graph in descending order of their degrees	(1) Calculate the degrees of the vertices in the graph	(1) Calculate the degrees of the vertices in the graph, and store them into the array degree[N]
Color the first node with the first color, and according to the sorted order of the vertices, color each vertex that isn't adjacent to the previously colored vertex with the same color	(2) Find the vertex with the maximum degree, k; (3) Is the colorPtr$_{th}$ color of the set of colors already used? {If k cannot use the colorPtr$_{th}$ color, then we change colorPtr} (4) Add k into the colorPtr$_{th}$ set of colored vertices.	(2) Find the vertex with the maximum degree in degree[], record the vertex in k, remove this vertex from degree[] (3) Is the colorPtr$_{th}$ color in the set of colors ColorSet[] already used? {!(k is not adjacent to all the vertices in ColorSet[colorPtr].node []), then colorPtr++} (4) Add k into the colorPtr$_{th}$ set of colored vertices ColorSet[colorPtr].node[];
Repeat step 2 with the next color on still uncolored vertices, until all vertices are colored	(5) Repeat steps (2) to (4), until all vertices are processed;	
Output the result	(6) Output the sets of similarly colored vertices.	

1. Design of data structure

When we put the vertices of the same color into one array node, the maximum length of the array would be N. The last index of the already existing vertices in the array is recorded by the tail pointer rear, as shown in Fig. 2.25. The set of colors ColorSet would have at most N colors, that is, the case where every node has a different color. Whether a color is used is marked with used. Suppose that 0 means "not used" and 1 means "used."

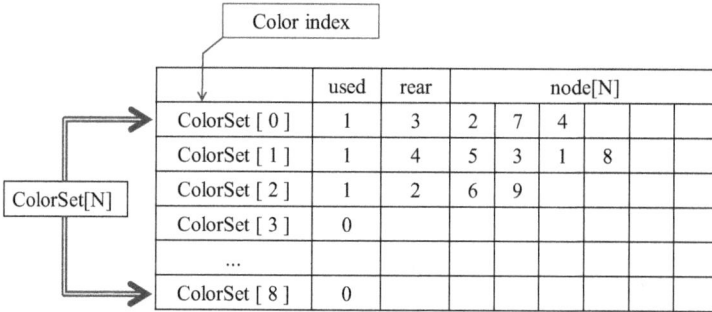

	used	rear	node[N]					
ColorSet [0]	1	3	2	7	4			
ColorSet [1]	1	4	5	3	1	8		
ColorSet [2]	1	2	6	9				
ColorSet [3]	0							
...								
ColorSet [8]	0							

Color index

ColorSet[N]

Fig. 2.25: Data structure design for the coloring of graph.

2. Program implementation

```c
#include <stdio.h>
#define TRUE 1
#define FALSE 0
#define N 9 // Number of vertices
int AdjMatrix[N][N]= // Adjacency matrix
  {{0,1,0,0,0,0,0,0,0},
   {1,0,0,0,1,1,0,1,1},
   {0,0,0,0,0,1,1,0,0},
   {0,0,0,0,1,0,0,0,1},
   {0,1,0,1,0,1,1,0,1},
   {0,1,1,0,1,0,1,0,0},
   {0,0,1,0,1,1,0,0,0},
   {0,1,0,0,0,0,0,0,0},
   {0,1,0,1,1,0,0,0,0}};
int degree[N]={0}; // Record the degrees of nodes
char *color[N]={"Red","Orange","Yellow","Green","Cyan"," Blue","Purple",
"Black","White"};
```

```
    struct ColorNode
    {
        int used; // Indicates whether the color is used. 0 indicates not used
        int rear; // The tail pointer for the set of vertices
        int node[N]; // The set of vertices colored the same
    } ColorSet[N]={{0,0,0,0}}; // ColorSet
/*================================================================
Functionality: Find the subscript of the node with the highest degree
Function input: Array of vertex degrees
Function output: The subscript of the vertex with the highest degree
==============================================================*/
int FindMax(int *a)
{
    int i,value,index;
    value=-1;
    index=0;
    for(i=0;i<N;i++)
    {
        if(value < a[i])
        {
            value=a[i];
            index=i;
        }
    }
    a[index]=-1;    // Clear the current maximum value
    return index;
}
/*================================================================
Functionality: Check whether vertex k can be added to the ith set of colored
vertices
Function input: The color i, the vertex k
Function output: 1 - Can be added
                 0 - Cannot be added
==============================================================*/
int judge(int i,int k)
{
    int p,q,m;
    p=0;
    q=ColorSet[i].rear;
    m=ColorSet[i].node[p];    // Node with subscript p in the ColorSet
    while (AdjMatrix[k][m]==0 && p!=ColorSet[i].rear)
    // k, p are not adjacent vertices and p is not the last vertex in the ColorSet
```

```
{
    p++;
    m=ColorSet[i].node[p];
}
    if (p==q)return 1;    // k can be added into the ColorSet
    return 0;             // k cannot be added into the ColorSet
}
/*=================================================
Functionality: Welsh_Powell coloring method for graph
Function input: None
Function output: None
Screen output: The sets of vertices colored the same
=================================================*/
void Welsh_Powell()
{
    int i,k;
    int colorPtr;    //Compute the degree of the vertices
    for (i=0; i<N; ++i)
    {
        for (int j=0; j<N; ++j)
        {
            if (i != j && AdjMatrix[i][j])
                degree[i]++;
        }
    }
    for (int j=0; j<N; ++j)
    {
        k=FindMax(degree);  // Find the vertex with the maximum degree k
        colorPtr=0;
        //The colorPtr_th color has already been used
        if( ColorSet[colorPtr].used==1)
        {
            while(!judge(colorPtr,k))
            // If k cannot be added to the colorPtr_th set of colored vertices
            colorPtr++;
        }
        // Add k into the ColorPtr_th set of colored vertices
        ColorSet[colorPtr].node[ColorSet[colorPtr].rear++]=k;
        if( ColorSet[colorPtr].used==0) ColorSet[colorPtr].used=1;
    }
    // Output the set of vertices colored the same
    for (j=0; j<N; ++j)
```

```
    {
      if (ColorSet[j].used==1)
      {
        printf("%s:",color[j]);
        for (i=0; i<ColorSet[j].rear; ++i)
        printf("%d ",ColorSet[j].node[i]+1);
        printf("\n");
      }
    }
}
int main()
{
    Welsh_Powell();
    return 0;
}
```

Results:

Red: 2 7 4

Orange: 5 3 1 8

Yellow: 6 9

Time complexity $O(n^2)$

2.5 Lookup of the vertices of graph – traversal of graph

2.5.1 Introductory example for lookup on graph – the search of web crawlers

"Web crawler" is an important component of a search engine. It traverses the web space, can scan websites within a certain IP range and follow links on the Internet from one webpage to another, from one website to another in order to collect webpage data. In the face of a huge amount of websites, how to efficiently visit all webpages without omission or repetition? We need a good strategy to carry out the traversal of graph.

We are already familiar with traversal methods of tree. Can we use the same traversal strategies we used or trees or graphs? First, we need to compare the structures of graph and tree, and analyze the features of graph. Only then we can adapt the corresponding processing methods.

Think and Discuss   !

Question 1: What are the differences between the traversal of graph and the traversal of tree?

Discussion: We can first review the traversal of tree. There are three parts of a tree: "Left subtree," "right subtree" and "root." As long as we traverse these three parts recursively, we can complete the traversal on the tree.

> We cannot divide a graph into corresponding parts as we divided the nodes of a tree. Therefore, we cannot use the method of traversing by parts. The feature of the graph is that any vertex could be connected to the rest of the vertices. Therefore, after visiting a certain vertex, it is possible for us to go back to it by following along a certain search path. If we can identify whether a vertex was visited already during the traversal process, we can fulfill the requirement of "visit only once."

> Question 2: How to avoid the same vertex of the graph from being visited multiple times during traversal?
> **Discussion:** In order to prevent a vertex from being visited multiple times, we can set a vector during the traversal process to mark the "visited" status of each vertex to tell whether a vertex was visited or not.

In actual applications, the parsing strategies of a web crawler include depth-first, breadth-first and best-first. Figure 2.26 displays the breadth-first and depth-first traversal results on a certain website.

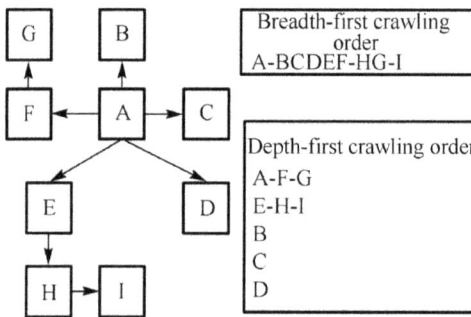

Fig. 2.26: Crawling strategy for webpages.

Similar to the traversal of tree, the traversal of graph starts from a certain vertex of the graph, and visit all the vertices in the graph. Also, each vertex is visited only once.

Many operations on graph, such as addition, modification or deletion of vertex or edge, solving the connectivity of the graph, topological ordering and calculation of the key path, are all based on the traversal of tree.

2.5.2 The breadth-first traversal of graph BFS

Breadth-first search (BFS) algorithm is a traversal strategy on connected graph. It's also one of the simplest search algorithms on graph. This algorithm is the prototype of a lot of important graph algorithms. The Dijkstra single-source shortest path algorithm and Prim's minimum spanning tree (MST) algorithm both adapt ideas similar to BFS.

2.5.2.1 Strategy for traversing a maze

The game of maze is familiar to everybody. Figure 2.27 shows a simple maze shape. If we want the computer to finish path exploration from entrance point 0 to exit point 6, and find the shortest path (we suppose that traversal of one node is counted as one step length), how should we design the algorithm?

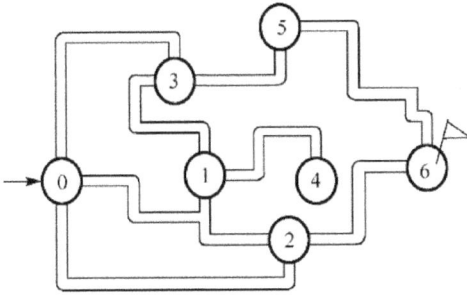

Fig. 2.27: A maze.

We can abstract the relations between the nodes in Fig. 2.27 and express them with graph, as shown in Fig. 2.28.

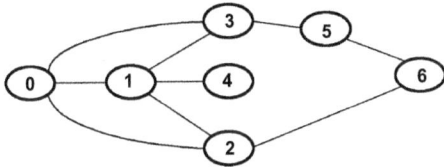

Fig. 2.28: Abstraction of the maze.

It is easy to observe that the shortest path here is V0→V2→V6, instead of V0→V3→ V5→V6. The way to find this path can be: first inspect the vertices directly connected to V0, V1, V2, V3. We realize that there is no V6. Then we further inspect the vertices directly connected to V1, V2, V3 and find them to be {V0, V2, V3, V4}, {V0, V1, V6} and {V0, V1, V5}. At this time, we can find V6 in the set of connected vertices to V2, which means we have found the shortest path from V0 to V6: V0→V2→V6. Although if we further search the set of connected vertices to V5, we can find another path V0→V3→V5→V6. Obviously, this is not the shortest path.

We will use sample Fig. 2.29 to illustrate this process, and the steps are as follows:

Step (a): Starting point V0 is marked gray, which means we are about to visit V0.

Step (b): V0 is marked black, which means we have already visited it. The directly connected vertices of V0, V1, V2, V3 are colored gray.

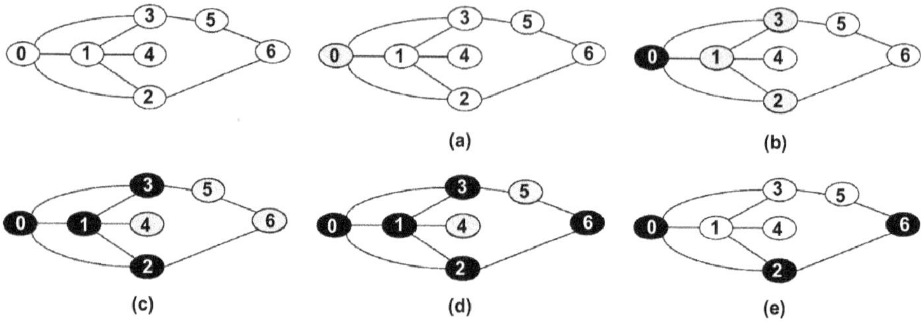

Fig. 2.29: Strategies for walking maze.

Step (c): Visit V1, V2, V3, the corresponding directly connected vertices are V0, V4, V6, V5, respectively. Since V0 is already visited, we only need to mark V4, V6 and V5 gray.

Step (d): We visited the target vertex V6. The search ends.

Step (e): The result of shortest path

The above searching process is a radiating search. We radiate from one vertex to all its directly connected vertices. In such a manner, we go on and radiate layer by layer, until the target vertex is radiated, that is when we Find the path from the beginning point to the ending point.

2.5.2.2 Depth-first traversal algorithm
1. Basic idea

For the above-mentioned maze-walking method, since its idea is to start from one vertex, and traverse the wider area around it in a radiating manner, we call it breadth-first traversal. The breadth-first traversal of graph (BFS algorithm) is a layered search process, similar to the layered traversal of trees. It also needs a queue to store the sequence of vertices traversed, so that the adjacent vertices of them can be visited in a dequeuing order.

2. Pseudocode description of BFS algorithm

Breadth-first traversal of graph starts from one vertex of the graph, v, as the starting point, from near to far, visits the vertices connected to v with paths one by one. During the search, how to generate the to-be-searched vertices from the already existing nodes can be ascertained according to the requirements of the question. Vertices on the same layer should have the same value for solving the problem, so we follow the principle of searching the earlier generated vertices first. Therefore, we design the list that stores the vertices as a queue. The detailed algorithm description is given in Table 2.4.

Table 2.4: BFS algorithm on graph.

Pseudocode description	Detailed description
(1) Initialization	– Initialize the array of visitation markers visited [N]=0. After a certain vertex is visited, we set the corresponding subscript element to 1. – Initialization of queue Q: Input the vertex to be visited v
(2) Queue the vertex v into the queue	Visit vertex v; visited[v]=1; Vertex v is queued into queue Q
(3) Continue execution if the queue is not empty, otherwise the algorithm ends	while (Queue Q is nonempty) v = Dequeued head element of queue Q
(4) Dequeue to obtain the queue head vertex v; visit the vertex v and mark v as visited	w = The first adjacent vertex of vertex v while (w exists) If w is not visited, then visit vertex w;
(5) Lookup the first adjacent vertex w of vertex v	visited[w] = 1; Enqueue vertex w into queue Q
(6) If the adjacent vertex w of v is not visited, then enqueue w	w = the next adjacent vertex of vertex v
(7) Continue searching for another new adjacent vertex w. Go to step (6) Until all the nonvisited adjacent vertexes of vertex v are processed. Go to step (3)	

– Implementation of BFS based on adjacency matrix

```
#include <stdio.h>
#define TRUE 1
#define FALSE 0
#define N 9 // Maximum number of vertices
int Visited[N];
int AdjMatrix[N][N]=// Adjacency matrix
 {{0,1,0,0,0,0,0,0,0},
  {1,0,0,0,1,1,0,1,1},
  {0,0,0,0,0,1,1,0,0},
  {0,0,0,0,1,0,0,0,1},
  {0,1,0,1,0,1,1,0,1},
  {0,1,1,0,1,0,1,0,0},
  {0,0,1,0,1,1,0,0,0},
  {0,1,0,0,0,0,0,0,0},
  {0,1,0,1,1,0,0,0,0}};
```

```
/*======================================================================
Functionality: Find the next adjacent vertex of vertex v after adjacent vertex i
Function input: Vertex v, adjacent vertex of v, i
Function output: The subscript of the next adjacent vertex after i; when there's
no adjacent vertex, return -1
=====================================================================*/
int FindVex(int v,int i)
{
    while (AdjMatrix[v][i]==0) i++;
    if (i<N) return i;
    else return -1;
}
/*===================================================================
Functionality: BFS traversal based on adjacency matrix
Function input: The beginning vertex for traversal of the graph v
Function output: None
Screen output: The BFS sequence of the graph
=================================================================*/
void GraphBFS(int v)
{
    int flag;
    SeqQueue struc;
    SeqQueue *sq;
    sq=&struc;
    int w,k;
    int count=0;

    initialize_SqQueue(sq);
    printf("BFS序列: ");
    printf("%d ",v);
    Visited[v]=1;
    count++;
    flag=Insert_SqQueue(sq,v);      // Queue the first vertex
    while (!Empty_SqQueue(sq))      // Queue Q is non-empty
    {
        if (count==N) break;
        k=Delete_SqQueue(sq);
        // Dequeue the head element of queue Q, return the subscript of the element
        v=sq->data[k];
        w=FindVex(v,0);       //w=The first adjacent vertex of vertex v;
        while(w!=-1)
```

```
    {
       if( Visited[w]!=1)          // If w is not visited
       {
          Visited[w]=1;            // Set the visitation marker of w
          printf("%d ",w);
          count++;
       }
       flag=Insert_SqQueue(sq,w);          // Enqueue vertex w into queue Q
       w=FindVex(v,w+1);           //w=the next adjacent vertex of vertex v
    }
  }
  printf("\n");
}
int main()
{
  GraphBFS(0);
  // Start traversal from vertex 1, the corresponding subscript is 0
  return 0;
}
```

The test graph is still Fig. 2.29 for the coloring problem. The result is as follows:

BFS Sequence: 1 2 5 6 8 9 4 7 3

 – BFS implementation based on adjacency list
Here only the function framework description is given. The readers can implement
the actual program by themselves.

```
/*==================================================
Functionality: BFS traversal based on adjacency list
Function input: The starting vertex v for the traversal
Function output: None
Screen output: The BFS sequence of the graph
==================================================*/
```

2.5.2.3 Discussion of the BFS algorithm
When we use BFS algorithm to solve problems, according to the conditions given
by the problem, starting from one node, we can generate one or multiple new
nodes. This is a process of expansion. The expansion is carried out according to the
degree of proximity to the starting vertex. First, we generate the first level of vertices, and simultaneously check if the target vertex is among the generated vertices.

If it is not there, then we expand all the first-level vertices one by one to obtain the second-level vertices, and check whether the second-level vertices contain the target vertex. Repeating the process, before we expand to obtain the $n + 1$ level of nodes, we must first consider all possible statuses of the nth level of nodes.

The relations between nodes in the search result are displayed in the form of a tree. In general, we call such a tree a solution tree.

During the expansion process, the solution tree has more vertices together with the increase in the number of levels. The quantity of search also expands rapidly. Suppose that the average number of adjacent vertices of a vertex is M. When the search begins, there is only one starting vertex in the queue. When the starting vertex is dequeued, we enqueue its adjacent vertices, then the queue will have M vertices. When the search at the next level continues to add elements into the queue, the number of vertices reaches square of M, as shown in Fig. 2.30. Once M is relatively large, the level of the solution tree is relatively deep, the queue will need massive amount of memory space; otherwise, data overflow will easily occur and cause the search to fail. Therefore, breadth-first search is suitable for the situations where the number of child vertices of nodes is not great, and the solution tree is not very deep.

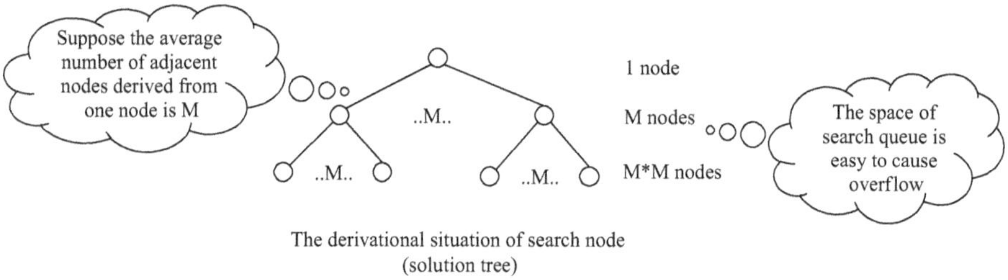

The derivational situation of search node
(solution tree)

Fig. 2.30: Derivative situations of breadth-first search.

2.5.3 Depth-first traversal of graph DFS

The depth-first search (DFS) of a graph is similar to the pre-order traversal of tree. The search strategy is to search the graph as "deeply" as possible.

2.5.3.1 Analysis of the DFS traversal method
Let us use an actual graph as an example to analyze the search process of DFS traversal. The illustration is shown in Fig. 2.31. Suppose that the starting vertex v = 1. The illustration of the search process is given in Table 2.5.

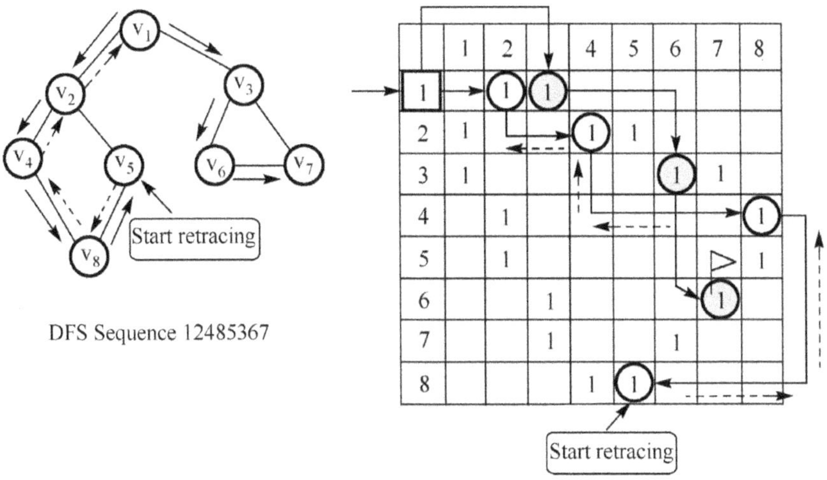

DFS Sequence 12485367

Fig. 2.31: Search path of DFS.

Table 2.5: DFS search process.

Step	Vertex to visit	The sequence of unvisited adjacent vertices	The sequence of already visited adjacent vertices	The turnback vertex when there is no adjacent vertex to visit
1	1	2 3		
2	2	4 5	1	
3	4	8	2	
4	8	5	4	
5	5	None	2 8	8
6	[8]	None	4 5	4
7	[4]	None	2 8	2
8	[2]	None	1 4 5	1
9	[1]	3	2	
10	3	6 7	1	
11	6	7	3	
12	7	None	3 6	

The "turnback vertex" in Table 2.4 is the point from which to retrace when v does not have any adjacent vertices to visit. The square brackets in the "visit vertex" item indicates turn back to this vertex, that is, the returning point of the recursion.

The order of vertex visit from steps 6 to 9 is a symmetry of the steps 1 to 4. The eventual DFS sequence of nodes is the order obtained from the "visit vertex" item of the table: 1 2 4 8 5 3 6 7.

From the above analysis, we can see that the search process of DFS traversal has the feature of iterative deepening and returning along the original path after satisfying certain conditions. Therefore, the DFS algorithm can be implemented by either recursion or stack.

2.5.3.2 Traversal algorithm for depth-first search
1. Analysis of DFS algorithm

Previously, we have analyzed the features of BFS. When there is a relatively large number of child nodes and too many levels in the solution tree, we need relatively large memory space for the queue. DFS deals with this shortcoming, since we only need to maintain one vertex per level during each search.

BFS can find the shortest path. Can DFS do it as well? The way of DFS is to go until the very bottom; thus, we obviously cannot know whether the current path being searched is the shortest or not. Therefore, we still need to continue exploring other paths to check whether it is the shortest path. This leads us to the shortcoming of DFS: it is hard to search for the optimal solution, and we can only check whether there is solution or not. Its advantage is that its memory cost is relatively little, which is better compared with BFS.

DFS algorithm is usually used when we only want a solution and in cases where there are a lot of repetitive vertices in the tree and it is hard to check for repetition.

> **Think and Discuss** Is there a unique DFS sequence?
> **Discussion:** Since we did not specify the order in which to visit adjacent vertices, the sequence from DFS is not unique.

2. DFS algorithm description: see Table 2.6.

Table 2.6: DFS algorithm on graph.

Top-level pseudocode	Elaboration of recursive solution	Elaboration of nonrecursive solution
Initialization	Construct a visited marker array visited[N], which contains initial values of 0. When a vertex is visited, then we set the variable at the corresponding subscript to 1.	
(1) Visit vertex v, and mark it as visited	(1) Input the vertex to visit Vi	(1) Initialize stack S; input the vertex to visit V

Table 2.6 (continued)

Top-level pseudocode	Elaboration of recursive solution	Elaboration of nonrecursive solution
	(2) Visit vertex v_i; visited[v_i]=1	(2) Visit vertex v: visited[v]=1; push vertex into stack S
(2) Select a vertex w from unvisited adjacent vertices of v, start the DFS from w	(3) Search in the ith row of the adjacency matrix. If v_i has an adjacent vertex v_j, and v_j is not visited, then we set i=j	(3) while（Stack S is not empty） x=S's stack-top element (without popping); if（there exists some unvisited adjacent vertex of x, w） { Visit w; visited[w] = 1; w is pushed onto the stack; } else x is popped
(3) Repeat steps 1 and 2 until all the vertices connected to v are visited	Repeat steps 1 to 3, until all vertices are visited	

Note: The above is DFS algorithm for connected graph. For DFS algorithm on unconnected graph, we should check whether there are still remaining vertices on the basis of the above algorithm. If there are still unvisited vertices, then we select an unvisited vertex as the new source vertex and repeat the DFS process, until all the vertices in the graph are visited.

3. Recursive implementation of DFS based on adjacency matrix

```
/*============================================================
Functionality: Recursive DFS traversal algorithm on graph based on adjacency
matrix
Function input: The starting vertex for traversal on graph v
Function output: None
Screen output: The DFS sequence of the graph
=============================================================*/
void GraphDFS(int i)
{  int j;
   printf("%d ",i+1);
   //i+1because the numbering of vertices in the graph starts from 1,
   while storage starts from 0
   Visited[i]=1;
   for (j=0; j<N; j++)
       if ((AdjMatrix[i][j]==1) && (!Visited[j]) )
          GraphDFS(j);   // Is called n times
}
```

Test program

The test graph is shown in Fig. 2.31. The DFS sequence is 1 2 4 8 5 3 6 7.

```
#include <stdio.h>
#define TRUE 1
#define FALSE 0
#define N 9 // Number of vertices
int Visited[N]={0};
int AdjMatrix[N][N]= // Adjacency matrix
{{0,1,1,0,0,0,0,0},
{1,0,0,1,1,0,0,0},
{1,0,0,0,0,1,1,0},
{0,1,0,0,0,0,0,1},
{0,1,0,0,0,0,0,1},
{0,0,1,0,0,0,1,0},
{0,0,1,0,0,1,0,0},
{0,0,0,1,1,0,0,0}};
int main()
{
    printf("DFS Sequence:");
    GraphDFS(0);
    return 0;
}
```

4. Recursive implementation of DFS based on adjacency list
The relations between various variables related to adjacency list in the program are shown in Fig. 2.32.

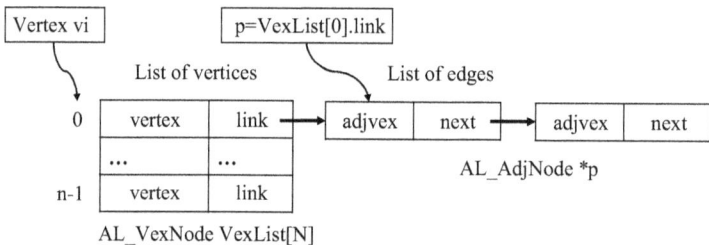

Fig. 2.32: Variables related to adjacency list.

```
#include <stdio.h>
#include <stdlib.h>

#define TRUE 1
#define FALSE 0
#define N 8 // Number of vertices
typedef int VexType;
typedef struct AdjNode     // Adjacency node struct
{
    int adjvex;      // Adjacent vertex
    AdjNode *next;      // Adjacent node pointer
} AL_AdjNode;
typedef struct      // Node structure for vertex in adjacency list
{
    VexType vertex;      // Vertex
    AdjNode *link;      // Head pointer of adjacent node
} AL_VexNode;
AL_VexNode VexList[N]={0,NULL};     // Vertex list

int Visited[N];
int AdjMatrix[N][N]=      // Adjacency matrix
{{0,1,1,0,0,0,0,0},
{1,0,0,1,1,0,0,0},
{1,0,0,0,0,1,1,0},
{0,1,0,0,0,0,0,1},
{0,1,0,0,0,0,0,1},
{0,0,1,0,0,0,1,0},
{0,0,1,0,0,1,0,0},
{0,0,0,1,1,0,0,0}};
/*=====================================================================
Functionality: DFS traversal recursive algorithm based on adjacency list
Function input: Starting vertex for the traversal of graph v
Function output: None
Screen output: The DFS sequence of the graph
=====================================================================*/
void GraphDFS_L(int vi)
// Starting from vi, traverse the graph with depth-first search.
// The graph is represented with adjacency list
{  AL_AdjNode *p;
```

```
    printf("%d ",VexList[vi].vertex+1);    // Visit the vertex vi
    Visited[vi]=1;    // Mark vi as visited
    p=VexList[vi].link;
    // Retrieve the head pointer for the edges of vi
    while( p!=NULL )
    // Search vi's adjacent vertices one by one
    {
        if (Visited[p->adjvex]==0)
        // Starting from unvisited adjacent vertices of vi, carry out
        depth-first search traversal
        GraphDFS_L(p->adjvex);
        p=p->next;
    }
}
int main()
{
    Create_AdjList();    // Construct the adjacency list
    printf("DFS Sequence:");
    DFSL(0);
    return 0;
}
```

Test result: Test sample is shown in Fig. 2.33. DFS sequence: 1 2 4 8 5 3 6 7.

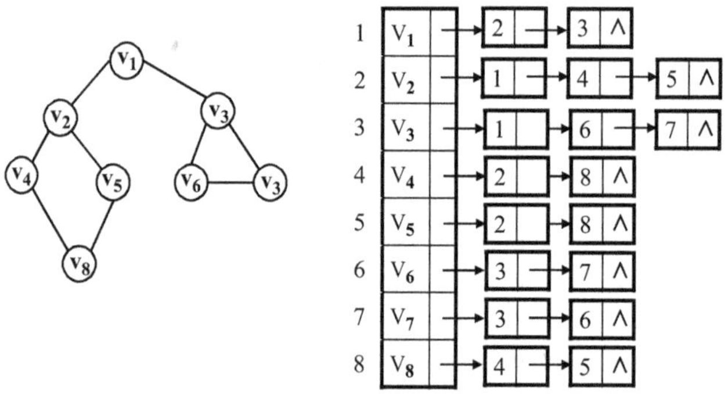

Fig. 2.33: Test for DFS based on adjacency list.

5. Nonrecursive program implementation based on adjacency list

```
/*================================================================
Functionality: DFS traversal of graph with non-recursive algorithm
Function input: Beginning vertex for traversal vi
Function output: None
Screen output: DFS sequence of the graph
================================================================*/
void DFS(int vi)
{
    SeqStack struc;
    SeqStack *s;
    int vj;
    int flag;
    s=&struc;
    initialize_SqStack( s );    // Initialization of stack
    Visited[vi]=1;              // Visit vertex vi
    printf("%d ",vi+1);
    flag=Push_SqStack(s,vi);   // Push vertex vi into stack
    while(!StackEmpty_SqStack(s))     // Stack is not empty
    {
        vj=0;
        flag=Get_SqStack(s, &vi);
        //vi=stack-top element (without popping)
        //Exists and finds unvisited adjacent vertex of vi, vj
        while(!(AdjMatrix[vi][vj]==1 && Visited[vj]==0)) vj++;
        if (vj<N) // Visit vj
        {
            printf("%d ",vj+1);
            Visited[vj]=1;
            flag=Push_SqStack(s,vj);            //vj is pushed onto stack
        }
        else
        {
            flag=Pop_SqStack(s, &vi);           //vi is popped out of stack
        }
    }
}
```

2.6 Classic application of graph – tree problem in graph

2.6.1 Introductory example of application of graph 1 – minimal cost for setting up communication network

In order to realize the high-speed communication between n cities, we need to set up communication cables between these n cities. We can set up a link between each pair of cities. The fee of the link is proportional to the length of the link. The distance between some cities is shown in Fig. 2.34. The cost and setup cost of fiber optic are very high. We will naturally consider this question: how to set it up so that each pair of cities can communicate with each other, and the total cost is kept at a minimum?

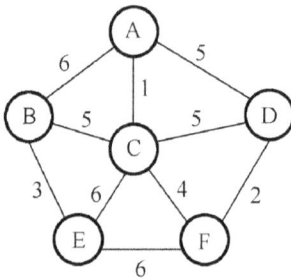

Fig. 2.34: Communication network with cost.

2.6.2 Introductory example of application of graph 2 – transmission of information in network

The rapid development of the Internet provides people with fast and convenient way to transmit information. People are increasingly reliant on computer network for work. If the network is disrupted, huge commercial loss would ensue. In the topological graph of Fig. 2.35, the switch mainly acts to connect networks. If the switch breaks down, then the several modules cannot work normally between each other, the client cannot visit the server, cannot connect to the Internet and cannot visit the printer. The switch becomes a single point of failure, that is, whenever one point breaks down, the whole network would not function normally. Therefore, in LAN network, in order to provide reliable network connection, there need to be redundant links. The so-called redundant links imply preparing two or more than two pathways. If one pathway breaks down, information can still go through other links, as shown in Fig. 2.36.

The aim of redundant linkages in the network is to eliminate network disruption caused by a single point of failure. But it will cause other problems. The initial switch does not have a learning mechanism, and it forwards each data frame it receives. Whether this method causes problems completely depends on the connection. If there is loop in the connection, then the data frame will circulate in the

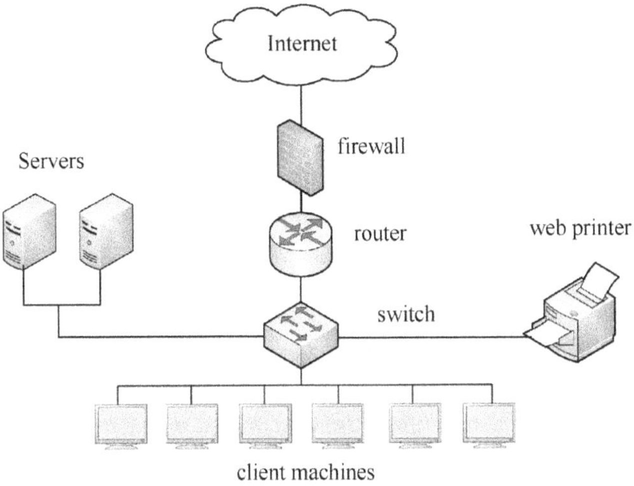

Fig. 2.35: Network structure without redundancy.

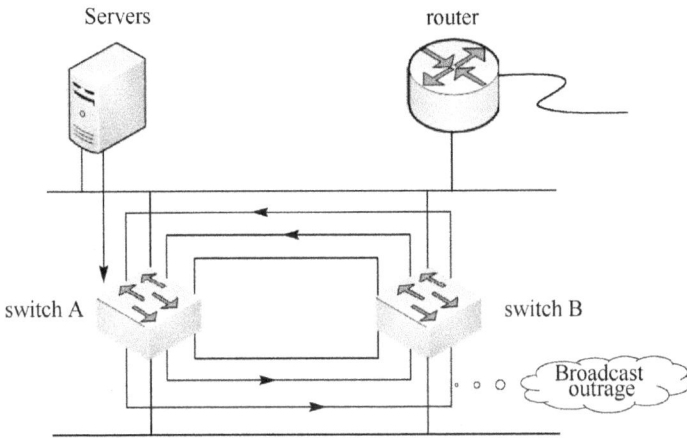

Fig. 2.36: Network structure with redundancy.

network forever, causing broadcast outrage. If there is no loop, then the data frame will automatically disappear when it reaches the end of the link. The existence of loop in the redundant linkage of Fig. 2.36 will cause the whole network to jam, and lead to network paralysis.

To avoid network jam, people avoid circularity in linkages with redundancy with software settings, that is, they cut off a certain linkage that might cause circularity and thus cut open the cycle. When there is some problem in the linkages, a noncircular linkage configuration can be automatically generated. For example, in the network of Fig. 2.37(a), there are two possible paths from A to C. When the path ABC is blocked,

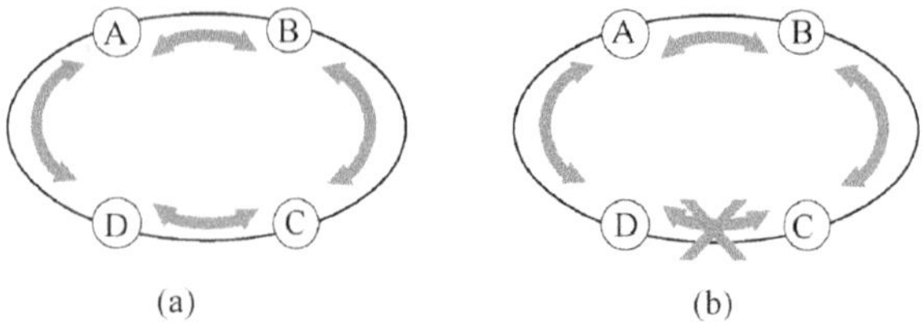

(a)   (b)

Fig. 2.37: Cycle in the network.

one can go through ADC. If at a certain moment, one can use software to stop the port from D to C, then the topology of the network will become Fig. 2.37(b). If there is some data frame, it must end at D or C, and thus will not be circularly forwarded.

From the above discussion, we can see that as long as there are no cycles in the topological structure of network, we can ensure the effective operation of network communication.

> **!**
> **Think and Discuss** How to obtain a noncircular structure from a graph?
> **Discussion:** If there is a path between any two vertices of a connected graph, but it does not have cycles, then graphs satisfying this description are actually trees in terms of graph theory. We will discuss problems related to trees from the perspective of graph.

> **i**
> **Knowledge ABC** STP
> STP represents spanning tree protocol. This protocol can be applied to the network. It implements path redundancy with a certain algorithm, and at the same time prunes the circular network into a tree-like network without cycles, and thus avoids the proliferation and infinite cycling of packets.
>
> The basic idea of STP is to generate "a tree." The root of the tree is a switch called root bridge. According to the differences in settings, different switches will be chosen as the root bridge. But at any given moment, there can only be one root bridge. Starting from the root bridge, a tree is formed level by level. The root bridge transmits configuration packets from time to time, while nonroot bridges receive the configuration packets and forward them. If a certain switch can receive the configuration packets from two or more ports, then there are more than one path from this switch to the root, and thus there is a cycle. At this time, the switch selects one port according to the configuration of the ports, and jam other ports, in order to eliminate the cycle. When a port cannot receive the configuration packets for a long period of time, the switch will consider the configurations for the port as expired, and the network topology potentially changed. At this time, it recomputes the network topology, and re-generates a tree.
>
> The SPT was invented by the famous engineer Radia Perlman of Sun Microsystems.

2.6.3 Spanning tree

If the subgraph of the connected graph G is a tree that contains all the vertices of G, then we call this subgraph a spanning tree of G.

A spanning tree is the minimum connected subgraph of a connected graph. The so-called minimum refers to the fact that if we add an edge in a tree, then a cycle forms; if we remove an edge, then it becomes a nonconnected graph.

If a spanning tree does not have a fixed root, then we call it a free tree. If we choose a vertex in the free tree as the root, then it becomes a normal tree. Starting from the root, specify a left-to-right order for the children of each vertex, then it becomes an ordered tree.

> **Think and Discuss** How to obtain a spanning tree of a graph?
>
> **Discussion:** The spanning tree of a graph contains all the vertices of the graph. We can review the traversal methods of graphs: the search path during the traversal leads to a process of avoiding repeated visit to vertices of the graph, that is, it is an acyclic path. Therefore, using different graph traversal methods, we obtain different spanning trees. Starting from different vertices, we can also potentially obtain different spanning trees. A concrete example of spanning tree is shown in Fig. 2.38, where figure G is the original figure. Depth-first spanning tree is the tree structure formed by connecting all the vertices according to the depth-first vertex search order. Similarly, breadth-first spanning tree is the tree structure formed by connecting all the vertices according to the breadth-first vertex search order.

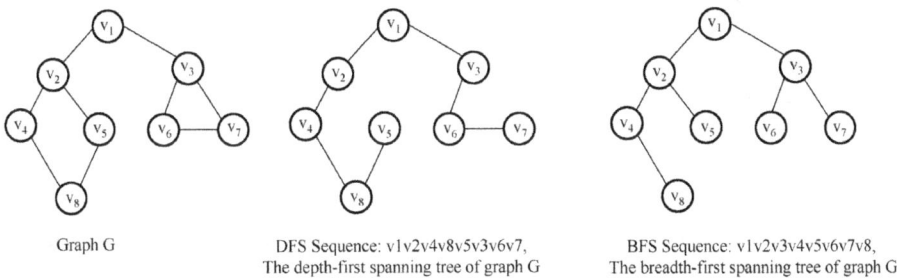

| Graph G | DFS Sequence: v1v2v4v8v5v3v6v7, The depth-first spanning tree of graph G | BFS Sequence: v1v2v3v4v5v6v7v8, The breadth-first spanning tree of graph G |

Fig. 2.38: Example of spanning tree.

2.6.4 Minimum spanning tree

In the introductory example 1, we can set up a link between any two cities. Then, between n cities (Fig. 2.39(a)), we can set up at most $n(n-1)/2$ links (Fig. 2.39(b)). How do we select $n-1$ links out of all these possible links (Fig. 2.39(c)), so that the total cost is minimized?

We can express n cities and the potential communication links between them with connected network, where the vertices of the network represent cities, and

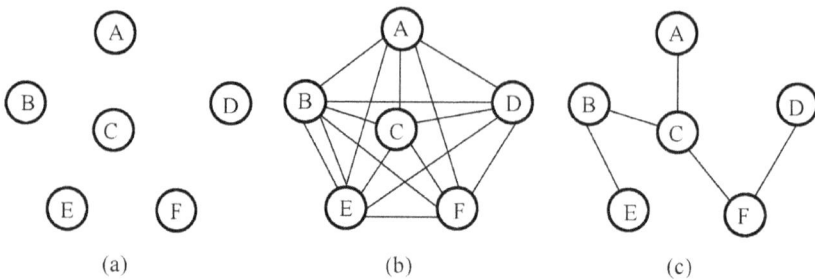

Fig. 2.39: Communication network between *n* cities.

edges represent links between two cities. The weights attached to edges represent the corresponding cost. For a connected network with *n* vertices, we can establish many different spanning trees, each of which can be a communication network. Now, we need to choose such a spanning tree, so that the total cost is minimized. This problem is to construct the minimum cost spanning tree of a connected network. The cost of a spanning tree is the sum of costs on all edges of the tree.

2.6.4.1 The concept of minimum spanning tree

The sum of weights of all edges of the spanning tree is called the weight of the spanning tree. The spanning tree with the minimum weight is called the MST.

Properties of MST The connected network spanning tree with *n* vertices has *n* vertices and *n*−1 edges.

To construct an MST, we need to solve the below two problems:
- Choose edges with smaller weights as much as possible, while avoiding cycles.
- Select *n*−1 appropriate edges to connect *n* vertices of the network.

2.6.4.2 Minimum spanning tree algorithm

The classical algorithms on MST include Prim's algorithm and Kruskal's algorithm. The two algorithms share a similar idea, that is, generate a "safe edge" that would not cause cycles each time. The difference between two algorithms is the method to generate the safe edge. The pseudocode is as follows.

Let T be the set of MSTs, and the set of MSTs should include *n* vertices and *n*−1 edges.

1. Initialize T = Empty set
2. while (T is not a spanning tree yet)
   3. find out an edge (u, v) that would not form a cycle for T and has minimum weight
   4. add edge (u, v) into T
5. return T

2.6.5 Algorithm for minimum spanning tree 1 – Prim's algorithm

The basic idea of Prim's algorithm: starting from a certain node u_0 in the connected network N = {V, E}, select the edge connected to it with the minimum weight (u_0, v), add its vertex to the set of spanning tree vertices U.

At each step that follows, select the edge with the minimum weight from the edges that have one vertex in U and another vertex not in U (u, v), add v into set U. Continue in this way, until all the vertices in the network are added to the set of spanning tree vertices U.

Detailed description of pseudocode:

Establish the list of candidate edge set, and record the weights from the beginning vertex u_0 to all the other vertices in it. Set u = u_0.

1. Select vertex u from the candidate edge set.
2. Select the shortest edge (u, v) from the candidate edge set.
3. Using v as the starting point, adjust the candidate edge set.

Adjustment method: when (u, x) > (v, x), substitute (u, x) with (v, x). x is a vertex other than u, v.

Repeat 1–3 until all the vertices are processed.

In the following, we use the communication network with weights from introductory example 1 to illustrate the execution steps of Prim's algorithm.

Set the beginning vertex u_0 = A, as shown in Fig. 2.40. The edges connected to A are (A, B) (A, C) and (A, D), where the edge with the lowest weight is (A, C).

–   For step 1 of Prim's algorithm, see Fig. 2.41. ∞ indicates that there is no path and no weight value.

    U_0 = A, list the weight values of A to all the other vertices (ending vertices) in the list of candidate edge set.

    (1) u = A.
    (2) Find the shortest edge (u, v) = (A, C) in the table of candidate edge set, v = C.
    (3) List the weight values from v to all ending points in the table of candidate edge set.
    (4) Compare the weights from u to the ending point and from v to the ending point x, select the edge with the smaller weight.

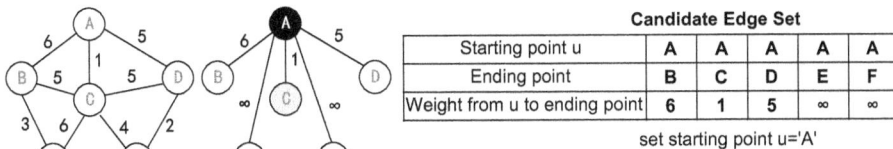

Candidate Edge Set

Starting point u	A	A	A	A	A
Ending point	B	C	D	E	F
Weight from u to ending point	6	1	5	∞	∞

set starting point u='A'

Fig. 2.40: Determining the starting point in Prim's algorithm.

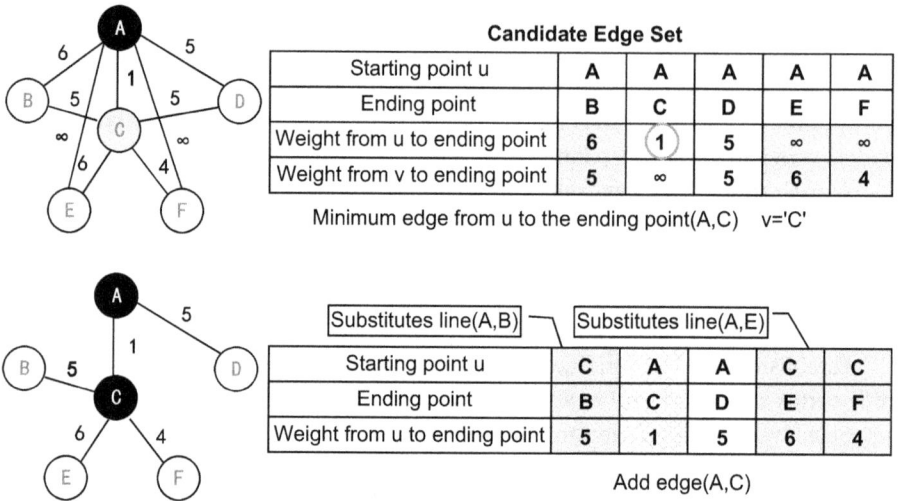

Candidate Edge Set

Starting point u	A	A	A	A	A
Ending point	B	C	D	E	F
Weight from u to ending point	6	(1)	5	∞	∞
Weight from v to ending point	5	∞	5	6	4

Minimum edge from u to the ending point(A,C)   v='C'

Substitutes line(A,B)		Substitutes line(A,E)		
Starting point u	C	A	A	C	C
Ending point	B	C	D	E	F
Weight from u to ending point	5	1	5	6	4

Add edge(A,C)

Fig. 2.41: Step 1 of Prim's algorithm.

If the ending point x = B, (A, B) = 6 > (C, B) = 5, then choose edge (C, B) to replace (A, B). The other substitutions include (C, E) substituting (A, E) and (C, F) substituting (A, F). Note that the edges that have the same weight will not be adjusted. After the whole candidate edge set is adjusted, we ascertain that edge (A, C) is added to U. We mark it in the table. The direct connections in the graph of the substituted edges will be removed, for example, the substitution of (A, B) line.

-   For Prim's algorithm step 2, see Fig. 2.42. The "/" in the table indicates that the edge from the beginning point to the ending point has been ascertained, and would not need be considered again.
    (1) Find the shortest edge in the candidate edge set table (u, v) = (C, F), v = F. Note that the previously already added edge (A, C) should not be considered again. All the algorithm steps will follow this principle.
    (2) List the weights from v to various ending points in the candidate edge set table. The edge adjustment is the substitution of (A, D) by (F, D). We ascertain that edge (C, F) is added to the set U.

-   For Prim's algorithm step 3, see Fig. 2.43.
    (1) Find the shortest edge in the table of set of candidate edges (u, v) = (F, D), v = D.
    (2) List the weights of v to all ending points in the table of set of candidate edges. We ascertain the addition of edge (F, D) to the set U.

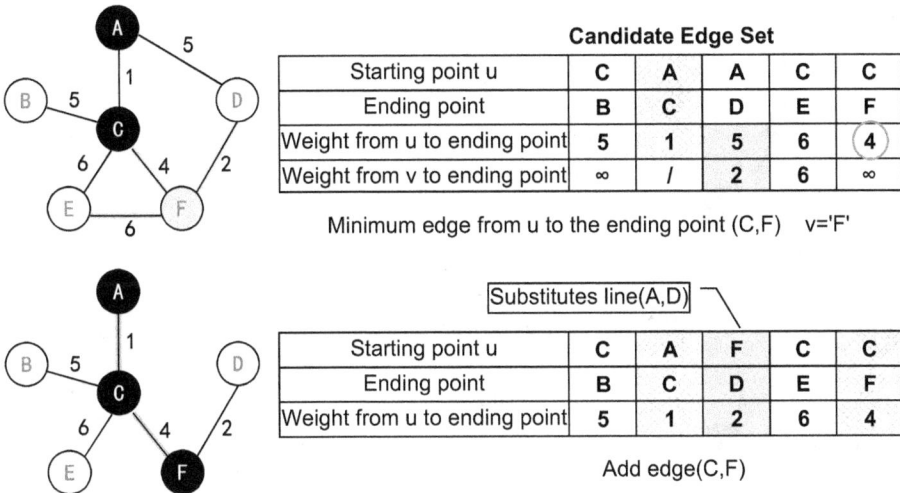

Candidate Edge Set

Starting point u	C	A	A	C	C
Ending point	B	C	D	E	F
Weight from u to ending point	5	1	5	6	4
Weight from v to ending point	∞	/	2	6	∞

Minimum edge from u to the ending point (C,F)   v='F'

Substitutes line(A,D)

Starting point u	C	A	F	C	C
Ending point	B	C	D	E	F
Weight from u to ending point	5	1	2	6	4

Add edge(C,F)

Fig. 2.42: Prim's algorithm step 2.

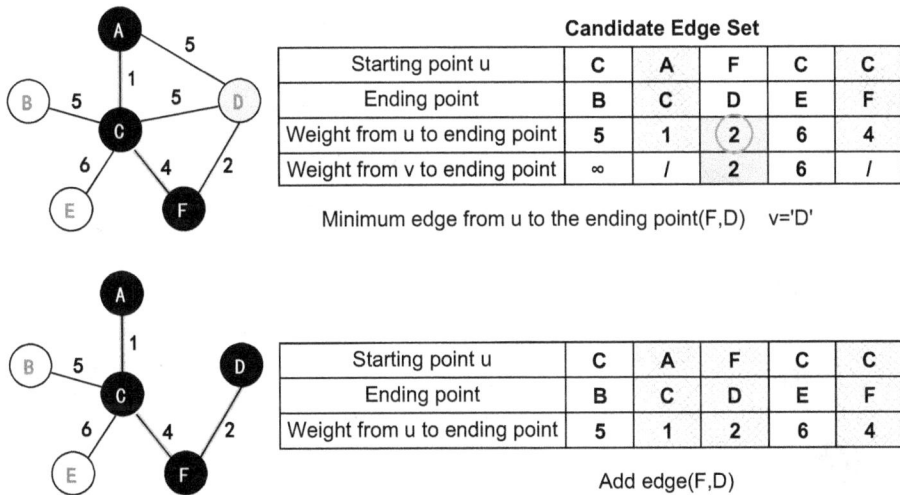

Candidate Edge Set

Starting point u	C	A	F	C	C
Ending point	B	C	D	E	F
Weight from u to ending point	5	1	2	6	4
Weight from v to ending point	∞	/	2	6	/

Minimum edge from u to the ending point(F,D)   v='D'

Starting point u	C	A	F	C	C
Ending point	B	C	D	E	F
Weight from u to ending point	5	1	2	6	4

Add edge(F,D)

Fig. 2.43: Prim's algorithm step 3.

- For Prim's algorithm step 4, see Fig. 2.44.
  (1) Find the shortest edge in the table of set of candidate edges (u, v) = (C, B), v = B.
  (2) List the weight values of v to various ending points, and the edge adjusted is edge (C, E) substituted by edge (B, E). We ascertain the addition of edge (C, B) to the set U.

Candidate Edge Set

Starting point u	C	A	F	C	C
Ending point	B	C	D	E	F
Weight from u to ending point	5	1	2	6	4
Weight from v to ending point	∞	/	/	3	/

Minimum edge from u to the ending point(C,B) v='B'

Substitutes line(C,E)

Starting point u	C	A	F	B	C
Ending point	B	C	D	E	F
Weight from u to ending point	5	1	2	3	4

Add edge(C,B)

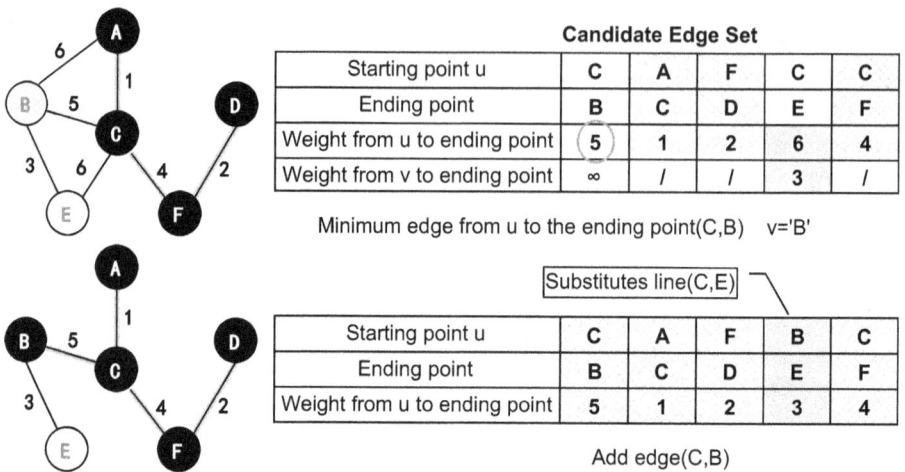

Fig. 2.44: Prim's algorithm step 4.

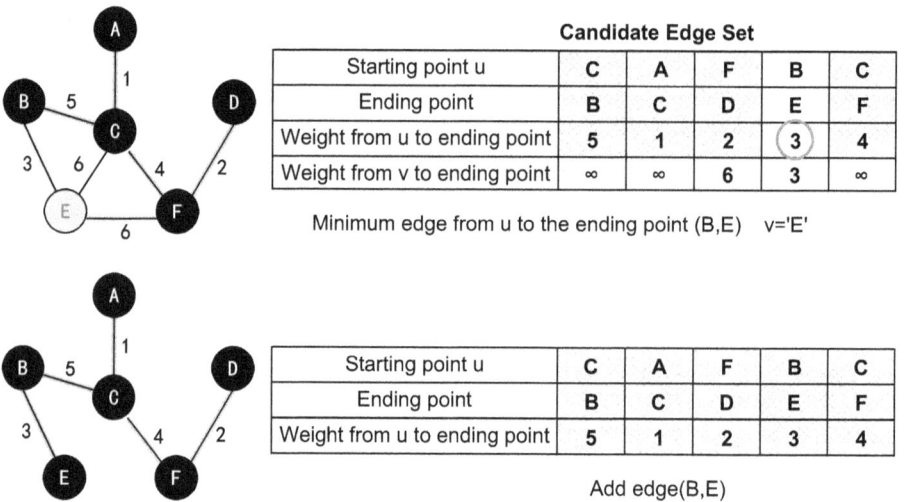

Candidate Edge Set

Starting point u	C	A	F	B	C
Ending point	B	C	D	E	F
Weight from u to ending point	5	1	2	3	4
Weight from v to ending point	∞	∞	6	3	∞

Minimum edge from u to the ending point (B,E) v='E'

Starting point u	C	A	F	B	C
Ending point	B	C	D	E	F
Weight from u to ending point	5	1	2	3	4

Add edge(B,E)

Fig. 2.45: Prim's algorithm step 5.

- For Prim's algorithm step 5, see Fig. 2.45.
  (1) Find the shortest edge in the table of set of candidate edges (u, v) = (B, E), v = E.
  (2) List the weights of v to all ending points in the table of set of candidate edges. We ascertain the addition of edge (B, E) to the set U.

Program implementation

```c
#include <stdio.h>
#define VERTEX_NUM 6        // Number of vertices of the graph
#define INF 32767        //INF represents ∞
typedef int InfoType;
InfoType AdjMatrix [VERTEX_NUM][ VERTEX_NUM];   //Adjacency matrix

void DispMat(InfoType AdjMatrix[][VERTEX_NUM]);   //Output the adjacency matrix
void prim(InfoType AdjMatrix[][VERTEX_NUM],int v);
int main()
{
    int A[VERTEX_NUM][VERTEX_NUM]=
    {{INF,6,1,5,INF,INF},
    {6,INF,5,INF,3,INF},
    {1,5,INF,5,6,4},
    {5,INF,5,INF,INF,2},
    {INF,3,6,INF,INF,6},
    {INF,INF,4,2,6,INF}};   //Initialize the adjacency matrix

    printf("Adjacency matrix of the graph:\n");
    DispMat(A);
    printf("\n");
    printf("Result of Prim algorithm:\n");
    prim(A,0);
    printf("\n");
    return 0;
}
/*============================================================================
Functionality: Construct minimum spanning tree with Prim algorithm
Function input: The adjacency matrix, the index of the starting vertex
Function output: None
Screen output: Adjacency matrix, various edges of the minimum spanning tree
============================================================================*/
// U--- Set of vertices of the spanning tree
void prim(int AdjMatrix[][VERTEX_NUM], int v)
{
    int i,j,k;
    struct set
    {
        int starNode[VERTEX_NUM];      // Starting node
        int endNode[VERTEX_NUM];      // ending node
```

```
    int value[VERTEX_NUM];        // Weights
} edge Set;    // Set of candidate edges

int Visited[VERTEX_NUM]={0};
// Visited marker array, records the nodes already added to U
int min;    // Records the minimum weight in the set of candidate edges
for (i=0;i<VERTEX_NUM;i++)    // Initialize the set of candidate edges
{
    if (i!=v)
    {
        edge Set.starNode[i]=v;       // At initialization, v is the starting node
        edge Set.endNode[i]=i;        // Assign ending node
        edge Set.value[i]=AdjMatrix[v][i];    // Assign weight value
    }
}
// Add the starting node v to the set of candidate edges
edge Set.starNode[v]=v;
edge Set.endNode[v]=v;
edge Set.value[v]=INF;

for (i=1;i<VERTEX_NUM;i++)
{
    min=INF;
    // Find vertex k in the candidate edge set which
    // hasn't been added to U and has minimum value
    for (j=0;j<VERTEX_NUM;j++)
    {
        if (edge Set.value[j]<min && Visited[j]==0)
        {
            min=edge Set.value[j];
            k=j;       //k records the vertex with minimum value
        }
    }
    Visited[k]=1;       //k is added to U
    if(min!=INF)
    printf("Edge (%c,%c) has weight value of:%d\n",edge Set.starNode[k]
    +'A',k+'A',min);
    // Due to the addition of vertex k,
    // adjust the value and startNode of candidate edge set
```

```
    for (j=0;j<VERTEX_NUM;j++)
    {
       if (AdjMatrix[j][k]<edge Set.value[j]&& j!=v)
       {
          edge Set.value[j]=AdjMatrix[j][k];
          edge Set.starNode[j]=k;
       }
    }
  }
}
/*======================================
Functionality: Print the adjacency matrix
Function input: Adjacency matrix
Function output: None
Screen output: The adjacency matrix
======================================*/
void DispMat(InfoType AdjMatrix[][VERTEX_NUM])
{
   int i,j;
   for (i=0;i<VERTEX_NUM;i++)
   {
      for (j=0;j<VERTEX_NUM;j++)
      if (AdjMatrix[i][j]==INF)
      printf("%3s","∞");
      else
      printf("%3d",AdjMatrix[i][j]);
      printf("\n");
   }
}
```

Test result

The adjacency matrix of the graph:

```
∞   6   1   5   ∞   ∞
6   ∞   5   ∞   3   ∞
1   5   ∞   5   6   4
5   ∞   5   ∞   ∞   2
∞   3   6   ∞   ∞   6
∞   ∞   4   2   6   ∞
```

Result from Prim's algorithm:

edge(A, C) with weight: 1   edge(C, F) with weight: 4   edge(F, D) with weight: 2
edge(C, B) with weight: 5   edge(B, E) with weight: 3

2.6.6 Algorithm for minimum spanning tree 2 – Kruskal's algorithm

2.6.6.1 Idea of Kruskal's algorithm

Prim's algorithm starts from a certain vertex and gradually find the smallest weight on each vertex in order to construct the MST. Now we can change a way of thinking: we start from edges, since the weights are actually attached to edges. It is natural to directly find the edges with the smallest weights and construct the spanning tree. This is the essence of Kruskal's algorithm.

Suppose that the set of spanning trees of graph G, T only has all the vertices of the graph but has no edges in the beginning. Each time, from the edge set of graph G, E, we choose the edge that has the smallest weight and would not cause a cycle to add to set T, until all vertices are covered, as shown in Fig. 2.46.

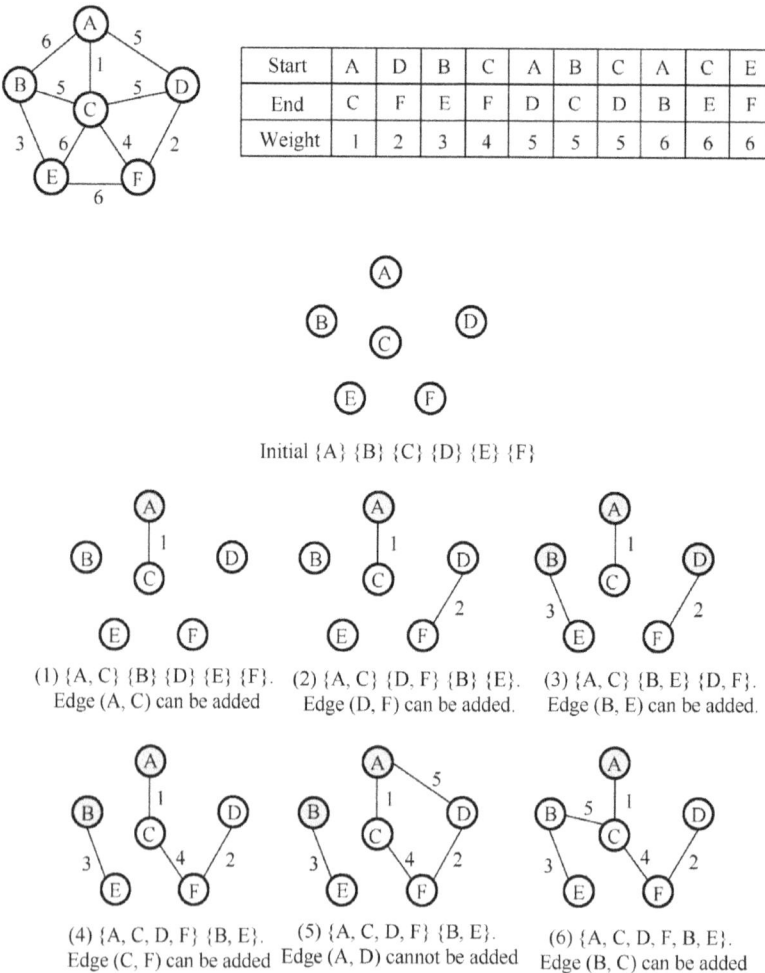

Start	A	D	B	C	A	B	C	A	C	E
End	C	F	E	F	D	C	D	B	E	F
Weight	1	2	3	4	5	5	5	6	6	6

Initial {A} {B} {C} {D} {E} {F}

(1) {A, C} {B} {D} {E} {F}.
Edge (A, C) can be added

(2) {A, C} {D, F} {B} {E}.
Edge (D, F) can be added.

(3) {A, C} {B, E} {D, F}.
Edge (B, E) can be added.

(4) {A, C, D, F} {B, E}.
Edge (C, F) can be added

(5) {A, C, D, F} {B, E}.
Edge (A, D) cannot be added

(6) {A, C, D, F, B, E}.
Edge (B, C) can be added

Fig. 2.46: Steps of Kruskal's algorithm.

In the steps of Kruskal's algorithm in Fig. 2.46, the initial spanning tree T is a forest with N nodes. Select an edge with the smallest weight from the edge set of the graph E, if the two vertices of this edge belong to different trees (connected component), then we combine the two trees that these two vertices, respectively, belong to one tree. Otherwise, if the two vertices of this edge already belong to the same tree, then this edge is not selectable. We should select the edge with next smallest weight and try again. Continue in this manner, until there is only one tree left, that is, the sub-graph has N–1 edges. For example, in step 1, the two vertices of edge (A, C) belong to two trees {A} and {C}, respectively, then we can combine these two trees into one tree {A, C}. In step 5, for edge (A, D), since both vertices A and D are in the connected component {A, C, D, F}, that is, in the same tree, adding this edge will cause cycle; thus, we discard this edge.

The description of Kruskal's algorithm is as follows.

Suppose that an undirected connected network G = (V, E), where V is the vertex set of the graph, E is the edge set of the graph, T is the set of spanning tree and N is the number of vertices.

1. Sort the edge set E in ascending order according to the weight of each edge.
2. Initialize the MST set T = (V, ф)– that is, it has only vertices and no edges.
3. Get the edge with minimum weight in the current edge set, suppose that edge e = (u, v), if the two vertices connected by the current edge u and v are not in the same tree, then combine the trees that two vertices are located into one tree. At the same time, delete edge e from edge set E and add edge e into the set of MST T.
4. Repeat step 3 until the number of edges in the minimal spanning tree set T is N–1.

Think and Discuss What are the key issues of implementing Kruskal's algorithm?   **!**
Discussion: The key to Kruskal's algorithm should be the following:
 – How to check whether the two vertices u and v connected by edge e are not on the same tree?
 – How to combine the two trees of vertices u and v into one tree?

To check two vertices belong to one tree, the key should be to find the relation between the marker of this tree and the vertex. An obvious marker of the tree should be the root node of the tree. If two vertices belong to the same tree, then they should share the same root node. Therefore, the way to solve the first problem can be to look up the root nodes of nodes u and v, respectively. If their root nodes are the same, then we can believe they are on the same tree.

The way to solve the second issue is to connect the root nodes of u and v nodes as parent/child nodes. In this way, we compose two separate trees into one tree.

2.6.6.2 The data structure design in Kruskal's algorithm

Each subtree is a connected component, that is, a subset. According to the idea discussed above, in each combination step of subtrees we should mark their roots, that is, for each node in the subset we should record the field for its root node. If we use continuous storage, then we can design a static link pointing to its root node (subscript). We can store the number of members of elements of subsets of this tree in the parent field of the root node. In order to differentiate it from ordinary node, we can set the number of elements in the parent field of the root node as negative. The design for the structure to record information above the parent of various vertices of the spanning tree is shown in Fig. 2.47.

	parent [N]						
Corresponding node	A	B	C	D	E	F	
Corresponding subscript	0	1	2	3	4	5	
Initial array	-1	-1	-1	-1	-1	-1	Initially every node is a root
Changed Array	-2	-1	0	-1	-1	-1	Edge (A, C) is added

Now the subtree with root A has two nodes		C's root is A, recorded at the position with subscript 0

Fig. 2.47: The structure to record information about parent of vertex.

In the beginning, every node is a root, then the initial value in the array would be −1, to indicate that the current node is the root. There is only one node in the subtree with this root as the tree. When the edge (A, C) is added, the roots of the two subtrees of A, C are different; thus, we can combine these two subtrees (or subsets). At this time, if we select A as the root of the subtree, then C becomes a normal node of the tree. We need to change its root field from the original −1 to 0, which points to node A that has subscript 0.

Figure 2.48 gives the relations between various nodes in the subtree corresponding to the steps of Kruskal's algorithm in Fig. 2.46. L root represents the root of the subset with relatively more nodes; S root represents the root of the subset with relatively fewer nodes.

In the beginning, every node is a subtree, whose root is itself. In step 4, the edge (C, F) is added. At the status at the end of step 3, we can see that the parents of nodes C and F are A and D, respectively; thus we can add the edge (C, F). Now, C can also be the parent of D. The problem is, which node should be the node of F? If we still use D as the root, then the components {A, C} and {D, F} did not change; if we choose C as the root, since the root of C is A, then the root of F and the root of C should combine into A. In this way, subtrees {A, C} and {D, F} are combined into one subtree due to the addition of edge (C, F). See steps 3 and 4 in Fig. 2.46.

Node	parent [N]						Add edge v1 v2	L Root	S Root
	A	B	C	D	E	F			
Subscript	0	1	2	3	4	5			
Initial	−1	−1	−1	−1	−1	−1	None	None	None
Steps	−2	−1	0	−1	−1	−1	A--C	A	C
	−2	−1	0	−2	−1	3	D--F	D	F
	−2	−2	0	−2	1	3	B--E	B	E
	−4	−2	0	0	1	3	C--F	A	D
	−4	−2	0	0	1	3	A--D	A	A
	−6	0	0	0	1	3	B--C	A	B
	−6	0	0	0	1	3	C--D	A	A
	−6	0	0	0	1	3	A—B	A	A
	−6	0	0	0	1	3	C—E	A	A
	−6	0	0	0	1	3	E--F	A	A

Fig. 2.48: Relations between nodes in Kruskal's algorithm.

Therefore, in step 4 of Fig. 2.48, the parent of D should be changed to subscript 0, the parent field of A, due to the addition of subtree {D, F}, should be increased by 2; thus, its value is changed to −4.

Knowledge ABC The combination problem of trees and union-find sets

The problem of combination of trees belongs to the union-find set problem of union operation of subsets. To carry out the union operation on subsets with union-find sets, that is, to union the subsets containing two elements, we need to first ascertain the root nodes corresponding to the subsets containing the two elements, and put the root node link as the subtree of another tree. To reduce/avoid the appearance of deformed tree (nearly singly linked tree) during the union operation, we normally use the tree corresponding to the subset with relatively fewer elements as the subtree of the tree corresponding to the subset with relatively more elements.

Union-find set is a tree-structured data structure used to process the union and find operations on some disjoint sets. It is usually represented in usage as a forest. A set is to let each element to construct a single element set, that is, to union the sets belonging to the same group of elements according to a set order.

The main operations on union-find sets are:

1. Initialization: initialize the set containing each node as itself.
2. Find: find the set that contains the element, that is, find the root node.
3. Union: union the sets containing two elements into one set. Normally speaking, before the union operation we should first check whether two elements belong to the same set. This can be realized with find operation.

2.6.6.3 The program implementation of Kruskal's algorithm

1. Data structure description

Set the set with relatively more nodes as L and the set with relatively fewer nodes as S.
- Edge set array

The weights are stored in the edge set array Edge Set[] in ascending order.

```
Struct for edge set array
typedef struct //Edge set array unit structure
{ VexType start_vex;    // Starting point
  VexType end_vex;      // Ending point
  InfoType weight;    // The weight can be assigned
  int sigle;    // Marker for whether the current edge is added. 0 is the
  initial value. 1 means added
} EdgeStruct;
```

- Array for parent node

Record the root of each vertex. If it is negative, it refers to the number of nodes in the subset.

```
int parent[VERTEX_NUM];
```

2. Pseudocode description

See Tables 2.7 and 2.8.

Table 2.7: Description of Kruskal's algorithm.

Description of Kruskal's algorithm
Order the weights of edges in the edge set of the graph in ascending order
Check the edges in the edge set e = (u, v) according to the order, one by one
If adding this edge to the spanning tree will cause a cycle, then discard it
Else add this edge to the spanning tree
Until N−1 edges are selected

Table 2.8: Description of cycle checking algorithm.

Description of cycle checking algorithm	Detailed description of cycle checking algorithm
Initialize parent[N]	Set the initial value of parent[N] to −1
If there is an edge (u, v)	If there is an edge (u, V), find the roots of nodes u and v
Check whether the roots of u, v are the same	Check whether the roots of u, v are the same
If yes, cycle exists, discard edge (u, v)	If yes, cycle exists, discard edge (u, v)
If no, there is no cycle, union the sets containing u and v	If no
	(1) Select (u, v) as one edge of the spanning tree
	(2) If the subset containing v has more nodes than that containing u, set the root of v as the root of the subset containing u; otherwise set the root of u as the root of the subset containing v

3. Program for Kruskal's algorithm

```
/*===========================================================================
Functionality: Kruskal algorithm to get the minimum spanning tree of the graph
Function input: Edge set of the graph, the number of edges, the number of
vertices
Function output: None
===========================================================================*/
void Kruskal(EdgeStruct Edge Set[],int edge_num, int vertex_num )
{
    int parent[VERTEX_NUM];
    // Record the root of each vertex. If it's negative,
    // it indicates the number of vertices in this set
    int i,k;
    int num=0;
    int v1Root,v2Root;
    int LRoot, SRoot;
    //LRoot: The root of the bigger set; SRoot: The root of the smaller set
    char LVertex,SVertex;
    for (i=0;i<vertex_num;i++)    parent[i]=-1;
    i=0; k=0;
    while ( k<edge_num && num<vertex_num )
```

```
{
    // Find the root of start_vexd v1Root
    v1Root=(Edge Set[k].start_vex-'A');
    while (parent[v1Root]>=0) v1Root=parent[v1Root];
    // Find the root of end_vexd v2Root
    v2Root=(Edge Set[k].end_vex-'A');
    while (parent[v2Root]>=0) v2Root=parent[v2Root];
    // Union set S into set L
    if (parent[v1Root]<=parent[v2Root])
    {
        LRoot=v1Root;
        SRoot=v2Root;
        LVertex= Edge Set[k].start_vex;
        SVertex= Edge Set[k].end_vex;
    }
    else
    {
        LRoot=v2Root;
        SRoot=v1Root;
        LVertex= Edge Set[k].end_vex;
        SVertex= Edge Set[k].start_vex;
    }
    printf("%c--%c ",Edge Set[k].start_vex,Edge Set[k].end_vex);
    printf("v1Root=%c v2Root=%c\n",LRoot+'A',SRoot+'A');
    // If start_vex and end_vex have different roots, then union set S into set L
    if (v1Root!=v2Root)
    {
        parent[LRoot]+=parent[SRoot];
        // The number of elements of subset L is combined with the number of
        elements of subset S
        parent[SRoot]=LRoot;
        // The root of node S is changed to the root of node L
        Edge Set[k].sigle=1;
        num++;
    }
    for (i=0;i<vertex_num;i++) printf("%4d",parent[i]);
    printf("\n");
    k++;
}
}
```

Program testing: test case is shown in Fig. 2.49.

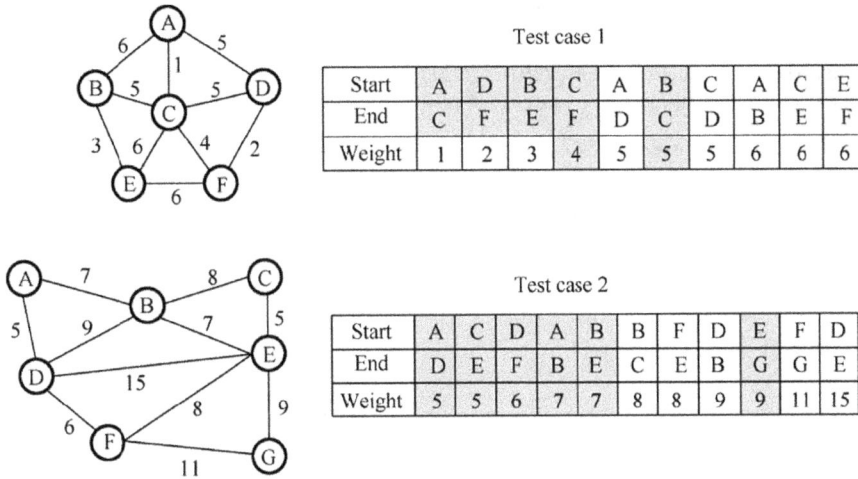

Test case 1

Start	A	D	B	C	A	B	C	A	C	E
End	C	F	E	F	D	C	D	B	E	F
Weight	1	2	3	4	5	5	5	6	6	6

Test case 2

Start	A	C	D	A	B	B	F	D	E	F	D
End	D	E	F	B	E	C	E	B	G	G	E
Weight	5	5	6	7	7	8	8	9	9	11	15

Fig. 2.49: Test case for Kruskal's algorithm.

```c
#include <stdio.h>
#define VERTEX_NUM 6 // The number of vertices for test data 1
#define EDGE_NUM 10 // The number of edges for test data 1
//#define VERTEX_NUM 7 // The number of vertices for test data 2
//#define EDGE_NUM 11 // The number of edges for test data 2

typedef char VexType;
typedef int InfoType;
typedef struct // Unit struct for the edge set array
{   VexType start_vex; // Starting vertex
    VexType end_vex; // Ending vertex
    InfoType weight; // The weight can be assigned
    int sigle;
} EdgeStruct;

void Kruskal(EdgeStruct Edge Set[],int edge_num, int vertex_num);
int main()
{
    EdgeStruct Edge Set[EDGE_NUM // Edge set array]
    //==========Test data 1===============
    n={{'A','C',1,0},{'D','F',2,0},{'B','E',3,0},{'C','F',4,0},
    {'A','D',5,0},{'B','C',5,0},{'C','D',5,0},{'A','B',6,0},
    {'C','E',6,0},{'E','F',6,0}};
```

```
//=========Test data 2=============
//={{'A','D',5,0},{'C','E',5,0},{'D','F',6,0},{'A','B',7,0},
//{'B','E',7,0},{'B','C',8,0},{'F','E',8,0},{'D','B',9,0},
//{'E','G',9,0},{'F','G',11,0},{'D','E',15,0}};
Kruskal(Edge Set,EDGE_NUM,VERTEX_NUM);
for (int i=0; i<EDGE_NUM; i++)
    if (Edge Set[i].sigle==1)
        printf("%c--%c %d\n",Edge Set[i].start_vex,Edge Set[i].end_vex,
        Edge Set[i].weight);
    return 0;
}
```

Test results

Test results of test case 1	Test results of test case 2
A--C v1Root=A v2Root=C	A--D v1Root=A v2Root=D
-2 -1 0 -1 -1 -1	-2 -1 -1 0 -1 -1 -1
D--F v1Root=D v2Root=F	C--E v1Root=C v2Root=E
-2 -1 0 -2 -1 3	-2 -1 -2 0 2 -1 -1
B--E v1Root=B v2Root=E	D--F v1Root=A v2Root=F
-2 -2 0 -2 1 3	-3 -1 -2 0 2 0 -1
C--F v1Root=A v2Root=D	A--B v1Root=A v2Root=B
-4 -2 0 0 1 3	-4 0 -2 0 2 0 -1
A--D v1Root=A v2Root=A	B--E v1Root=A v2Root=C
-4 -2 0 0 1 3	-6 0 0 0 2 0 -1
B--C v1Root=A v2Root=B	B--C v1Root=A v2Root=A
-6 0 0 0 1 3	-6 0 0 0 2 0 -1
C--D v1Root=A v2Root=A	F--E v1Root=A v2Root=A
-6 0 0 0 1 3	-6 0 0 0 2 0 -1
A--B v1Root=A v2Root=A	D--B v1Root=A v2Root=A
-6 0 0 0 1 3	-6 0 0 0 2 0 -1
C--E v1Root=A v2Root=A	E--G v1Root=A v2Root=G
-6 0 0 0 1 3	-7 0 0 0 2 0 0
E--F v1Root=A v2Root=A	F--G v1Root=A v2Root=A
-6 0 0 0 1 3	-7 0 0 0 2 0 0
A-C 1	D--E v1Root=A v2Root=A
D-F 2	-7 0 0 0 2 0 0
B-E 3	A-D 5
C-F 4	C-E 5
B-C 5	D-F 6
	A-B 7
	B-E 7
	E-G 9

2.6.7 Summary of spanning tree algorithm

The basic ideas of Prim's algorithm and Kruskal's algorithm are similar. Prim's algorithm focuses on vertices, and Kruskal's algorithm focuses on edges. The aim is to let the sum of weights of all the edges in the spanning tree to be minimum. We need to ensure the weight of each edge in the spanning tree to be minimum.

Kruskal's algorithm is more efficient than Prim's algorithm, since Kruskal's algorithm only needs one sorting of weight values, while Prim's algorithms need to perform multiple sorts. However, each run of Prim's algorithm would not necessarily cover all the edges of a connected graph.

Prim's algorithm and Kruskal's algorithm are suitable for dense graph and sparse graph, respectively.

2.7 Classical application of graph – shortest path problems

2.7.1 Introduction to shortest path problems

2.7.1.1 Effective transmission of data in the Internet

The Internet is used to transmit data. For the data to be transmitted from the origin to the target, more than one intermediary points will be passed, and there will potentially be multiple transmission paths. What mechanisms are used to transmit data and how to choose the best transmission paths are problems to be solved in network communication.

> **Knowledge ABC** Data packet interchange technology
>
> Most computer networks are not able to transmit arbitrarily long data. Hence, a network system is to divide the data (also called message) to be transmitted into pieces, and send them piece by piece. Such pieces are called data packets. A message will normally be divided into thousands of packets. The transmission of these packets is independent of each other, and can follow different paths (also called routes) to arrive at the destination, just as there are multiple transportation routes between two places. Once they arrive at the destination, the original message is constructed again. To ensure the correctness of transmission and construction, each data packet needs to contain information such as source, destination, data packet serial number and data content.
>
> The concept of data packet proposed by the French Louis Pouzin appeared the earliest in the CYCLADES packaging network plan in 1970. This packeted data transmission technology is called data packet exchange. IP (Internet Protocol) network is a typical packet exchange network; the corresponding international standard is X.25.
>
> During the transmission of data packets, there must be at least one data transmission pathway between the two sides of communication. Such pathways might need to go through multiple intermediate node devices – router. A router acts like a "traffic police"; it will

automatically select and set routes according to the conditions of information pathways, sending the received data packets along the best paths and according to the order of arrival shown in Fig. 2.50.

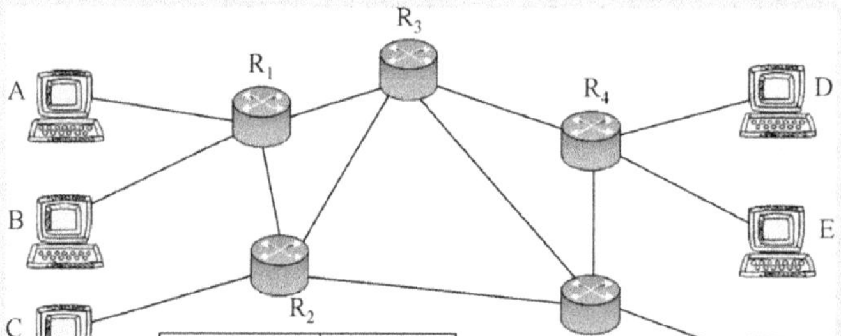

Routing table of R2

Fig. 2.50: Data packet transmission in the Internet.

After the router at each node receives the packets, according to the destination addresses in the packets, it looks up the corresponding next stop and sends them out. A data packet reaches its destination after multiple such transmissions. The process of selecting the path for data packets from the starting site to the destination site is called "routing" and the corresponding algorithm is called "routing algorithm."

In the network, in order to efficiently transport the data packets to their destinations, the shortest path from the source site to the destination site would be needed (there are various methods for measuring path length, e.g., distance, pathway bandwidth, communication throughput, communication cost, queue length and transmission delay). For actual networks, their topological structure is a network graph. Each vertex in the graph represents a router, and each arc represents the connection relation between routers (link pathway). The number on the arc (weight) represents the cost of the linkage (e.g., transmission delay), as shown in Fig. 2.51.

For the problem of selecting the best routing path between a pair of given routers, we can convert it to the problem of finding the shortest path between these vertices in the graph.

Routing algorithms are algorithms to improve the functionalities of routing protocols and reduce the cost that comes with routing as much as possible. When the software to implement routing algorithms must be run on devices with limited physical resources, its efficiency is especially important.

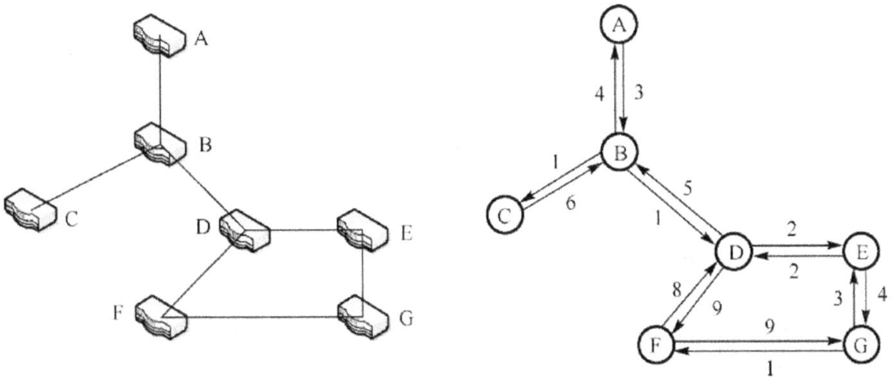

Fig. 2.51: Topological model of network.

2.7.1.2 Algorithms related to the shortest path problem and its related applications
Shortest path problem is a classical problem in the study of graph theory. It aims at
finding the shortest path between two vertices, that is, for all paths that begin from
a certain vertex and reach at another vertex, we need to find the one that has the
smallest sum of the weights on all its edges – the shortest path.

The calculation of shortest path can be divided into single-source shortest path
and all-pair shortest path according to the number of source vertices. The shortest
path problem involving single source tries to calculate the shortest paths from one
certain vertex to all the other vertices in a directed weighted graph. All-pair shortest
path problem is to calculate the shortest paths between all pairs of vertices in directed or undirected weighted graph. Classical algorithms for single-source shortest
path include the Dijkstra algorithm and the Bellman–Ford algorithm, where the
Dijkstra algorithm mainly applies to single-source shortest path problems where all
weights are nonnegative, while the Bellman–Ford algorithm can also be applied to
problems where the weights can be negative. Algorithms for all-pair shortest path
mainly include the Floyd–Warshall algorithm and the Johnson algorithm.

Shortest path calculation can be divided into static shortest path calculation
and dynamic shortest path calculation, depending on whether the data for calculation change. Static shortest path algorithms calculate the shortest path assuming
the external environment doesn't change. They mainly include Dijkstra's algorithm,
A* (A star) algorithm. Dynamic shortest path calculation refers to the calculation of
shortest path when the external environment constantly changes, that is, under un-
predictable circumstances, for example, the constant moving of enemies and hur-
dles in a game. A typical algorithm is D* algorithm.

Shortest path problem is a classical problem in the study of graph theory and
network technology. It is always widely applied in areas such as geographical information, computer network and traffic query. The other important applications now

include routing algorithms for computer networks, robot pathfinding, traffic navigation, artificial intelligence and game design. The core pathfinding algorithm for American Mars explorers uses D* (D star) algorithm.

This chapter only introduces the classical Dijkstra's algorithm and Floyd–Warshall's algorithm.

2.7.2 Single-source shortest path algorithm – Dijkstra's algorithm

Dijkstra's algorithm is the most typical shortest path routing algorithm, which is used to calculate the shortest paths from one vertex to all the other vertices in a directed graph. The algorithm uses BFS strategy. It solves single-source shortest path problems in directed graphs with nonnegative weights. The algorithm eventually obtains a shortest path tree. Although nowadays we already have better solutions for finding the shortest path, many systems still use Dijkstra's algorithm out of stability requirement.

The Dijkstra algorithm was proposed by the Dutch computer scientist Turing laureate Edgar Wybe Dijkstra in 1959.

2.7.2.1 Problem description
Given a weighted directed graph G and a source vertex v, obtain the shortest paths from v to the rest of the vertices in G. The weights in the graph are nonnegative.

2.7.2.2 Problem analysis
To obtain the shortest paths from the source vertex v to the rest of the vertices in G, one possible method is to list all paths and calculate the length of each path, then choose the shortest one. We will illustrate how to effectively solve this type of problem in as follows.

In a network, the following are the possibilities regarding whether there is a path between any two vertices:
- No path
- Only one path
- n paths exist

There are only two possibilities for the shortest path from a vertex v to an arbitrary vertex v_i. The first is that there is only one path from v to v_i, then it will be the shortest path; the second is that there are multiple paths from v to v_i, then we need to ascertain which is the shortest among these n paths.

In Fig. 2.52, suppose that the source vertex is v_0, then by listing the path situations from v_0 to the other vertices, we can get a path table.

By observing and analyzing the table, we can discover the following two properties:

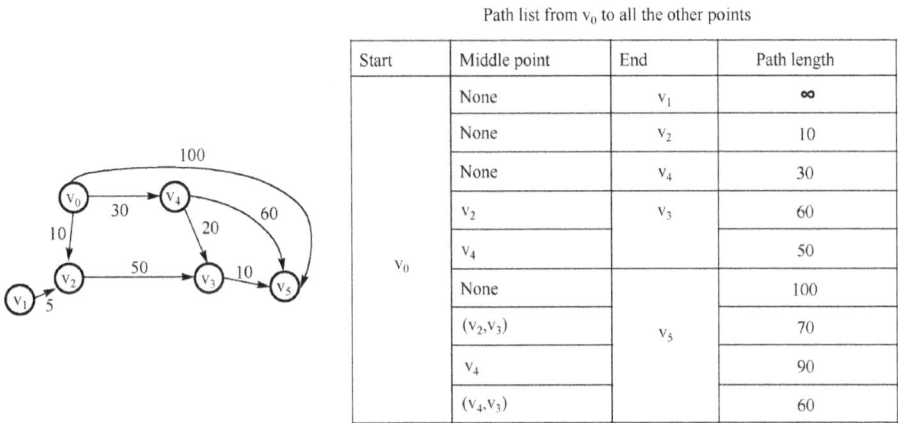

Path list from v_0 to all the other points

Start	Middle point	End	Path length
v_0	None	v_1	∞
	None	v_2	10
	None	v_4	30
	v_2	v_3	60
	v_4		50
	None		100
	(v_2,v_3)	v_5	70
	v_4		90
	(v_4,v_3)		60

Fig. 2.52: Analysis of single-source shortest path.

- To obtain the shortest path from the source vertex to various vertices is to start from the closest vertices from the source vertex (i.e., no intermediary vertices), ascertain the shortest paths to each vertex one by one. For example, if we first list the paths from v_0 to v_1, v_2, v_4, which do not have any intermediary vertex, then these paths are already the shortest paths.
- If there are multiple paths from the source vertex to a vertex, and there are intermediary vertices on its shortest path, then the shortest paths from the source vertex to these intermediary vertices should have already been generated before this. For example, from v_0 to v_3, one can either go through intermediary vertex v_2 or v_4 with the shortest path of 50. Before ascertaining, the shortest paths from v_0 to v_2 and v_4 are already ascertained. As another example, the shortest path from v_0 to v_5 is 60. The intermediary vertices between v_0 and v_5 include v_4, v_3 and v_2, v_3. Then if the shortest path from v_0 to v_5 goes through v_3, then the shortest path from v_0 to v_3 should have been already ascertained before this.

2.7.2.3 Description of the idea of Dijkstra's algorithm

The calculation for Dijkstra's shortest distance is to start from source vertex v_0, list all its adjacent vertices, find a shortest vertex among them as the new expansion vertex u, update the distances from v_0 to all the vertices adjacent to u.

Consider using Fig. 2.53 as an example to illustrate the concrete calculation process as follows:

Figure 2.53(a): Suppose that the source vertex is v_0, the value of source vertex v_0 is 0; the shortest paths from v_0 to all the other vertices are all ∞. (The value of the path is recorded in the circle on each vertex.)

Figure 2.53(b): There are three direct paths starting from v_0, which arrive at vertices v_2, v_4, v_5 respectively and have lengths of 10, 30, 100. We modify the values in the corresponding vertex circles. The shortest path among them corresponds to vertex v_2.

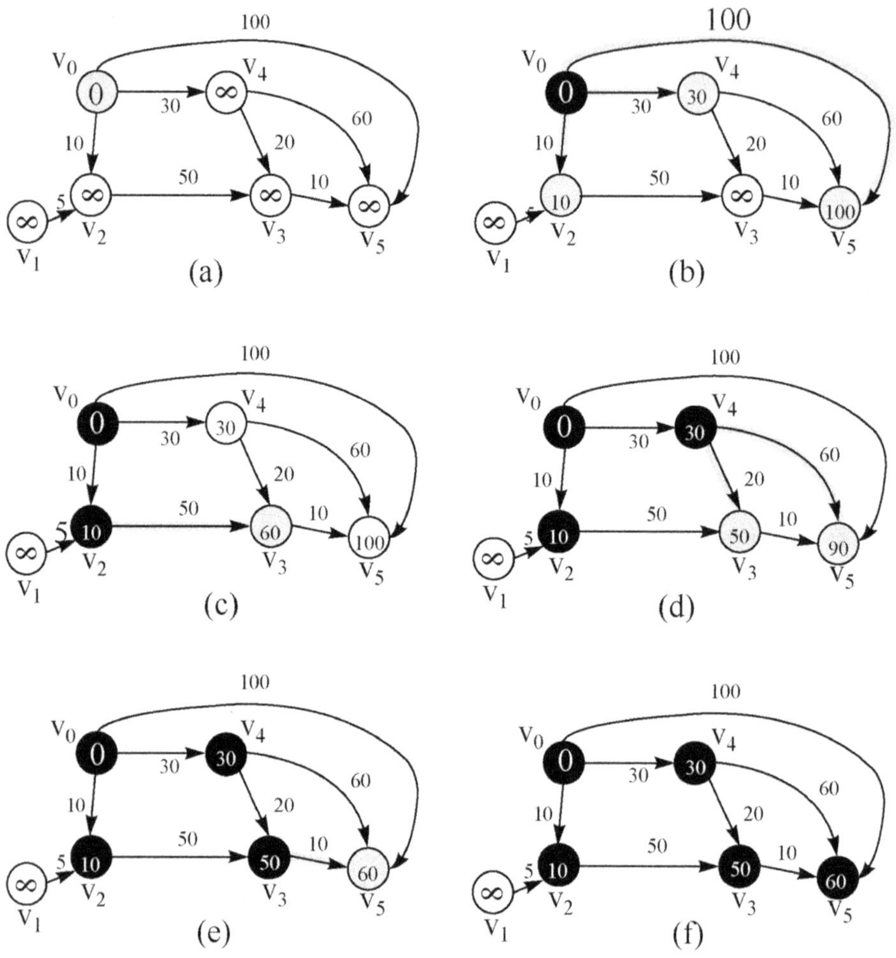

Fig. 2.53: Dijkstra's solution process illustration.

Figure 2.53(c): Ascertain the new expansion vertex u = v_2 (v_2 is colored deep). Use u as the intermediary vertex.

Dist(v_0, v_3) = Dist(v_0, v_2) + Dist(v_2, v_3) = 60, which is smaller than the original value of ∞. Thus, we modify it.

At this time, among the paths from v_0 to v_3, v_4, v_5 (we have already ascertained the shortest distance from v_0 to v_2; thus, we do not consider v_2 anymore), the shortest path corresponds to the vertex v_4, with a distance of 30.

Figure 2.53(d): u = v_4, we can modify the values of the vertices v_3, v_5.

Figure 2.53(e): u = v_3, modify v_5.

Figure 2.53(f): u = v_5, all the vertices with connection to v_0 have been processed.

Record the intermediary results of the above computation process, as shown in Fig. 2.54, where S is the set of vertices for which we have already obtained the shortest path; Dist[] is the array for shortest path and Dist[x] represents the path length from v_0 to vertex x; Path[] is the intermediary vertex path array and Path[x] records the intermediary vertex u on the path from v_0 to vertex x. Note: (1) x is the numbering of the vertex. (2) If there are multiple intermediary vertices between v_0 and vertex x, then Path[x] records the value of the most recently modified intermediary vertex.

Cycle	Set of vertices with shortest path	Distance from source vertex to various vertices Dist[]						Path of the intermediary vertex Path[]					
	{s}	0	1	2	3	4	5	0	1	2	3	4	5
1	{0}	0	∞	10	∞	30	100	0	−1	0	−1	0	0
2	{0,2}	0	∞	10	60	30	100	0	−1	0	2	0	0
3	{0,2,4}	0	∞	10	50	30	90	0	−1	0	4	0	4
4	{0,2,4,3}	0	∞	10	50	30	60	0	−1	0	4	0	3
5	{0,2,4,3,5}	0	∞	10	50	30	60	0	−1	0	4	0	3

Fig. 2.54: Table for the solution process of Dijkstra.

2.7.2.4 Concrete steps of Dijkstra's algorithm
1. Find the vertex u corresponding to the smallest value in Dist; add u to S.
2. Use u as the intermediary vertex and calculate the distances from the source vertex v_0 to all the other vertices. If it is smaller than the original distance, then modify Dist and Path arrays. In terms of concrete data structure, it can be expressed as

If Dist[x] > Dist[u] + AdjMatrix[u][x], then Dist[x] = Dist[u] + AdjMatrix[u][x].

(Note: AdjMatrix[][] is the adjacency matrix)
3. Repeat steps 1 and 2 until all the vertices with paths to v_0 are added to S.

Think and Discuss Why does Dijkstra's algorithm restrict weight values to be nonnegative? **Discussion:** When all the edges have positive weights, there cannot be another closer, nonexpanded vertex u; thus, the distance between v_0 to a vertex will not be changed once ascertained. If there exist edges with negative weights, then closer distances will be generated during expansion, which might disrupt the property that already updated vertices will not have their distances changed. Therefore, only when all the edges have nonnegative weights, we can ensure the correctness of the algorithm. Thus, graphs on which Dijkstra's algorithm is used to obtain the shortest path cannot have edges with negative weights.

2.7.2.5 Code implementation of the algorithm

```
#include <stdio.h>
#define N 20 // The maximum number of vertices of the graph
#define MAX 32767
typedef struct // The struct for the adjacency matrix of the graph
{
    int AdjMatrix[N][N];    // Adjacency matrix AdjMatrix
    int VexNum,ArcNum;      // Number of vertices, number of arcs
    //int vexs[N];    // Store vertex information -
                      // e.g. the next vertex of this vertex
} AM_Graph;
void DisplayAM(AM_Graph g); // Output adjacency matrix
void Dijkstra(AM_Graph g,int v0);
//Dijkstra algorithm -
// the shortest paths form vertex v0 to the rest of the vertices
void DisplayPath(int dist[],int path[],int s[],int n,int v0);
// Calculate the shortest paths from path
void PPath(int path[],int i,int v0);


/*========================================================================
Dijkstra algorithm
Functionality of the function:  the shortest paths form source vertex to the
rest of the vertices
Function input:  The adjacency matrix of the graph, source vertex v0
Function output:  None
========================================================================*/
void Dijkstra(AM_Graph g,int v0)
{
    int i,j;
    int Dist[N];
    //Shortest distance array, records the shortest distance from v0 to vertex j
    int Path[N];
    //Intermediary vertex array, records the previous vertex of vertex j
    int S[N];
    // The set of vertices with shortest paths calculated.
    // Values: 1: The vertex is in the set; 0: The vertex is not in the set
    int MinDis;    // The minimum distance to v0
    int u;         // The vertex with shortest distance to v0
    for (i=0;i<g.VexNum;i++)
```

```
{
    Dist[i]=g.AdjMatrix[v0][i]; // Initialize the distance
    if (g.AdjMatrix[v0][i]<MAX)        // If there is a path from v0 to i
    Path[i]=v0;          // The predecessor of i is v0
    else Path[i]=-1;     // There is no predecessor to i, we mark -1
    S[i]=0;
    //S[]=0, which indicates vertex i is not in the set S
}
S[v0]=1;
//The initialization, add source vertex v0 into the set S
for (i=0;i<g.VexNum;i++)
{
    MinDis=MAX;
    // During initialization, set the minimal distance to v0 as MAX
    u=-1;
    //u is -1, means there is no corresponding vertex

    // Find the minimum value in Dist and its corresponding vertex u
    for (j=0; j<g.VexNum; j++)
        if (S[j]==0 && Dist[j]<MinDis)
        {
            MinDis=Dist[j];
            u=j;
        }

    if(MinDis!=MAX)  S[u]=1;       // Add vertex u into the set S
    else break;

    // Use u as the intermediary vertex,
    // look at the distance from v0 to other vertices
    for (j=0;j<g.VexNum;j++)
    {
        // Select vertex j which is not in set S and is connected to u
        if (S[j]==0 && g.AdjMatrix[u][j]<MAX)
        {
            // If [distance from v0 to j] >
            // [distance from v0 to u + distance from u to j]
            if (Dist[j]>Dist[u]+g.AdjMatrix[u][j])
```

```
                {
                    // Modify distance from v0 to j
                    Dist[j]=Dist[u]+g.AdjMatrix[u][j];
                    Path[j]=u;          // Modify j's predecessor vertex
                }
            }
        }
    }
    printf("Output the shortest path:\n");
    DisplayPath(Dist,Path,S,g.VexNum,v0);    //Output the shortest path
}
/*======================================================================
Functionality: Output the shortest paths from the source vertex to other
vertices
Function input: Array of shortest distances, array of paths, set of vertices
with shortest paths, number of vertices, source vertex
Function output: None
======================================================================*/
void DisplayPath(int Dist[],int Path[],int S[],int n,int v0)
{
    int i;
    for (i=0;i<n;i++)
    if (S[i]==1 && i!=v0)
    // Only the vertices in the set S have paths to output
    {
        printf("The shortest path from %d to %d has a length of:%d",v0,i,Dist[i]);
        printf("\t The path is:%d-",v0);
        PPath(Path,i,v0);
        printf("%d\n",i);
    }
    else
    printf("There isn't a path from %d to %d\n",v0,i);
}
/*======================================================================
Functionality: Print the shortest path from the source vertex to the designated
vertex
Function input: Array of paths, ending vertex, source vertex
Function output: None
Screen output: Shortest path
======================================================================*/
```

```
void PPath(int Path[],int i,int v0)
{
   int k=Path[i];
   if (k==v0) return;
   else PPath(Path,k,v0);
   printf("%d-",k);
}
/*=====================================
Functionality: Output the adjacency Matrix
Function input: Adjacency matrix
Function output: None
Screen output: Adjacency matrix
=====================================*/
void DisplayAM(AM_Graph g)
{
   int i,j;
   for (i=0;i<g.VexNum;i++)
   {
      for (j=0; j<g.VexNum; j++)
      {
         if (g.AdjMatrix[i][j]==MAX) printf("%4s","∞");
         else printf("%4d",g.AdjMatrix[i][j]);
      }
      printf("\n");
   }
}

int main()
{
   int A[N][6]={{MAX,MAX,10, MAX,30, 100},
               {MAX,MAX,5, MAX,MAX,MAX},
               {MAX,MAX,MAX,50, MAX,MAX},
               {MAX,MAX,MAX,MAX,MAX,10 },
               {MAX,MAX,MAX,20, MAX,60 },
               {MAX,MAX,MAX,MAX,MAX,MAX}};
   AM_Graph g;    // Define the adjacency matrix g
   g.VexNum=6;
   g.ArcNum=8;
   for (int i=0;i<g.VexNum;i++)    // Give value to the adjacency matrix
   for (int j=0;j<g.VexNum;j++)
   g.AdjMatrix[i][j]=A[i][j];
```

```
    printf("The adjacency matrix of the directed graph G:\n");
    DisplayAM(g);    // Output the adjacency matrix
    int v0=1;          // Set the initial vertex
    Dijkstra(g,v0);
    return 0;
}
```

Test result: test cases are shown in Fig. 2.52.
The adjacency matrix of the directed graph G:

∞	∞	10	∞	30	100
∞	∞	5	∞	∞	∞
∞	∞	∞	50	∞	∞
∞	∞	∞	∞	∞	10
∞	∞	∞	20	∞	60
∞	∞	∞	∞	∞	∞

Output the shortest path:
There is no path from 1 to 0
There is no path from 1 to 1
The shortest path from 1 to 2 has length 5, the path is 1–2
The shortest path from 1 to 3 has length 55, the path is 1–2–3
There is no path from 1 to 4
The shortest path from 1 to 5 has length 65, the path is 1–2–3–5

2.7.3 All-pair shortest path algorithm – Floyd's algorithm

To obtain the shortest distance between all vertex pairs in the weighted graph, following the idea from the Dijkstra algorithm introduced previously, we can use one vertex as the source vertex each time, and repeatedly execute the Dijkstra algorithm n times. In this way we can obtain the shortest path between each pair of vertices. Besides directly using the Dijkstra algorithm, in 1962 Robert W. Floyd proposed another solution. Its idea is to use the vertices in the graph as the intermediary vertex one by one, and view the distances between all pairs of vertices u, v. If it is shorter than the original distance, then update. Floyd's algorithm is also called vertex insertion method. It can normally be used in any graph, including directed graph and graphs with negative weights.

In the following we illustrate the method to update the distances between various pairs of vertices with a concrete example, as shown in Fig. 2.55. First, we set up the Dist matrix and the Path matrix to record the distances between the vertices and the corresponding intermediary vertex.

(1) Initialize the matrix

Dist$^{(-1)}$

	0	1	2	3
0	0	1	∞	4
1	∞	0	9	2
2	3	5	0	8
3	∞	∞	6	0

Path$^{(-1)}$

	0	1	2	3
0	0	0	0	0
1	0	1	1	1
2	2	2	2	2
3	0	0	3	3

(2) Update the matrix using vertex 0 as the middle point

Dist$^{(0)}$

	0	1	2	3
0	0	1	∞	4
1	∞	0	9	2
2	3	4	0	7
3	∞	∞	6	0

Path$^{(0)}$

	0	1	2	3
0	0	0	0	0
1	0	1	1	1
2	2	0	2	0
3	0	0	3	3

(3) Update the matrix using vertex 1 as the middle point

Dist$^{(1)}$

	0	1	2	3
0	0	1	10	3
1	∞	0	9	2
2	3	4	0	6
3	∞	∞	6	0

Path$^{(1)}$

	0	1	2	3
0	0	0	1	1
1	0	1	1	1
2	2	0	2	1
3	0	0	3	3

(4) Update the matrix using vertex 2 as the middle point

Dist$^{(2)}$

	0	1	2	3
0	0	1	10	3
1	12	0	9	2
2	3	4	0	6
3	9	10	6	0

Path$^{(2)}$

	0	1	2	3
0	0	0	1	1
1	2	1	1	1
2	2	0	2	1
2	2	0	3	3

(5) Update the matrix using vertex 3 as the middle point

Dist$^{(3)}$

	0	1	2	3
0	0	1	9	3
1	11	0	8	2
2	3	4	0	6
3	9	10	6	0

Path$^{(3)}$

	0	1	2	3
0	0	0	3	1
1	2	1	3	1
2	2	0	2	1
3	2	0	3	3

Length	End -> Start	Length	End -> Start
0	0→0	3	2→0
1	0→1	4	2→0→1
9	0→1→3→2	0	2→2
3	0→1→3	6	2→0→1→3
11	1→3→2→0	9	3→2→0
0	1→1	10	3→2→0→1
8	1→3→2	6	3→2
2	1→3	0	3→3

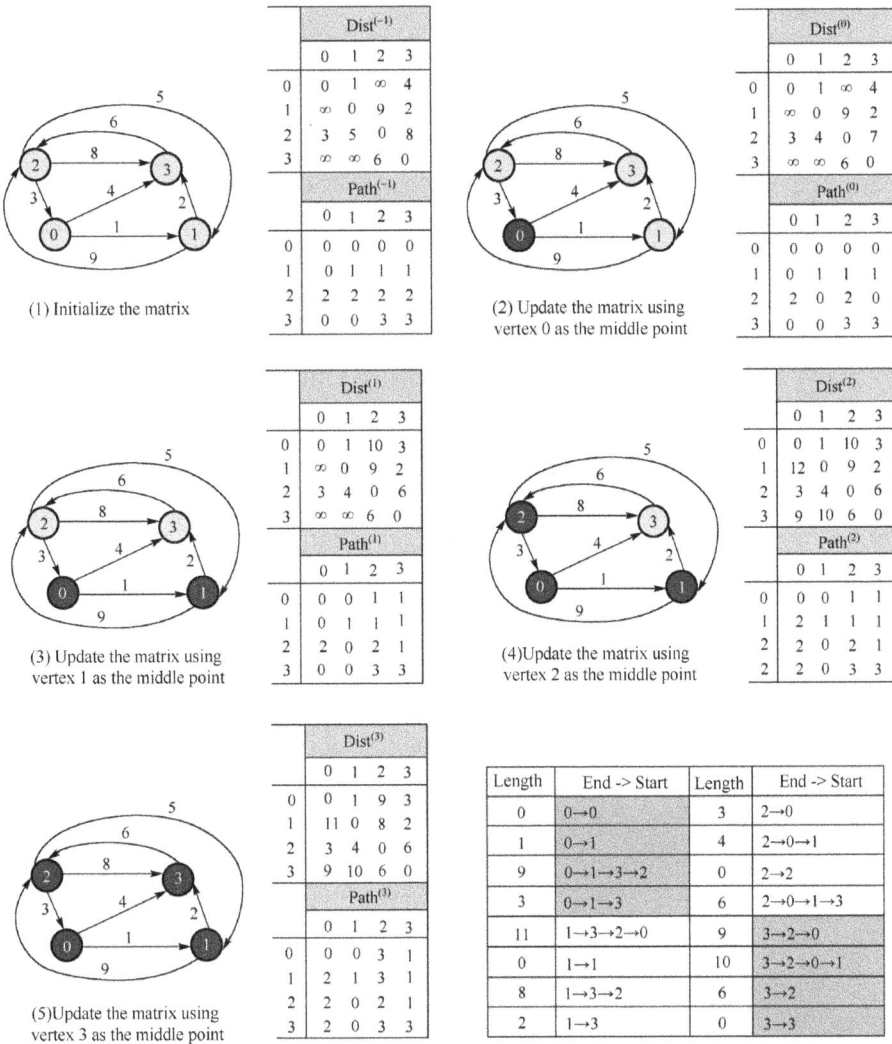

Fig. 2.55: Analysis of Floyd's algorithm.

Step 2 of the algorithmic analysis of Floyd's algorithm is shown in Fig. 2.55.

1. Path(−1) matrix: From vertex 2 to vertex 1, through intermediary vertex 2, noted as Path(2, 1) = 2;
   Dist(−1) Matrix: The path from vertex 2 to vertex 1 has length 5, noted as Dist(2, 1) = 5.
2. Using vertex 0 as the intermediary vertex
   Because Dist(2, 1) = Dist(2, 0) + Dist(0, 1) = 3 + 1 = 4 < 5

Therefore, we have

In Dist(0) matrix: Dist(2, 1) = 4

In Path(0) matrix: Path(2, 1) = 0

In a similar manner: Dist(2, 3) = 7; Path(2, 3) = 0

The principles for the changes on Dist and Path matrices in steps 3–5 are the same, and we would not repeat them here. Ultimately, we list the values of the shortest paths between all pairs of vertices and the list of vertices each path goes through.

2.7.3.1 Description of Floyd's algorithm

1. Set Dist(u, v) as the distance from vertex u to vertex v with the intermediary vertex w;
2. If there exists Dist(u, w) + Dist(w, v) < Dist(u, v), then Dist(u, v) = Dist(u, w) + Dist (w, v);
3. After iterating all vertices, Dist(u, v) will record the value of the shortest path from u to v.

2.7.3.2 Program implementation

```
#include<stdio.h>
#define N 100
//Maximum number of vertices in the graph
#define MAX 32767
typedef struct
// Struct for adjacency matrix for the graph
{
    int AdjMatrix[N][N];    // Adjacency matrix
    int VexNum,ArcNum;      // Number of vertices, arcs
    //int vexs[N];
    // Store vertex information, e.g. the next vertex of this vertex
} AM_Graph;
void DisplayAM(AM_Graph g);    // Output the adjacency matrix
void Floyd(AM_Graph g);
// Floyd algorithm - Calculate the shortest path between all pairs of vertices
void DisplayPath(int A[][N],int Path[][N],int n);    // Output the paths
void PPath(int Path[][N],int i,int j);
int main()
{
    int A[N][4]={ {0,1,MAX,4},
```

```
               {MAX,0,9,2},
               {3,5,0,8},
               {MAX,MAX,6,0 },
};
   AM_Graph g;        // Define the adjacency matrix
   g.VexNum=4;
   g.ArcNum=8;        // 4 vertices, 8 edges
   // Assign values to the adjacency matrix
   for (int i=0;i<g.VexNum;i++)
   for (int j=0;j<g.VexNum;j++)
     g.AdjMatrix[i][j]=A[i][j];
   printf("The adjacency matrix of the direct graph G:\n");
   DisplayAM(g);       // Output the adjacency matrix
   Floyd(g);
   // Call the algorithm and output the distance between each pair of vertices
   return 0;
}
/*=========================================================================
Floyd algorithm
Functionality of the function: Calculate the shortest paths between each pair
of vertices in the graph
Function input: Adjacency matrix
Function output: None
=======================================================================*/
void Floyd(AM_Graph g)
{
   int i,j,k;
   int Dist[N][N],Path[N][N];
   // Dist matrix records the shortest distances between each pair of vertices,
   // path matrix records the intermediary vertices of paths
   // Initialize Dist and Path matrices
   for (i=0;i<g.VexNum;i++)
   {
     for (j=0;j<g.VexNum;j++)
     {
        Dist[i][j]=g.AdjMatrix[i][j];
        // The initial status of the Dist matrix is the adjacency matrix
        Path[i][j]=-1;
        // When there isn't any intermediary vertex, it is -1
     }
   }
```

```
    // Use various vertices in the graph as the intermediary vertex k,
    // traverse Dist matrix
    for (k=0; k<g.VexNum; k++)
    {
        // Query from Dist matrix whether the distances to other vertices through
        // vertex k have improved (i.e. decreased)
        for (i=0;i<g.VexNum;i++)
        for (j=0;j<g.VexNum;j++)
        {
            // The distance between i and j through k has decreased
            if (Dist[i][j]>(Dist[i][k]+Dist[k][j]))
            {
                Dist[i][j]=Dist[i][k]+Dist[k][j];
                // Modify the distance between vertices i and j
                Path[i][j]=k;
                // Record the intermediary vertex k
            }
        }
    }
    printf("\nOutput the shortest paths:\n");
    DisplayPath(Dist,Path,g.VexNum);
    // Output the shortest paths
}
/*=================================================================
Functionality: Output the shortest paths between each pair of vertices
Function input: Adjacency matrix, array of paths, number of vertices
Function output: None
=================================================================*/
void DisplayPath(int A[][N], int Path[][N], int n)
{
    int i,j;
    for (i=0;i<n;i++)
    or (j=0;j<n;j++)
    if(A[i][j]==MAX)
    {
        if(i!=j)
        // To reach itself the distance would be 0. We should exclude that situation
        printf("There isn't a path from %d to %d\n",i,j);
    }
    else
    {
        printf("The length of path from %d to %d is: %d",i,j,A[i][j]);
```

```
      printf("\t The path is");printf("%d-",i);PPath(Path,i,j);
      printf("%d\n",j);
  }
}
/*===============================================================
Functionality: Print out the path between two designated vertices
Function input: Array of paths, vertex 1, vertex 2
Function output: None
Screen output: Path between the vertices
==============================================================*/
void PPath(int Path[][N],int i,int j)
{
   int k=Path[i][j];
   if (k==-1) return;
   PPath(Path,i,k);
   printf("%d-",k);
   PPath(Path,k,j);
}
/*=====================================
Functionality: Output the adjacency matrix
Function input: Adjacency matrix
Function output: None
Screen output: Adjacency matrix
====================================*/
void DisplayAM(AM_Graph g)
// Output the adjacency matrix
{
   int i,j;
   for (i=0;i<g.VexNum;i++)
   {
      for (j=0;j<g.VexNum;j++)
         if (g.AdjMatrix[i][j]==MAX)
            printf("%4s","∞");
         else printf("%4d",g.AdjMatrix[i][j]);
      printf("\n");
   }
}
```

Test result: test the graph with shape shown in Fig. 2.55.

The adjacency matrix of the directed graph G:

```
0    1    ∞    4
∞    0    9    2
3    5    0    8
∞    ∞    6    0
```

Output the shortest path:

From 0 to 0, the length of path is 0   The path is 0–0
From 0 to 1, the length of path is 1   The path is 0–1
From 0 to 2, the length of path is 9   The path is 0–1–3–2
From 0 to 3, the length of path is 3   The path is 0–1–3
From 1 to 0, the length of path is 11 The path is 1–3–2–0
From 1 to 1, the length of path is 0   The path is 1–1
From 1 to 2, the length of path is 8   The path is 1–3–2
From 1 to 3, the length of path is 2   The path is 1–3
From 2 to 0, the length of path is 3   The path is 2–0
From 2 to 1, the length of path is 4   The path is 2–0–1
From 2 to 2, the length of path is 0   The path is 2–2
From 2 to 3, the length of path is 6   The path is 2–0–1–3
From 3 to 0, the length of path is 9   The path is 3–2–0
From 3 to 1, the length of path is 10 The path is 3–2–0–1
From 3 to 2, the length of path is 6   The path is 3–2
From 3 to 3, the length of path is 0   The path is 3–3

2.7.3.3 Time complexity analysis

Each vertex needs to be compared with the other $n-1$ vertices in terms of length of edge; thus, its time complexity is $O(n^3)$.

2.7.4 Summary of shortest path problems

Shortest path problems can have multiple algorithms, which are suitable for different scenarios. The Dijkstra algorithm is suitable for solving single-source shortest path problems without negative weights, and its time complexity is $O(V^2)$ (V is the number of vertices). Floyd's algorithm is a dynamic planning algorithm, which is best applied to dense graphs. The weights of the edges can be either positive or negative. This algorithm is simple and effective. Since the three loops are compact, applied to dense graphs, it is more efficient than the Dijkstra algorithm, which must be executed V times if we want to obtain the all-pair shortest paths. Its advantages are ease of comprehension, ability to calculate the shortest distance between any two vertices, and ease of code writing. Its disadvantage is that it has a relatively high time complexity and is unsuitable to calculate massive amounts of data. Floyd's algorithm cannot straightfowardly reflect the ordering of the shortest paths between all pairs of vertices.

2.8 Classical application of graph – problems of moving vertices and moving edges

2.8.1 Introduction to the problem of ordering of moving vertices of graph

In general, complicated projects can be divided into various operations/steps/sub-projects. Some operations can be finished independently, but most are related to other operations, and can only be started after other operations are finished. Only after we finish these operations in a certain order, we can correctly finish the whole project. Such actual examples abound in our daily lives. Let us see some concrete examples in the following.

2.8.1.1 Problem of assembly order

When we buy some goods online, we might need to finish the final assembly by ourselves, for example, board furniture and bicycles. Actually, many of our daily activities are also a process of assembly, for example, putting on clothes.

Professor Brown wakes up in the morning. He must first put on some clothes before he can put on some other clothes (e.g., wear socks before putting on shoes). Some other clothes can instead be put on in an arbitrary order (e.g., socks and trousers), as shown in Fig. 2.56. If we treat the wearing of various attires as vertices, the order of these activities as directed edges, then we can draw the "assembling of attires" order in Fig. 2.56(a). Figure 2.56(b) is a cleaned-up version of Fig. 2.56(a), where we form a sequence of vertices in the horizontal direction, so that all directed edges in the graph point from left to right. "Assembly order problem" is then converted into the ordering of vertices for directed graph.

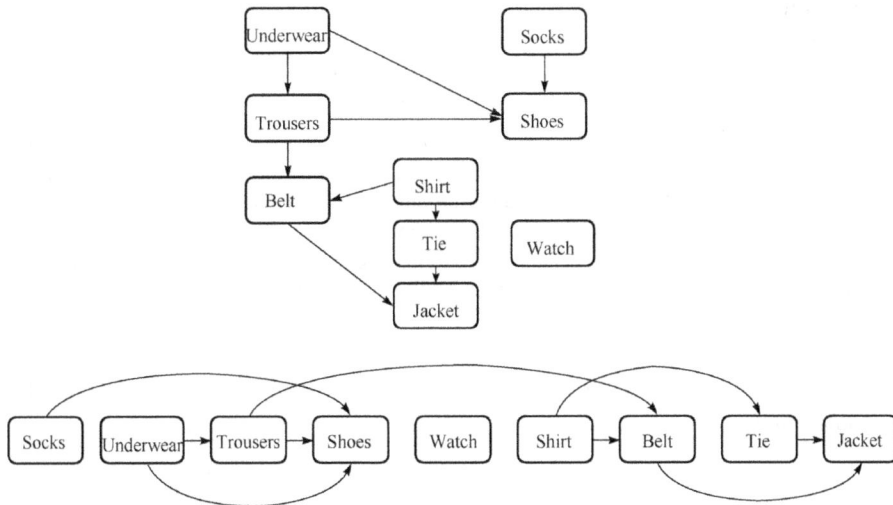

Fig. 2.56: The problem of attire assembly of Professor Brown.

2.8.1.2 Problem of prerequisite courses

Some students have viewed the introduction to the "machine learning" Massive Open Online Course, and are interested in studying it. However, they do not know whether their foundational knowledge is sufficient. In such an occasion, they would need to look up the prerequisites for this course. They need to study the related courses in a certain order, based on the corresponding knowledge system. Some courses are basic courses, which can be taken independent of other courses; some other courses need prerequisite courses as foundations. To build a study plan, one would need to ascertain the order in which to take the courses. Figure 2.57 shows some courses in computer science and their relations.

Course code	Course name	Prerequisites
C1	Introduction to computing	None
C2	Principles of microprocessor	C1, C3
C3	Principles of computer architecture	C1
C4	Program design	C1, C6
C5	Data structures	C3, C4, C6
C6	Advanced mathematics	None

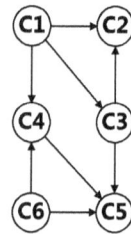

Fig. 2.57: Some courses of computer science and their relations.

We can abstract a course into a vertex, then the dependency relations between courses are the directed edges that connect the vertices. In this manner, we can express the courses and their related relations with a directed graph. The problem of ascertaining the order of taking various courses is converted into the problem of ordering the vertices of the directed graph. Since this is an ordering based on vertices of a topological structure, it is called topological ordering.

Think and Discuss Can all directed graphs be topologically ordered?
Discussion: If we add a relation in the description of course relations, so that one needs to study data structures (C5) first before studying introduction to computer science (C1), would not that be illogical? In that case, we cannot perform topological ordering, since if there is a mutual dependency between C1 and C5, we cannot ascertain the order of the two vertices. In a directed graph, this situation is described as a cycle, as shown in Fig. 2.58.

The necessary and sufficient condition for a directed graph to be topologically ordered is that it is a directed acyclic graph (DAG). The feature of such a graph is that it would not contain any cycles.

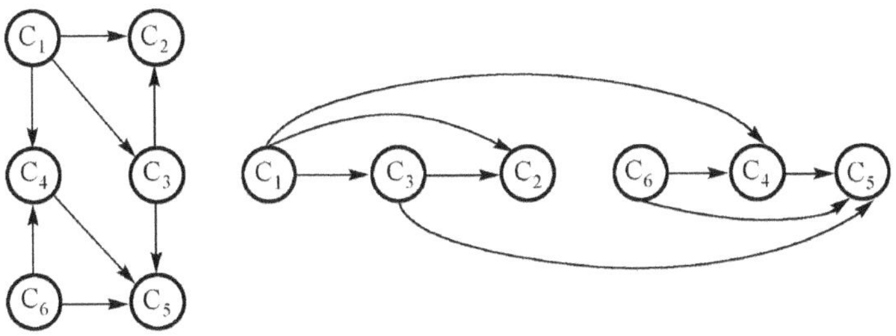

Fig. 2.58: Problem of course selection.

2.8.1.3 Analysis of source code

The mainstream compilers we use daily need to support multiple source files. For example, when we put function definitions in the corresponding files, whenever we want to use those functions, we would need the files that define such functions (e.g., C), or refer to the namespace containing the type (e.g., Java), or refer to this file as a unit (e.g., Object Pascal).

Compilers that support multiple source files need to first compile the file referred to by this section of source code before compiling this source code file itself. For example, there are two header files A.h and B.h, with the following contents.

```
// Header file A.h
int next(void)
{
...
}
int last(void)
{
...
}
int news( int i)
{
...
}
// Header file B.h
int reset(void)
{
    ...
}
```

The source file D.cpp uses functions defined in A.h and B.h, with the following contents:

```
int main()
```

```
{
  ...
  i=reset();
  for (j=1; j<4; j++)
  {
    printf("%d\t",next());
    printf("%d\t",last());
    printf("%d\n",news(i+j));
  }
  return 0;
}
```

Whenever we compile D.cpp, if the files in A.h and B.h have not been precompiled, the compiler would not be able to identify functions such as reset, and whether the parameters for calling reset are correct. In this situation, it would need to first analyze what files D.cpp refers to, and whether those files further refers to other files, and compile the files at the top of the list and so on.

To analyze the dependency relations of source code, we can view all the source code files as vertices. If a vertex (source code file) refers to another vertex, then we add an outgoing edge from the current vertex to the vertex being referred to. After we have added the outgoing edges of all vertices, normally those vertices would form a DAG. If there is a cycle, then the source code files contain erroneous referring, which would cause compilation to fail.

For example, the following are the source code files A, B, C, D, E, with the following reference relations:

- A refers to B,C
- B refers to D,E
- C refers to B,E
- D refers to E
- E does not refer to any other file

From the above reference relations, we can draw Fig. 2.59. The compilation order should be E D B C A.

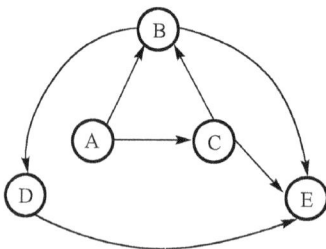

Fig. 2.59: Reference relations between source code files.

2.8.2 AOV network and topological ordering – ordering of moving vertices

2.8.2.1 AOV network model and topological ordering

A directed graph where the vertices represent movement and edges represent the ordering of the movements is called an activity on vertex (AOV) network.

To order all the vertices in an AOV network into a linear sequence, so that this sequence keeps the original priority relations of the vertices, and establish artificial ordering between vertices originally without any ordering relation, this sorting process is called topological ordering.

AOV network has important applications in fields such as product assembly planning, software development task assignment and source code analysis. In recent years, the application of AOV network has gradually expanded to newly emergent fields such as arrangement of grid workflow, GIS applications and knowledge self-organization.

2.8.2.2 Definition of topological ordering

Topological ordering: the operation of obtaining a total order relation on a set from a partial order relation on the set is called topological ordering.

Topological sequence: in an AOV network, if there is no cycle, then all activities can be ordered into a linear sequence, so that each activity's predecessor activity is arranged before this activity. We call this sequence a topological sequence.

Partial order and total order are concepts from discrete mathematics. Straightforwardly speaking, partial order means some of the elements of the set can be compared with each other, while total order means all the elements in the set can be compared with each other. Here, the meaning of "comparison" is comparison of size, and can also be the ordering in a sequence. For example, for the two directed graphs in Fig. 2.60, if the arc(x, y) in the graph indicates $x \leq y$, then the symbol "$\leq$" represents x's ordering in the sequence that is prior to y. (a) is a partial order as the ordering of v_2 and v_3 cannot be ascertained. (b) is a total order as the ordering of all vertices is ascertained. If we artificially add an arc that indicates $v_2 \leq v_3$ in the directed graph of (a), then (a) would also be a total order, and this total order is

Arc<x,y>represents x<y

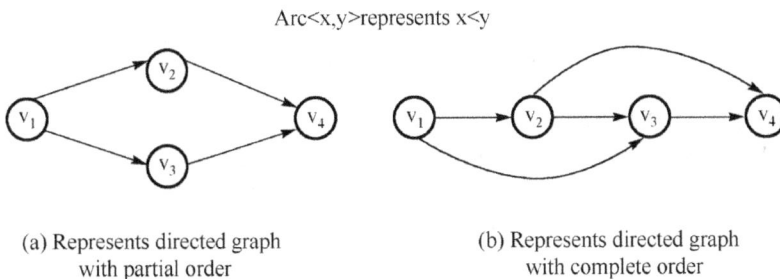

(a) Represents directed graph
with partial order

(b) Represents directed graph
with complete order

Fig. 2.60: Partial order and total order.

called topological order. The operation to obtain topological order from partial order is topological ordering.

To illustrate the notions of partial order and total order with the previous example of prerequisite courses, suppose that after taking the course "introduction to computer science," one can take "advanced mathematics" or "principles of computer architecture," then there is no fixed ordering between these two courses. It is fine to take either one before the other. On the contrary, the course "data structures" must be taken after finishing the course "program design basics." In all selectable courses, the relation between any two courses is either fixed (i.e., with a clear ordering), or unfixed (i.e., without a clear ordering). In the set "courses," only some courses have fixed ordering relations. Then, this "courses" set is a partially ordered set.

If we specify that one must take "advanced mathematics" before taking "principles of computer architecture," then there exists a certain ordering between them as well. If we give a certain ordering to all the uncertain relations in the "courses" set, then the original partial order relation becomes a total order relation. We can see that total order is a special case of partial order.

> **Think and Discuss** Is the result of topological ordering unique?
> **Discussion:** Take the example of the ordering of the several integers stored in the array.
>
> If all the integers in the array are different, then the ordering of all elements according to the size is certain, that is, this sequence satisfies total order relation. For a structure with total order, the result after its linearization (ordering) is certain to be unique.
>
> If there are integers of the same value in the array, the relation between elements with the same value cannot be determined. Therefore, they can appear in arbitrary order in the final ordering result.
>
> For elements with the same value, if we specify that the one that appears earlier is bigger than the one that appears later, then we can convert the partial order to a total order, obtaining a unique result.
>
> Expanding it to topological ordering, the condition for the result to be unique is that all vertices have a complete order relation among them. If there is no complete order relation, then the result of the topological ordering would not be unique.
>
> For Fig. 2.61, there are multiple topological ordering sequences. We list two of them: (C1, C3, C2, C6, C4, C5) and (C6, C1, C3, C2, C4, C5).

Topological ordering has the following characteristics:
- If we list all the vertices in the graph into a row according to the topological ordering, then all the directed edges in the graph will point from left to right.
- If there is a directed cycle in the graph, then it is impossible for the vertices to satisfy topological ordering.
- The topological sequence of a DAG usually suggests that a certain plan is viable.

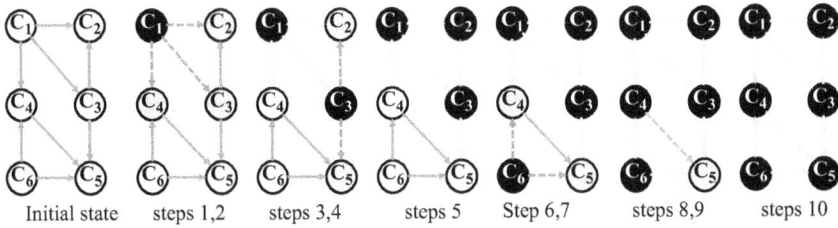

| Initial state | steps 1,2 | steps 3,4 | steps 5 | Step 6,7 | steps 8,9 | steps 10 |

Step	In-degree					
	c_1	c_2	c_3	c_4	c_5	c_6
1	0	2	1	2	3	0
2		1	0	1	3	0
3		1	0	1	3	0
4			0	1	2	0
5			0	1	2	0
6				1	2	0
7				0	1	
8				0	1	
9					0	
10					0	

Steps of topological ordering
(1) Find the vertex v with in-degree 0, fetch it
(2) Delete the arc <v, u> connected with v
(3) Decrease the in-degree of u, an adjacent vertex of v.
Topological ordering sequence 1:$C_1C_3C_2C_6C_4C_5$
Topological ordering sequence 2: $C_6C_1C_4C_3C_5C_2$

Fig. 2.61: Analysis of topological ordering algorithm.

2.8.2.3 Topological ordering algorithm

To topologically order DAGs, where is the starting point?

1. Algorithm analysis

First let us use course selection as an example. In Fig. 2.57, the course that is taken in the beginning should not have any prerequisite courses. For example, one might first choose "introduction to computer science" C1 or "advanced mathematics" C6 as the beginning vertex. If we choose C1 first, then afterward we may choose C3 or C6. Observing the selected vertex, its feature is that it has an indegree of 0. The reason to choose C3 is that after removing C1, its indegree becomes 0 from the original 1. The process of step-by-step ordering is shown in Fig. 2.61.

Since in this type of topological ordering, each step always outputs the vertex without any predecessor (i.e., with indegree of 0), it is also called predecessor-less vertex first topological ordering algorithm.

2. Algorithm description

See Table 2.9 for the pseudocode description of the algorithm. The detailed step 2 uses adjacency list as the storage structure.

Table 2.9: Algorithm description of the topological sorting algorithm.

Top-level pseudocode	Elaboration 1	Elaboration 2
(1) Select a vertex without predecessor v in the directed graph and output it; (2) Delete the vertex v and all its outgoing edges in the directed graph Repeat the above two steps until all the vertices are processed. If you can't find any vertex without a predecessor, then there is a cycle in the graph.	Find out the degrees of the vertices in the graph	Find out the degrees of the vertices in the graph Push all the vertices with indegree 0 in the adjacency list onto the stack
	When (there is still vertex v with indegree 0 in the graph) Output v; Delete v and all its outgoing edges in the graph;	When the stack is not empty Pop and output the vertex *i* at the top of the stack; At the *i*th row of singly linked list of the adjacency list, find all the immediate successor vertices of *i*, and decrease the indegree of k by 1. If the indegree of vertex k is 0, then push k onto the stack
	If (the number of vertices output < the total number of vertices in the graph) Then there is a cycle in the graph and the ordering fails	If the number of vertices output when the stack is empty is not *n*, then there is cycle in the directed graph. Otherwise the topological ordering is complete.

3. Data structures description
   – Structure of adjacency list: see Fig. 2.62.
   – Array of indegrees int indegree[] – stores the indegree of each vertex

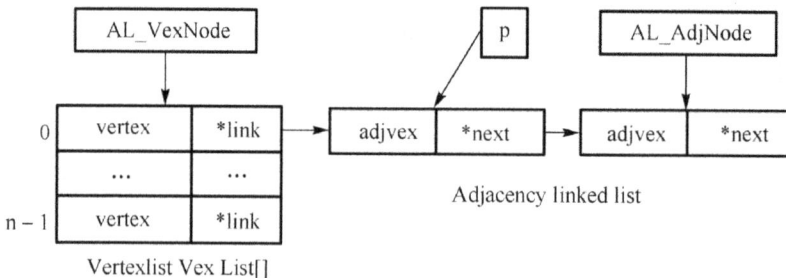

Fig. 2.62: Topological ordering data structure based on adjacency list.

Array of sequence of topological ordering int v[] – stores the results of topological ordering (indexes of vertices).
   Stack of vertices SeqStack S – stores vertices with indegree of 0.

4. Algorithm implementation

```
/*============================================================================
Functionality: Topological ordering which prioritizes vertices without
predecessors
Function inputs: Adjacency list G, (topological ordering sequence array v[])
Function output: TURE—G has no cycle, FALSE—G has cycles
Topological ordering sequence array v[]
============================================================================*/
int TopologicalSort(AL_Graph G,int v[])
{
  int i,k,count=0;
  int indegree[VERTEX_NUM];   // indegree array
  SeqStack S;   // vertices stack
  AL_AdjNode *p;   // Pointer of adjacent node
  FindInDegree(G,indegree);   // Get the vertex and the indegree
  initialize_SqStack(&S);   // Initialize the stack
  // Push the vertex with indegree of 0 onto the stack S
  for(i=0;i<G.VexNum; ++i)   //  for(i=G.VexNum-1;i>=0; i--)
  // We might start scanning from the reverse direction of indegree,
  // to obtain the other set of topological ordering results
  if(!indegree[i]) Push_SqStack(&S,i);   // Stack S is not empty
  while(!StackEmpty_SqStack(&S))
  { Pop_SqStack(&S,&i);
    v[count]=i; ++count;      // Record the index of the vertex 0, i
    // Scan the row of linked list with index i in the adjacency list
    for(p=G.VexList[i].link; p; p=p->next)
    { k=p->adjvex;   // Find the successor vertex of i, k
      if(!(--indegree[k])) Push_SqStack(&S,k);
      // Decrease the degree of k by 1,
      // push the vertex with indegree of 0 onto the stack
    }
  }
  if(count<G.VexNum) return FALSE;
  else return TRUE;
}

#include <stdio.h>
#include <stdlib.h>
#include "SqStack.h"
// Functions for the operations of the sequential stack
#include "AdjList.h"
```

```
// Functions for the construction of adjacency list
/*===============================================================
Functionality: Obtain the indegrees of the graph
Function inputs: Adjacency list G, (indegree array indegree[])
Function output: Indegree array indegree[]
===============================================================*/
void FindInDegree(AL_Graph G, int indegree[])
{
  int i;
  AL_AdjNode *p;
  for(i=0;i<G.VexNum;i++)
    indegree[i]=0;
  for(i=0;i<G.VexNum;i++)
  {
    p=G.VexList[i].link;
    while(p)
    {
      indegree[p->adjvex]++;
      p=p->next;
    }
  }
}
int main()
{
  AL_Graph G;
  int a[VERTEX_NUM];
  G=Create_AdjList();
  TopologicalSort(G,a);
  printf("Sequence of topological    ordering:\n");
  for (int i=0; i<    VERTEX_NUM; i++)
  printf("c%d ",a[i]+1);
  printf("\n ");
  return 0;
}
```

Sample figure for the test is Fig. 2.61. The test results are:
Sequence 1 of topological ordering: c1 c3 c2 c6 c4 c5
Sequence 2 of topological ordering: c6 c1 c4 c3 c5 c2

5. Complexity analysis
For a directed graph with n vertices and e edges, the time complexity of obtaining the indegree of all vertices is $O(e)$; the time complexity of constructing the stack of

vertices with indegrees of 0 is $O(n)$. During the topological ordering process, if the directed graph does not have cycles, then we perform a push and pop on every vertex. The operation to decrease indegree by 1 is executed e times in total in the while statement. Therefore, the total time complexity is $O(n + e)$.

6. Other topological ordering algorithms

Besides the previously introduced method that prioritizes vertices without predecessors, there are also topological ordering that prioritizes vertices without successors and topological ordering, which orders DAG with depth-first traversal. In the following we only introduce the algorithm ideas.

 – Topological ordering that prioritizes vertices without successors

Each step of this method outputs the vertex without successors (i.e., with outdegree of 0) currently. For a DAG, the sequence output with this method is a reverse topological sequence. Therefore, we need to set up a stack T to store the sequence of vertices output, then we can obtain the topological ordering. Whenever we output a vertex, we only need to perform push operation. Once the ordering is finished, we only need to pop the vertices in the stack one by one to obtain the topological ordering.

Algorithm description

```
NonSuccFirstTopSort(G)
{ // Prioritize vertices without successors
  while(G has vertices with outdegrees 0)
  {
    Find a vertex v with outdegree 0 from G and output v;
    Delete v and all indegrees of v from G
  }
  if(Number of vertices output < Total number of vertices in the graph)
  Error("There are directed cycles in G, the ordering fails!");
}
```

Note: We can use a reverse adjacency list as the storage structure for G. We can set a vector of out degrees or add an outdegree field in the vertex node of the reverse adjacency list to store the current outdegrees of various vertices. We can set a stack or queue to store all vertices with outdegree of 0 temporarily. Except for adding a stack or vector T to store the vertex sequence output, this algorithm is totally similar to NonPreFirstTopSort.

 – Topologically order DAG using depth-first traversal

Whenever a DFS starting from a certain vertex v finishes, all the successors of v must have been visited (image has been deleted). At this moment, v is equivalent to a vertex without successors. Therefore, if we output the vertex v just before DFS algorithm returns, we can obtain the reverse topological sequence of the DAG.

Algorithm description

```
void DFSTopSort(G, i, T)
{
  // Invoke this algorithm in DisTraverse, i is the starting point of the
  search, T is the stack
  int j;
  visited[i]=TRUE;      // Visit i
  for(all the adjacent vertices of i, j)    //i.e. <i, j>∈E(G)
     if(!visited[j]) DFSTopSort(G, j, T);   // Similar to DFS algorithm
       Push(&T, i);      // The search starting from i is finished, output i
}
```

The first vertex being output must be a vertex without successors (with outdegree 0). It should be the last vertex of the topological sequence. If we want to obtain the normal topological sequence, we can similarly add a stack T to store vertices output, and initialize T at the beginning of DFSTraverse algorithm.

As long as we change the invocation of DFS in DFSTraverse to invocation to DFSTopSort, we can obtain the topological sequence T. The concrete algorithm can be easily obtained from the above pseudocode.

If G is a DAG, then the topological ordering realized with DFS traversal is completely similar to NonSuccFirstTopSort algorithm. If C has directed cycles, then the former would not function normally.

2.8.3 AOE network and critical path – longest activity edge

AOE (Activity on Edge) network is a weighted directed graph, where vertices represent events, edges represent activities and weights represent the duration of the activity.

Critical path method (CPM) is an analysis method for longest path based on AOE network.

2.8.3.1 Problem background

Topological ordering mainly solves the problem of whether a project can be carried out in order. Besides that, people are usually also concerned about the progress of the project, for example, how much time would it take at least to finish the whole project, and what activities are keys to the project progress.

For example, for the course selection problem introduced in topological ordering, we can obtain the order of each course via topological ordering. If we want to know

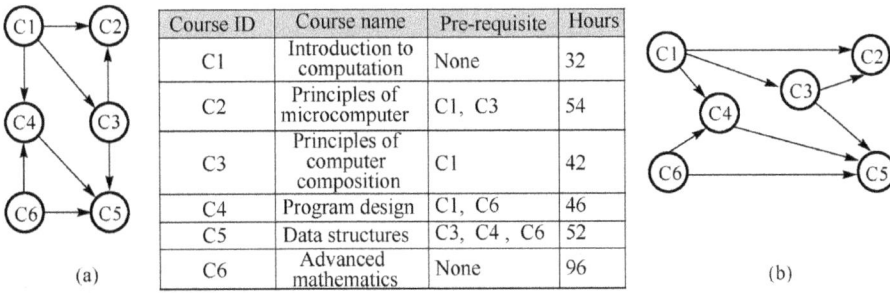

Fig. 2.63: The shortest time problem on course selection.

within how much time at least will we be able to finish those courses, we need to add a parameter – the number of hours of each course, as shown in Fig. 2.63. In this way, the problem is converted into obtaining the shortest time on a flowchart, that is, how much time we need at least to finish the whole project.

To facilitate problem description, we topologically transform Fig. 2.63(a) into Fig. 2.63(b). In an AOV network, vertices represent activities. It only describes the restraints between activities. The parameter of course hours is the duration of an activity. We can view it as the weights of the activities. How should we represent this newly added information in the graph?

Think and Discuss How should we describe the AOV network after the activity weights have been added?

Discussion: Reviewing the concept of graph, the marker for weight is normally on the edges of the graph, which makes it straightforward and easy to comprehend. Therefore, in this problem, we need to convert the vertices that represent activities in AOV network into "activity edges." This step is easy to realize. The following problem is, what should be the connecting vertices for such "activity edges" be?

From the concrete meanings of "activity edges," its connecting vertices should be the beginning and end of this activity, respectively, as shown in Fig. 2.64.

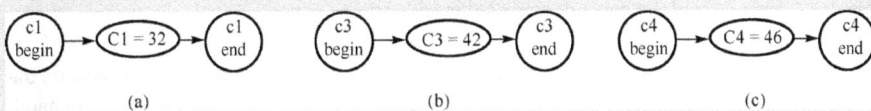

Fig. 2.64: Information description in the conversion between vertices and edges.

After we have analyzed the conversion from a vertex to edge that represents an "activity," the next step should be to connect these "activity edges" together. We can connect them via the fact that the end of one activity is the beginning of another. For example, logically speaking, the vertex where C1 ends is the vertex where C3 starts. Therefore, these two vertices can be combined into one vertex. To facilitate description, we give it a vertex index v2, as shown in Fig. 2.65. It suffices for each vertex to have a different index.

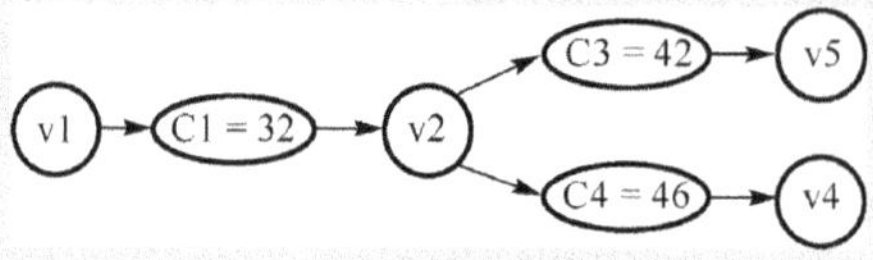

Fig. 2.65: Connections of "activity edges".

Following the above idea, referring to Fig. 2.63, and using "course hours" as the weights of the activity edges, the complete network construction is shown in Fig. 2.66, where "source vertex" and "sink vertex" represent the beginning and end of the whole project, respectively. Observing this figure, we can discover that some activity edges are still similar to the vertices, such as C2, C4 and C5. We can divide them into multiple edges according to the logical relations between edges and vertices and the situations of connection, as shown in Fig. 2.67.

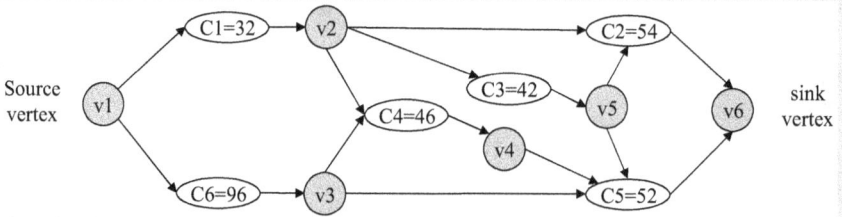

Fig. 2.66: Initial form of the network represented with activity edges.

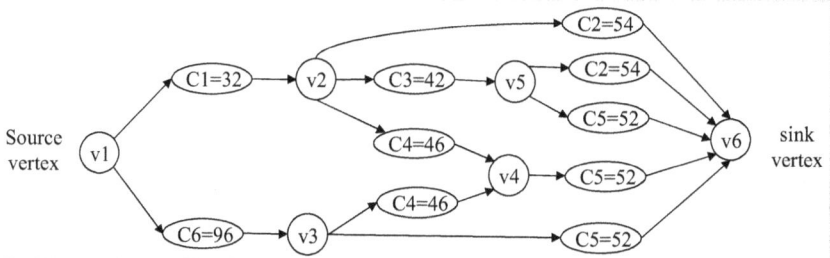

Fig. 2.67: Final form of the network represented with activity edges.

From the final form of the network, we can clearly observe that in the whole course network, the longest paths are C6, C4, C5. Under the condition that multiple courses can be learned in parallel, we need at least 96 + 46 + 52 = 194 course hours to finish the study of the courses in the table. This provides reliable foundation for the writing of teaching plans.

2.8.3.2 The concept of AOE network

AOE network is a weighted directed graph that represents the project flow, where vertices represent events and directed edges represent activities. The weights on the edges represent the duration of the activities. The activities (also called step or work) refer to any kind of activities that cost time or resources, for example, the initial design, technical design and manufacturing of a new product. According to needs, the steps can be divided more roughly or with more details. The events in the AOE network are marks of the beginning or end of such steps.

Properties of AOE network:

1. Only after something represented by a vertex occurs, the activities represented by all the edges starting from that vertex can begin.
2. Only after the activities represented by all edges entering a certain vertex end, the event represented by this vertex can happen.

Since a project has a general beginning and end, in normal circumstances, AOE network has only one source vertex and one sink vertex.

Knowledge ABC Project management and CPM

CPM is a project planning and management method based on mathematical calculations. It is one of the network graph planning methods. CPM decomposes a project into multiple independent activities and ascertains the duration of each activity. It then connects the activities based on logical relations (End – Begin, End – End, Begin – Begin and Begin – End), so that the duration of the project and the features of each activity (earliest/latest time, time difference) can be calculated. After we load resources onto the activities of CPM, we can also analyze the resource needs and distributions of the project. The CPM is an important analytical tool in modern project management.

2.8.3.3 Analysis on the calculation method of critical activities

Using the AOE network to model a project, on the basis that there is no conflict stemming from the mutual restrictions between activities, we can analyze how much time it takes to finish the whole project, or problems such as what activities should we speed up in order to shorten the time needed to finish the project.

Term Explanation Critical path and critical activities

In an AOE network, some activities are carried out sequentially, while some activities are carried out in parallel. There might be multiple directed paths from the source vertex to various vertices, or even from the source vertex to the sink vertex; these paths will also differ in lengths. Although the time needed to finish activities on different paths is different, it is always the case that only when you finish all the activities on all the paths, you can finish the whole project. Therefore, the time needed for finishing the whole project depends on the length of the longest path from the source vertex to the sink vertex, that is, on the sum of the continuous time of all the activities on this path. This longest path is called critical path. Since some activities in the AOE network can be carried out simultaneously, the time required to finish the whole project should be the length of the longest path from the source vertex to the sink vertex. The length of the critical path is the shortest time needed by the whole project, and the activities on the critical path are the critical activities

1. Algorithm analysis of critical activities

Let us analyze the way to obtain critical path with a concrete example of AOE network. For example, a certain research project consists of 10 activities, as shown in Fig. 2.68. According to the restriction relations between the various activities in the table, we can draw the network representation of the project, as shown in Fig. 2.69. We want to calculate the longest path from the source vertex to the sink vertex. There are two types of scenarios depending on the types of the vertices: one is vertices with indegree 1, and the other is vertices with indegree bigger than 1.

Activity	Content	Amount of work	Pre-requisite activities
a1	Prepare the materials	3	/
a2	Preliminary study	2	/
a3	Site selection	1	a1
a4	Design investigation plan	8	a1
a5	Contact investigation site	3	a2
a6	Train personnel	7	a2
a7	Prepare the table	4	a3,a5
a8	On-site investigation	2	a3,a5
a9	Write investigation report	9	a4,a7
a10	Meeting and summary	6	a6,a8

Fig. 2.68: Details of the research activity.

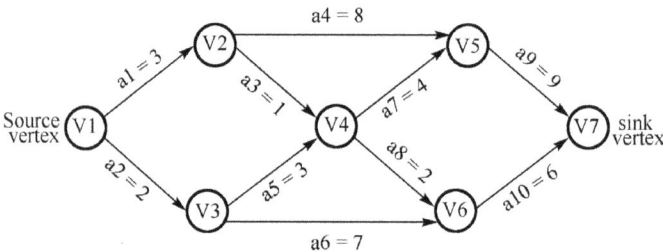

Fig. 2.69: Network for the project.

- Vertices with indegree 1: the moment the activity on the outgoing edge of this vertex starts is the moment the activity on the incoming edge of this vertex ends. For example, for vertex V2, activity a3 starts at the time a1 finishes.
- Vertex with indegree bigger than 1: the moment this vertex starts its activities on its outgoing edges is the moment the activities on all the incoming edges of this vertex end. We need to take the largest value among all paths of incoming edges.

For example, for vertex V4, activity a7 must start when the two paths before it are a1, a3 and a2, a5 at both ends. Therefore, for the moment V4 starts, we need to choose the biggest among path values 4 and 5, which is 5. To calculate the values according to the beginning time of events at all vertices, as shown in Fig. 2.70, we can get that the sink vertex has a value of 20, whose meaning is the longest path of the whole AOE network.

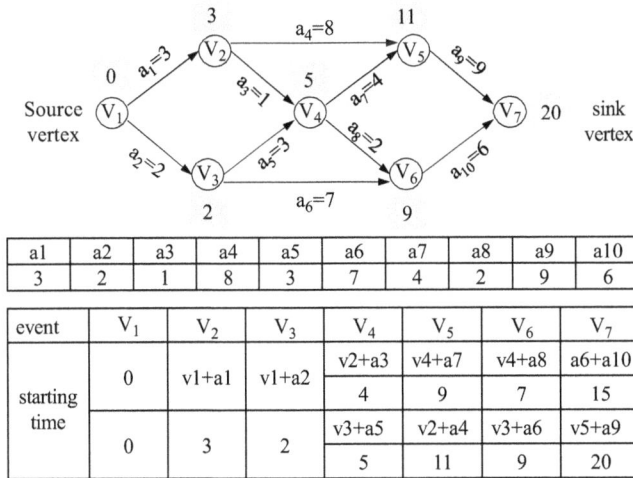

al	a2	a3	a4	a5	a6	a7	a8	a9	a10
3	2	1	8	3	7	4	2	9	6

event	V_1	V_2	V_3	V_4	V_5	V_6	V_7
starting time	0	v1+a1	v1+a2	v2+a3	v4+a7	v4+a8	a6+a10
				4	9	7	15
	0	3	2	v3+a5	v2+a4	v3+a6	v5+a9
				5	11	9	20

Fig. 2.70: The starting time of various vertex events from source vertex to sink vertex.

Think and Discuss If we only know the value of the longest path, can we determine the critical events?

Discussion: From the longest path value of 20 in Fig. 2.70, we can observe that the longest path is v1v2v5v7. But this is only a result of observation. Only using the information from various points in this table, we cannot "calculate" the corresponding vertices for the critical events. We need further information to ascertain the characteristic data of the critical events.

We can try to deduce from the other direction. Using vertex v6 as an example, since the value of vertex v7 is 20, even if vertex v6 starts at 20−a10 = 14, the overall progress would not be impacted. Therefore, the starting moment of v6, 9, in Fig. 2.70, is the earliest possible starting time. The moment 14 reverse deduction from the sink vertex is the latest possible starting time.

Let us look again at the latest starting time of vertex v5, 20 − a9 = 11. This moment is the same as its earliest starting time. Therefore, we can use reverse deduction to calculate the latest possible starting moments of all events, and then perform the analysis.

To reverse-deduct the latest possible starting moment of all events from the sink vertex to the source vertex, see Fig. 2.71, where ee[k] represents the earliest possible moment for the event to occur (earliest event) and le[k] represents the latest possible moment for the event to occur (latest event).

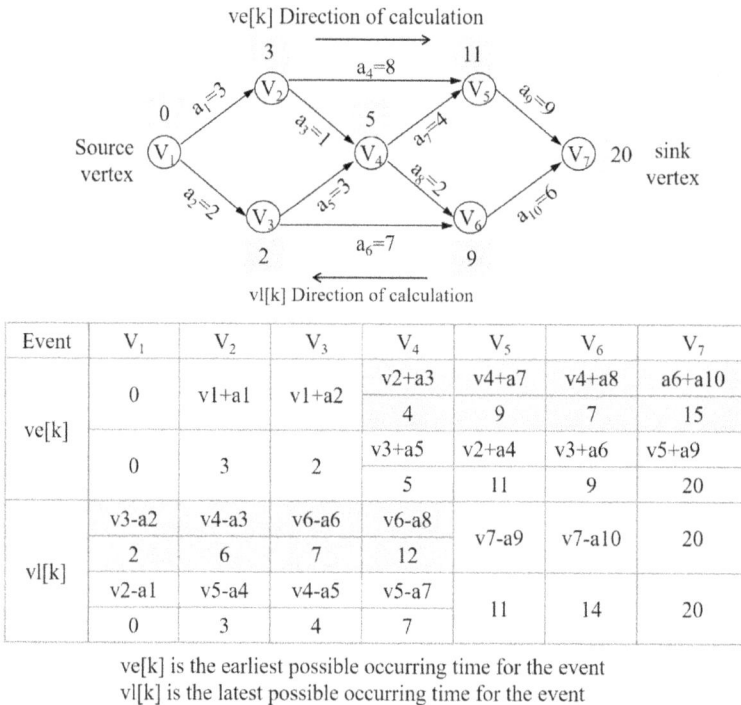

Event	V_1	V_2	V_3	V_4	V_5	V_6	V_7
ve[k]	0	v1+a1	v1+a2	v2+a3	v4+a7	v4+a8	a6+a10
				4	9	7	15
	0	3	2	v3+a5	v2+a4	v3+a6	v5+a9
				5	11	9	20
vl[k]	v3-a2	v4-a3	v6-a6	v6-a8	v7-a9	v7-a10	20
	2	6	7	12			
	v2-a1	v5-a4	v4-a5	v5-a7	11	14	20
	0	3	4	7			

ve[k] is the earliest possible occurring time for the event
vl[k] is the latest possible occurring time for the event

Fig. 2.71: The earliest and latest possible starting times of the events at all vertices.

Observing the situations of the vertices, we can still divide them into two types: one is with outdegree of 1 and another is with outdegree greater than 1.

- Vertex with outdegree 1: the latest moment for the activities on the outgoing edges of this vertex to start is the time on its arc head vertex minus the activity time on the edge, such as vertex v6. The latest beginning moment for activity a10 = time of V7 − a10 = 20 − 6 = 14.
- Vertex with outdegree greater than 1: the latest starting moment for the activities on all outgoing edges is the smallest one out of the time obtained by arc head vertex minus activity time. For example, vertex v4 has two outgoing edges a7 and a8, with the corresponding arc tail vertices v5 and v6; therefore, the latest starting time corresponding to v4 has two values, 7 and 12. If we take the value 12, then among the two paths from v4 to the sink vertex, the time going through the longest path (v4, v5, v7) to the sink vertex will be 25, which

is bigger than the previously deduced longest path; if we take the value 7, then its value will be 20, which is the same as the value of the longest path. Therefore, for a vertex with outdegree higher than 1, the latest starting time needs to be selected from the smallest values.

Conclusion
Source vertex → Sink vertex: for the vertices with indegree greater than 1, the earliest beginning moments need to be selected from the biggest values.
   Sink vertex → Source vertex: for the vertices with outdegree greater than 1, the latest beginning moments need to be selected from the smallest values.

Think and Discuss How to ascertain critical events?   !
Discussion: We already know that the longest path in Fig. 2.69 passes through the vertices v1, v2, v5, v7. If we observe the ee[k] and le[k] of the same vertex, then find the difference between the two values is always 0. On the contrary, vertices not on the longest path will always have a nonzero difference. Therefore, this condition of "the time difference between the latest and earliest starting times is 0" can be used as the condition to check whether the vertex is a critical event, as shown in Fig. 2.72.

Event	v1	v2	v3	v4	v5	v6	v7
ve[k]	0	3	2	5	11	9	20
vl[k]	0	3	4	7	11	14	20
vl[k]−ve[k]	0	0	2	2	0	5	0

Fig. 2.72: Analysis table of the event.

Looking at the actual meaning of this difference, the situation where the earliest and latest possible starting times are the same indicates that this event can be either moved earlier or later, but can only be started at that very moment. The situation where the times are different indicates that this event can be started any time between the two.

In the following, we can use a similar method to analyze and ascertain critical activities, as shown in Fig. 2.73, where ee[i] is the earliest starting time of the activity, which is the same as the corresponding event: el[i] is the latest possible starting time for the activity, which is obtained by subtracting the latest possible starting time of the vertex at the arc tail with the corresponding activity time. The activities whose differences of el[i] and ee[i] are 0 are the critical activities a1, a4, a9.

   The ultimate results:
   Critical activities: a1, a4, a9
   Critical path: v1→v2→v5→v7
   Shortest time: a1 + a4 + a9 = 20

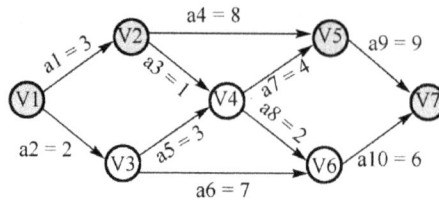

Activity	a1	a2	a3	a4	a5	a6	a7	a8	a9	a10
ee[i]	v1	v1	v2	v2	v3	v3	v4	v4	v5	v6
	0	0	3	3	2	2	5	5	11	9
el[i]	v2–a1	v3–a2	v4–a3	v5–a4	v4–a5	v6–a6	v5–a7	v6–a8	v7–a9	v7–a10
	0	2	6	3	4	7	7	12	11	14
el[i]–ee[i]	0	2	3	0	2	5	2	7	0	5

Earliest starting time for the activity ee[i]	Latest starting time for the activity el[i]

Fig. 2.73: Table for the analysis of activities.

2. Features of critical paths
   - The duration of the activities along the critical path ascertain the duration of the project. The sum of all the durations of the activities on the critical path is the duration of the project.
   - Any activity along the critical path is a critical activity. Delay of any of such activities will cause the delay of the whole project.
   - The time cost of the critical path is the shortest possible time for the project to finish. If we shorten the total time cost of the critical path, we will shorten the duration of the project; on the contrary, we will lengthen the duration of the project. However, if we only shorten the time needed by activities on noncritical paths within a certain extent, we would not impact the duration of the whole project.
   - The activities on the critical path are activities with the smallest amount of total differences. If we change the time cost of a certain activity there, we might change the critical path.
   - There might be multiple critical paths, whose time cost are the same and equal to the total duration of the project.

3. The generalized method for calculating critical path
   - Set: the earliest possible starting time for events is ve(i); latest possible starting time for events is vl(i).
   - Set: the earliest possible starting time for activities is ee(i); latest possible starting time for activities is el(i).
   Definition: activities whose ee(i) = el(i) are called critical activities.

- Let arc <j, k> represent activity ai, and denote its duration as dut(<j, k>), for example, the relations between events and activities in Fig. 2.74.

$$\text{Then } ee(i) = ve(j)$$

$$El(i) = vl(k) - dut(< j, k >)$$

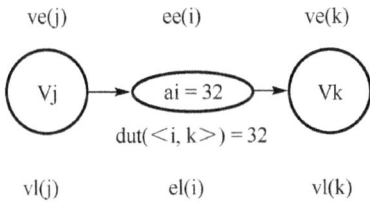

Fig. 2.74: Relation between events and activities.

- The method to calculate the earliest possible starting time of events is ve(j) and latest possible starting time for events is vl(i). The relations between various starting times of events are shown in Fig. 2.75.
  (1) To start from ve(1) = 0, deducing from the source vertex to the sink vertex. Source vertex → Sink vertex: The vertices with indegree >1 should select the largest value
  $ve(j) = Max\{ ve(i) + dut(<i, j>) \}$
  where T is the set of arcs with j as arc heads, $<i, j> \in T, 2 \le j \le n$.
  (2) To start from vl(n) = ve(n), deducing from the sink vertex to the source vertex Sink vertex → Source vertex: The vertices with indegree >1 should select the smallest value
  $vl(i) = Min\{ vl(j) - dut(<i, j>) \}$
  where S is the set of arcs with i as arc tails, $<i, j> \in S, 1 \le i \le n-1$.

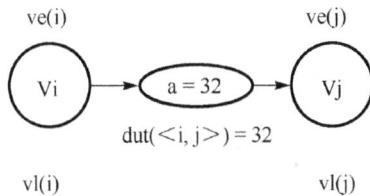

Fig. 2.75: Relations between events.

The two above deduction formulas are carried out with the premises of topological order and reverse topological order.

4. Algorithms to obtain the critical path
   - Pseudocode description
Pseudocode description 1
   (1) Topologically order the graph. If it fails then exit.

(2) Following the topological order, search activity edges sequentially and calculate the earliest possible starting times of events.
(3) Following the topological order, search activity edges reversely and calculate the latest possible starting times of events.
(4) Calculate the earliest possible starting times and latest possible starting
times of various .activities.
(5) Obtain the critical events and critical activities.

Detailed description of pseudocode 1
(1) Topologically order the graph and construct topological sequence S and reverse sequence T. If it fails then exits.
(2) Following the order of topological sequence S, calculate the earliest possible time for all vertices (events), with the formula

$$ve(j) = Max\{ve(i) + dut(<i, j>) \}$$

(3) Following the order of the reverse topological sequence T, calculate the latest possible time for all vertices (events), with the formula

$$vl(i) = Min\{vl(j) - dut(<i, j>)\}$$

(4) Calculate the earliest possible time for each activity <i, j> ee(i, j) and latest
possible time el(i, j).
If it satisfies ee(i, j) = el(i, j), then it is a critical path and we can output it.
Detailed description of pseudocode 2
TopoSort function: Topologically sort the graph and obtain the earliest possible
time for events.
(1) Use adjacency list as storage structure.
(2) Push all the vertices with indegree 0 in the adjacency list into the stack S.
(3) While stack S is nonempty, pop the stack top element Vi and push Vi to
stack T.
(4) Search for the direct successor of Vi, Vk in the adjacency list. Decrease the
indegree of Vk by 1. If the indegree of Vk is 0, then push it into stack S.
(5) Calculate the earliest possible time for the adjacent vertex of vi, k:

$$if(ve[i] + dut > ve[k]), ve[k] = ve[i] + dut$$

(6) Repeat the above steps 3–5, until the stack S is empty.
(7) If the number of vertices output when the stack is empty is not the number of vertices of the graph, then there is a cycle in the directed graph.
Otherwise, the topological ordering is finished.

CriticalPath function gets the critical path and the topological ordering of the critical activities. The result of the reverse ordering will be stored in the stack T:

  (1) Initialize the latest possible time of vertex events v1 as the earliest possible time of the sink vertex, ve.
  (2) Get the latest possible starting time of each vertex according to the order of stack T.
  (3) Iterate each activity in order.

The earliest possible starting time of activity <i, k> ee = The earliest possible starting time of vertex i, ve[i].

The latest possible starting time of activity <i, k> el = The latest possible starting time of vertex k vl[k] – dut.

If(ee = = el)then < i, k > is a key activity.

− Data structure design
  (1) The graph is represented with adjacency table, as shown in Fig. 2.76. We can add a data item indegree in the vertex structure of the adjacency table to store the indegree value of each vertex in the AOV network. We can add a data item weight to store the weights of the edges in the adjacent vertex structure. The data structures and functions to construct adjacency list are in the file AdjList.h.
  (2) Construct sequential stacks S and T to store the topological ordering sequence and reverse topological ordering sequence, respectively. The data structure and functions for sequential stack are in the file SqStack.h.
  (3) The array for the earliest possible starting time of events int ve[VERTEX_NUM];

The array for the earliest possible starting time of events int vl[VERTEX_NUM];

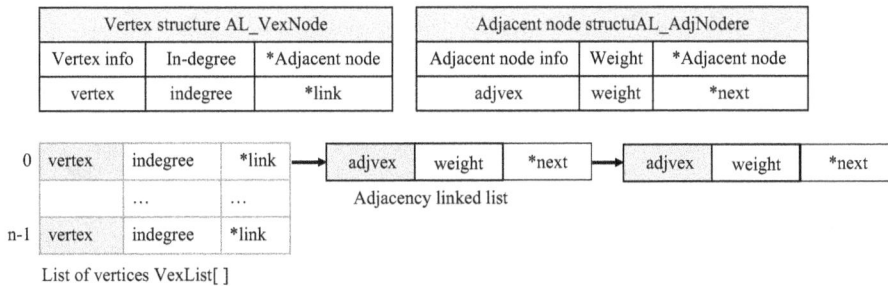

Fig. 2.76: The adjacency table storage of graph for critical path algorithm.

Program implementation

```c
#include <stdio.h>
#include <stdlib.h>
#define TRUE 1
#define FALSE 0
#include "AdjList.h"
#include "SqStack.h"
/*===============================================================
Functionality: Topological ordering and reverse topological ordering,
obtain the earliest starting time for all vertex events
Function input: Adjacency table of the graph, (Sequential stack S), (Array
for the earliest starting time of events)
Function output: Sequential stack (Its content is the reverse topological
sequence), (Array for the earliest starting time of events)
===============================================================*/
void TopoSort(AL_Graph &G, SeqStack &T, int ve[])
{
    SeqStack S;
    initialize_SqStack(&S);

    AL_AdjNode *p;
    int count=0;
    int i;
    int k;
    int dut;

    for(i=0;i<G.VexNum;i++)
        G.VexList[i].indegree=0;
    // The degree of the vertex is initialized to 0
    // Calculate the indegrees of various vertices
    for(i=0;i<G.VexNum;i++)
    {
        p=G.VexList[i].link;
        while(p)
        {
            G.VexList[p->adjvex].indegree++;
            p=p->next;
        }
    }
    // Push the vertex with indegree 0 into the stack
    for(i=0;i<G.VexNum;i++)
```

```
{
  // Push the vertex with indegree 0 into the stack at
  // position i of VexList array
  if(G.VexList[i].indegree==0)
  Push_SqStack(&S,i);
}
// Initialize the earliest starting time of the event as 0
or(i=0;i<G.VexNum;i++) ve[i]=0;
printf("Topological ordering sequence:");
while(S.top!=-1)
{
  Pop_SqStack(&S, &i);   // Pop the element with indegree 0
  printf("%d ",G.VexList[i]);
  // Output vertex with indegree 0
  // Popping the topological sequence in S in order and push them
  // into T in order, we obtain the reverse topological sequence
  Push_SqStack(&T,i);
  count++;
  // Vertex counter is increased by 1
  p=G.VexList[i].link;
  // Point p to the first edge table node of the vertex with indegree 0
  while(p)
  {
    dut=p->weight;
    k=p->adjvex;
    // Decrease the indegree of the adjacent vertex of
    // the vertex with indegree 0
    G.VexList[k].indegree--;
    // If the vertex has indegree 0 after its indegree is decreased by 1,
    //push it onto the stack
    if(G.VexList[k].indegree==0) Push_SqStack(&S,k);
    // Through a while loop, calculate the earliest starting time of all
    // adjacent vertices of the vertex
    if(ve[i]+dut>ve[k]) ve[k]=ve[i]+dut;
    p=p->next;
  }
}
printf("\n");
if(count<G.VexNum)
  printf("Network G has circuits!\n");
}
```

```
/*==============================================================
Functionality: Obtain the critical path and critical activities
Function input: Adjacency table of the graph
Function Output:  None
Screen output: Critical path and critical activities
==============================================================*/
void CriticalPath(AL_Graph &G)
{
    int i,j,k,dut;
    int ee,el;
    // Earliest starting time of the activity and latest starting time
    int ve[VERTEX_NUM];
    // Earliest starting time of the events
    int vl[VERTEX_NUM];
    // Latest starting time of the events
    AL_AdjNode *p;
    SeqStack T;

    initialize_SqStack(&T);
    TopoSort(G,T,ve);
    // Topological ordering. The reverse ordering order is stored in stack T
    // Initialize the latest starting time of the vertex event as the earliest
    // starting time of the sink vertex
    for(i=0;i<G.VexNum;i++) vl[i]=ve[G.VexNum-1];
    //while loop, obtain the latest starting time of each vertex according to
    // the order of stack T
    while(T.top!=-1)
    {
        // According to the order of events in stack T, iteratively calculate the
        // latest starting time of the events at all vertices
        for(Pop_SqStack(&T,&j),p=G.VexList[j].link; p ;  p=p->next)
        {
            k=p->adjvex;
            dut=p->weight;
            if(vl[k]-dut<vl[j])
            vl[j]=vl[k]-dut;
            // Calculate the latest starting time of vertex events
        }
    }
    printf("Starting point, ending point, earliest starting time, latest
    starting time, critical activities\n");
```

```
int totaltime=0;
//Iterate through each activity <i,k>
for(i=0;i<G.VexNum;i++)
{
  for(p=G.VexList[i].link;p;p=p->next)
  {
    k=p->adjvex;
    dut=p->weight;
    ee=ve[i];
    // Earliest starting time of activity <i, k>
    el=vl[k]-dut;
    // Latest starting time of activity <i, k>
    // To accommodate the fact that the vertex numbering in the test case
    // starts from 1, we add 1 to the vertex numbering being output.
    printf("%4d %4d",i+1,k+1);
    printf("%12d %12d",ee,el);
    if(ee==el)
    // Critical activity
    {
      printf(" (%2d,%2d)",G.VexList[i].vertex+1,G.VexList[k].vertex+1);
      totaltime+=dut;
    }
    printf("\n");
  }
}
printf("The length of the critical path is: %d\n",totaltime);
}
int main()
{
  AL_Graph G;
  G=Create_AdjList();
  CriticalPath(G);
  return 0;
}
```

Test result: the graph serving as test case is Fig. 2.69.
Topological ordering sequence: 0 2 1 3 5 4 6.
Starting point, ending point, earliest starting time, latest starting time, critical activities.

1	2	0	0 (1, 2)
2	4	3	6
2	5	3	3 (2, 5)

3	4	2	4
3	6	2	7
4	5	5	7
4	6	5	12
5	7	11	11 (5, 7)
6	7	9	14

The length of the critical path will be 20.

2.8.4 Summary of activity vertices and activity edges

The problems to be solved by AOV network are: first, is there any cycle within the network, that is, whether the flow of the project makes sense; second, if there is no cycle in the network, then what are the orders for the arrangement of various child processes. These two problems can be solved with topological ordering.

AOV network is a network where the vertices represent activities. It only describes the constraints between the activities. On the contrary, AOE network uses edges to represent activities, and the weights on the edges represent the durations of the activities. We use deduction and reverse deduction methods to calculate the earliest possible starting time, latest possible starting time, earliest finishing time and latest finishing time of each activity and calculate the time difference of each activity. Then we find out the path composed of all the activities with time difference of 0, which will be the critical path.

2.9 Chapter summary

The connections between main contents of this chapter are shown in Fig. 2.77.
1. Storage methods of graph
   - Adjacency matrix
   - Adjacency linked list

2. Traversal of graphs
Traversal of graphs is an extension of traversal of trees. It is the operation of visiting the vertices in the graph in order and only once, and is also the process of linearizing the network structure according to the certain rule.

Since there might be cycles in the graph, in order to distinguish whether a vertex is visited, and avoid repeated visits to vertices, we should mark each vertex visited during the traversal.

Depth-first: the depth-first traversal of graph is similar to the pre-order traversal of tree. The sequence obtained according to the order of visiting the vertices is called DFS sequence.

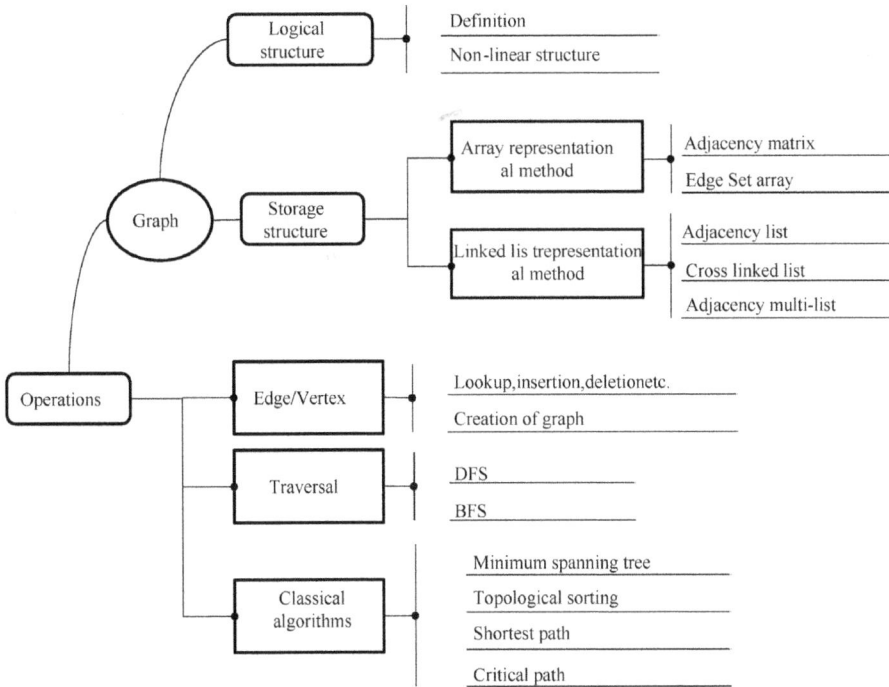

Fig. 2.77: The connections between various concepts of graph.

Breadth-first: the breadth-first traversal of graph is similar to the layered traversal of tree. The sequence obtained according to the order of visiting the vertices is called BFS sequence.

3. Applications of graph
   – MST: the spanning tree with minimal weights (for undirected graphs). The corresponding algorithms include Prim's algorithm and Kruskal's algorithm.
   – Shortest path (weighted directed graph). Single-source shortest path problem: obtain the shortest paths from a certain source vertex to the rest of the vertices; All-pair shortest path problem: convert to single-source shortest path problem for different vertex pairs in the graph. The corresponding algorithms include the Dijkstra algorithm and the Floyd algorithm.
   – Topological ordering (DAG): to line up all the vertices in the graph into a linear sequence, which satisfies the condition that the arc tail is before the arc head.
   – Critical path (weighted directed graph): the weights on the edges represent the duration needed to complete this activity. The critical path is the longest path in the graph.

The storage methods for the graph include adjacency matrix and adjacency list.

The traversal on the graph involves running either BFS or DFS.

A spanning tree has paths between its vertices and no cycles.

The minimal spanning tree needs to be compared with weights attached.

The critical path to obtain the longest distance between the source vertex and the sink vertex in a DAG.

The shortest path is to obtain the shortest distances between vertices in a network.

Topological ordering of a activity network should derive total ordering from partial ordering.

The applications of graph under different circumstances should be well remembered.

2.10 Exercises

2.10.1 Multiple-choice questions

1. In a graph, the sum of the degrees of all the vertices is ( ) times the number of edges.
    (A) 1/2   (B) 1   (C) 2   (D) 4

2. In a directed graph, the sum of the indegrees is ( ) times the sum of outdegrees of all the edges.
    (A) 1/2   (B) 1   (C) 2   (D) 4

3. An undirected graph with n vertices has at most ( ) edges.
    (A) n (B) $n(n-1)$ (C) $n(n-1)/2$ (D) $2n$

4. An undirected connected graph with eight vertices has at least ( ) edges.
    (A) 5   (B) 6   (C) 7   (D) 8

5. For an undirected graph with n vertices, if we represent it with an adjacency matrix, then the size of this matrix is ( ).
    (A) n (B) $(n-1)^2$ (C) $n-1$ (D) n^2

6. When we perform breadth-first traversal on a graph represented with an adjacency list, usually we use ( ) to implement the algorithm.
    (A) stack   (B) queue   (C) ordering   (D) sorting

7. When we perform depth-first traversal on a graph represented with an adjacency list, usually we use ( ) to implement the algorithm.
    (A) stack   (B) queue   (C) ordering   (D) sorting

8. If we can visit all the vertices of an undirected graph by performing a DFS starting at any vertex, then this graph must be ( ).
(A) complete graph   (B) connected graph   (C) cyclic graph   (D) a tree

9. To store a weighted directed graph G with adjacency matrix A, then the indegree of vertex i would be ( ) in A.
(A) sum of noninfinite elements at row i
(B) sum of number of noninfinite elements at the ith column
(C) number of noninfinite and nonzero elements at row i
(D) sum of noninfinite and nonzero elements at the ith row and the ith column

10. Depth-first traversal on graphs stored with adjacency list is similar to ( ) of binary trees.
(A) in-order traversal
(B) pre-order traversal
(C) post-order traversal
(D) layered traversal

11. The adjacency matrix of an undirected graph is a ( ).
(A) symmetric matrix
(B) zero matrix
(C) upper triangular matrix
(D) diagonal matrix

12. Adjacency list is a ( ) of graphs.
(A) sequential storage structure
(B) linked storage structure
(C) indexed storage structure
(D) hashed storage structure

13. To define the path from v_i to v_j in an undirected graph is a ( ) from v_i to v_j.
(A) vertex sequence
(B) edge sequence
(C) sum of weights
(D) number of edges

14. In the reverse adjacency list of a directed graph, the adjacency list for each vertex links all the ( ) for the adjacent vertices of this vertex.
(A) incoming edges
(B) outgoing edges
(C) incoming edges and outgoing edges
(D) neither outgoing edges nor incoming edges

15. Let G1=(V1, E1) and G2=(V2, E2) be two graphs, if V1 V2,E1 E2, then we say ( ).
    (A) G1 is a subgraph of G2
    (B) G2 is a subgraph of G1
    (C) G1 is a connected component of G2
    (D) G2 is a connected component of G1

16. We know the adjacency matrix representation of a directed graph. To delete all the edges outgoing from the ith vertex, we should ( ).
    (A) delete the ith row of the adjacency matrix
    (B) set all elements in the ith row of the adjacency matrix to 0
    (C) delete the ith column of the adjacency matrix
    (D) set all elements in the ith column of the adjacency matrix to 0

17. The topological ordering of any directed graph ( ).
    (A) might not exist
    (B) has one possibility
    (C) definitely has multiple possibilities
    (D) has one or multiple possibilities

18. The incorrect statement about graph traversal below is ( ).
    (A) the DFS of connected graph can be a recursive process
    (B) the search of adjacent vertices in BFS has the characteristic of "first in first out"
    (C) unconnected graph cannot be searched via DFS
    (D) the traversal of graph requires each vertex to be visited just once

19. The BFS algorithm on graphs stored with adjacency lists is similar to ( ) of binary trees.
    (A) pre-order traversal
    (B) in-order traversal
    (C) post-order traversal
    (D) layered traversal

20. Critical path is ( ) in the AVE network.
    (A) the longest path from source vertex to sink vertex
    (B) the shortest path from source vertex to sink vertex
    (C) the longest cycle
    (D) the shortest cycle

21. The ( ) below can be used to check whether there is a cycle in the graph.
    (A) breadth-first traversal
    (B) topological ordering
    (C) obtaining shortest path
    (D) obtaining critical path

22. A result sequence for topological ordering of the below directed graph (Fig. 2.78) is ( ).
    (A) 125634   (B) 516234   (C) 123456   (D) 521643

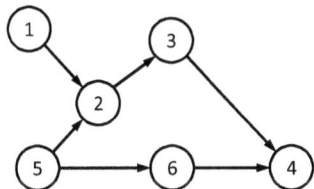

Fig. 2.78: A directed graph.

23. There are ( ) types of MSTs for any undirected connected graph.
    (A) only one
    (B) one or multiple
    (C) definitely multiple
    (D) it may not exist

2.10.2 Long-form questions

1. The table of participation for a tracks competition is in Table 2.10. Suppose matches A to F represent a data element, respectively. If two matches cannot be held simultaneously, then they are connected (restraining condition). (1) According to this table and the restraining condition, draw the corresponding graph structure model, and draw the adjacency list structure of this graph. (2) Write the element sequence obtained by BFS starting from element A.

Table 2.10: Table of participation for a tracks competition.

Name	Match
ZHAO	A B C
QIAN	C D
SUN	C E F
LI	D F A
ZHOU	B F

2. The number of strongly connected components in Fig. 2.79 is ___

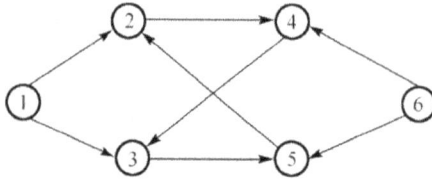

Fig. 2.79: A directed graph.

3. Given an undirected graph as shown in Fig. 2.80, draw its adjacency list, write out the sequence of vertices of edges obtained by DFS and BFS that start from vertex 1.

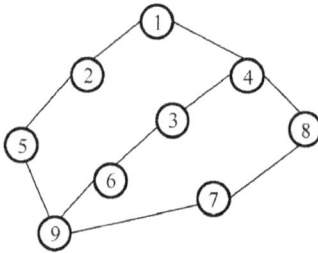

Fig. 2.80: An undirected graph.

4. For the connected network shown in Fig. 2.81, construct its MST with Prim's algorithm and Kruskal's algorithm, respectively.

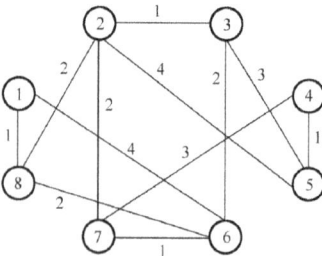

Fig. 2.81: A connected network.

5. According to the search paths of depth-first traversal and breadth-first traversal, give the construction algorithms for the spanning trees of graphs. Try to realize the union-search set with linked list.

6. For the directed graph shown in Fig. 2.82, try to use the Dijkstra algorithm to obtain the shortest paths from vertex v1 to all the other vertices, and write out the situations at each step of algorithm execution.

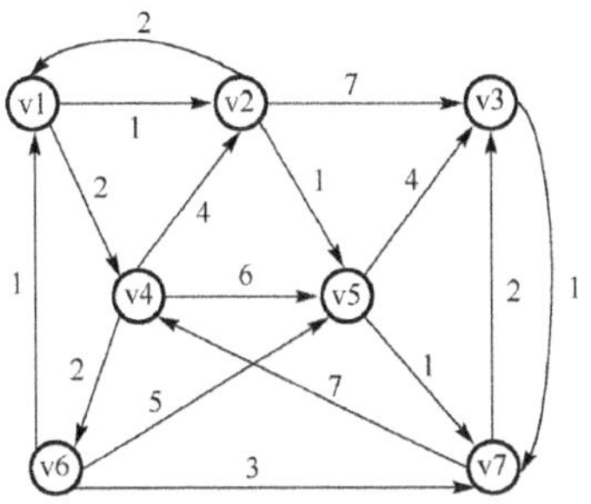

Fig. 2.82: A directed graph.

7. Figure 2.83 is the adjacency list representation of weighted directed graph
G. Draw:
   (1) This graph in the form of normal graph figure
   (2) The vertex sequence obtained via DFS starting from vertex v1
   (3) The vertex sequence obtained via BFS starting from vertex v1
   (4) The critical path from v1 to v8

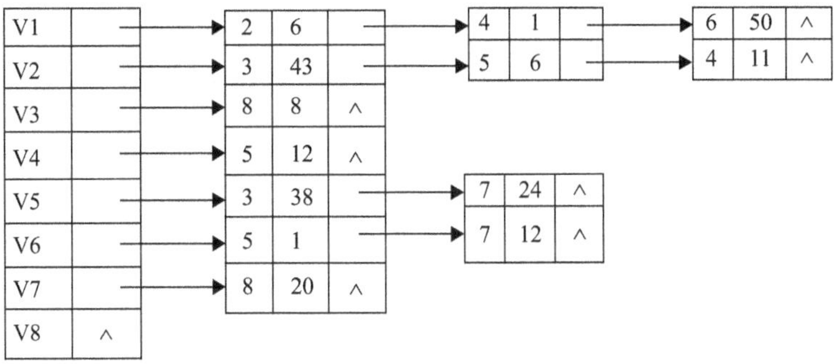

Fig. 2.83: The adjacency list representation of weighted directed graph.

8. For the AOV network shown in Fig. 2.84, obtain the changing process of indegree
field when topological ordering is performed with predecessorless-vertex-first topological ordering algorithm. Write out all possible topological ordering sequences.

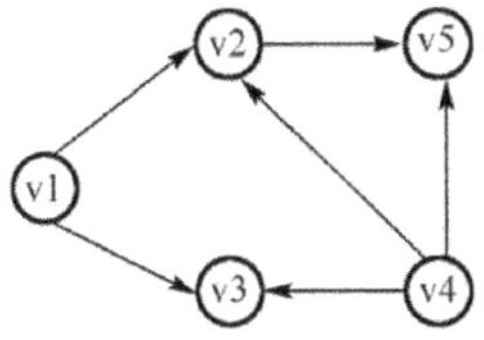

Fig. 2.84: An AOV network.

9. Try to solve the single-source shortest path problem of DAG with topological ordering.
10. For the AOE network shown in Fig. 2.85, obtain:
    (1) The earliest possible and latest possible starting time
    (2) All critical paths
    (3) The shortest duration for the project
    (4) Can we accelerate the project by accelerating some activities?

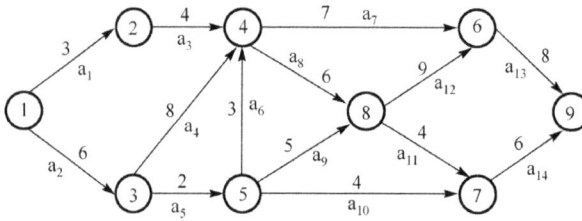

Fig. 2.85: An AOE network.

11. We already know the adjacency matrix of a directed network as follows. If we need to construct an entertainment center at one of the vertices, and require the longest distance from this vertex to other vertices to be the shortest, and the total distance to be as short as possible. Where should the entertainment center be located? Write out the solving process.

$$
\begin{array}{c}
V1 \\ V2 \\ V3 \\ V4 \\ V5 \\ V6
\end{array}
\begin{bmatrix}
0 & 2 & \infty & \infty & \infty & 3 \\
\infty & 0 & 3 & 2 & \infty & \infty \\
4 & \infty & 0 & \infty & 4 & \infty \\
1 & \infty & \infty & 0 & 1 & \infty \\
\infty & 1 & \infty & \infty & 0 & 3 \\
\infty & \infty & 2 & 5 & \infty & 0
\end{bmatrix}
$$

12. Table 2.11 lists the priority relations of projects and the required duration of each project. Please:
    (1) Draw the AOE network
    (2) List the earliest possible time and latest possible time of each event
    (3) Find the critical paths of this AOE network, and answer the shortest time needed to finish the whole project

Table 2.11: Relations between projects and the required duration of each project.

Project ID	Time needed	Prerequisite	Project ID	Time needed	Prerequisite
A	15	None	H	15	G, I
B	10	None	I	120	E
C	50	A	J	60	I
D	8	B	K	15	F, I
E	15	C, D	L	30	H, J, K
F	40	B	M	20	L
G	300	E			

13. We already know the sequence of edges and weights of vertices 1–6 (as shown in Table 2.12): the three numbers on each row represent the two endpoints and the weight of each edge. In total, there are 11 rows. Represent this undirected network with adjacency multilist, describe the data structure with C-like language and draw the illustration figure for the storage structure. They should conform to the order of algorithm that inserts at the head of edge node linked list and the order of input sequence. Write the DFS and BFS traversal vertex sequences starting from vertex 1, and the corresponding spanning trees. Calculate the MST starting from vertex 1 according to the Prim's algorithm, draw the calculation table and draw this tree.

Table 2.12: Sequence of edges and weights of vertices in a graph.

Numbering	Endpoint	Endpoint	Weight
1	1	2	5
2	1	3	8
3	1	4	3
4	2	4	6
5	2	3	2
6	3	4	4
7	3	5	1
8	3	6	10
9	4	5	7
10	4	6	11
11	5	6	15

2.10.3 Algorithm design questions

1. Basic programming exercise about graphs. In the different storage methods of graph:
   (1) Add a vertex
   (2) Add an edge (arc)
   (3) Lookup whether the graph contains a certain item
   (4) Lookup the designated item and return
   (5) Print each vertex and its adjacent vertices
2. Design an algorithm to check whether the undirected graph is connected. If yes return 1, if no return 0.
3. Using the storage structure of adjacency list, write an algorithm to check whether there is a simple path with length k between any two given vertices of an undirected graph.
4. Suppose that on average each person has 25 friends. According to the six-degree hypothesis, the connection between any two people can definitely be indirectly established via six people. If a person is connected via N people, then he would be your degree–n friend. Now try to check this six-degree theory with program.
5. Suppose that the following are competitions for a track event: A (high jump), B(long jump), C(javelin), D(shot put), E(100 m), F(200 m). Table 2.13 is the competition list for the participating athletes. How to arrange the competitions so that each competition can be carried out successfully (without conflict). Shorten the whole duration as much as possible.

Table 2.13: List of participating athletes for a competition.

Name	Competition 1	Competition 2	Competition 3
Alpha	A	B	E
Beta	C	D	
Gamma	C	E	F
Delta	D	F	A
Epsilon	B	F	

3 Data processing methods – sorting technologies

Main Contents
- The concept of sorting
- Common sorting methods

Learning Targets
- Familiarize yourselves with the features of various sorting methods
- Familiarize yourselves with various sorting processes and their principles.

!

3.1 Introduction

Nowadays, the smartphones store a huge amount of contact information. Most of them are displayed according to the surname of the contact person. We can imagine that if such name information is chaotic and unordered, as shown in Fig. 3.1, then to find the name of somebody we will need to compare the names one by one, which would be very inconvenient. Also we would need to spend relatively large amount of time when there are relatively more contact persons.

In general, we do not want to waste our time on finding certain contact people in the phone, an operation that is supposed to be easy. Therefore, the current mobile operating systems will automatically sort the contact people with a certain order (Chinese character, English character, number all that have their respective orderings), as shown in Fig. 3.2. In this way, looking up a contact person with a certain name according to the order will save us a lot of time. Then, how can we effectively order some originally chaotic records into sorted orders? This introduces an important problem in computer program design, that is, sorting problem.

3.1.1 The basic concepts of sorting

Sorting is an important operation in computer program design. Its functionality is to re-order an initially unordered arbitrary sequence into a sequence ordered according to the keywords. Apparently, to facilitate lookup, we would normally want a sequence to be nonincreasing (or nondecreasing) ordered. In this manner, we can use highly efficient lookup algorithms to perform the lookup operation. Besides, some basic operations in computer programs, such as constructing binary ordered tree, themselves are processes of sorting. Therefore, to learn and study sorting algorithms is an important aspect of computer science students.

https://doi.org/10.1515/9783110676075-003

Name	Telephone
Zhang Dongdong	135XXXXXXXX
An Xiaohong	136XXXXXXXX
LiXiaoming	138XXXXXXXX

Fig. 3.1: Unordered contact information.

Name	Telephone
An Xiaohong	136XXXXXXXX
Li Xiaoming	138XXXXXXXX
Zhang Dongdong	135XXXXXXXX

Fig. 3.2: Ordered contact information.

First, we will give out a concrete definition of sorting operation with a concrete example. Suppose that we have five contact people stored in our mobile phone unordered.

{An Xiaohong, Zhang Dongdong, Li Xiaoming, Wang Xiaohua, Li Daming}

We can denote the sequence of these five records as follows:

R = {R1 = An Xiaohong, R2 = Zhang Dongdong, R3 = Li Xiaoming, R4 = Wang Xiaohua, R5 = Li Daming}

In this way, we have established an unordered sequence R, where each name in the sequence R_i is called a **record**.

In the next step, we need to ascertain what **keyword** are we going to use for the sorting. Keyword refers to the basis used for sorting. The difference in the selection of keyword will cause differences in the sorting results. In this example, we choose the first character of the Pinyin of the name of the contact people as the basis for sorting. Therefore, the current keyword sequence is

$$K = \{K1 = A, K2 = Z, K3 = L, K4 = W, K5 = L\}$$

After we have the keyword sequence, lastly we need to ascertain the **sorting order** (nonincreasing, nondecreasing, ascending, descending). Here, if we choose the nondecreasing sorting method, then eventually we can obtain a sorted sequence

$$\overline{K} = \{K1, K3, K5, K4, K2\}$$

where $\overline{K}$ with a bar indicates the keyword sequence after sorting. After we have established the keyword sequence, we can then follow the clue of the corresponding record, and eventually obtain a nondecreasing ordered list of contact people:

$$\overline{R} = \{R1, R3, R5, R4, R2\}$$

$\overline{R}$ refers to an ordered sequence after ordering. Corresponding to the actual phone, it will be displayed as {An Xiaohong, Li, Xiaoming, Li Daming, Wang Xiaohua, Zhang Dongdong}, which satisfies our requirement for the order.

From the above concrete example, we can see that ordering is a process of adjusting an unordered sequence R according to its keyword K in a designated order, until we finally obtain the ordered sequence $\overline{R}$.

Knowledge ABC Order of records

Generally speaking, the order relation among a series of records include nonincreasing ($A \geq B \geq C$) and nondecreasing ($A \leq B \leq C$), ascending ($A < B < C$) and descending ($A > B > C$). In this book, for the ease of illustration, all examples and programs will use nondecreasing relation. To obtain other relations, the reader can modify the algorithm by themselves.

Illustrations of key issues in ordering:

3.1.1.1 The basis to establish the keyword K

During the sorting process, the keyword can be of some ascertained basis (using the above example, we can use the first name of the pinyin of the name of the contact people, or we can use the stroke of the name). It can also be a combination of multiple bases. It is flexible.

3.1.1.2 Processing principle in the case of repeated keyword

During the sorting process, several cases might occur, where the keywords K_i of the record are the same. For example, the records above {Li Xiaoming} and {Li Daming}, and the keywords K3 and K5 are the same. When such situations occur, we need to check the stability of the sorting algorithm. When there are two records with the same keyword, such as R3 and R5, if before the sorting R3 is before R5, and after the sorting R3 is still guaranteed to be before R5, then we say **the sorting algorithm is stable**. Otherwise, if it is possible that R5 gets after R3 after sorting, then we say **the sorting algorithm is unstable**.

3.1.1.3 The storage of sorted records

From the perspective of whether it involves external storage, we can divide sorting algorithms into **internal sorting** and **external sorting**. Internal sorting indicates the process of sorting the records within the computer's random access memory; the whole process does not involve operations on external storages. On the contrary, external sorting refers to situations where there are huge amounts of records to be sorted, and the memory cannot contain all the records at once. For such algorithms, operations on external storage must be relied upon during the sorting process.

This chapter mainly introduces internal sorting algorithms. Readers interested in external sorting algorithms can refer to the corresponding books.

3.1.2 Classification of sorting algorithms

In internal sorting, there are mainly two types of operations: comparison and moving. Comparison refers to the comparison between keywords. This is the most fundamental operation of sorting. Moving refers to the moving of a record from one location to another. There are various sorting algorithms. If we classify them according to the main operations during the sorting process, they can be divided into five categories: insertion sort, exchange sort, selection sort, merge sort and allocation sort. If we divide them according to the time complexity of the algorithm, there are three types:
1. Simple sorting algorithms, with complexity $O(n^2)$;
2. Advanced sorting algorithms, with complexity $O(n \log_2 n)$;
3. Radix sort, with complexity $O(d(n + k))$.

(Note: n is the number of elements to be sorted; it has d digits, and each digit has k possible values.)

For each type of algorithm, this book will introduce several classical algorithms. When learning sorting algorithms, besides comprehending the idea of the algorithm itself, the more important is to understand the principles based on the algorithm during sorting, in order to develop and create better algorithms. For the sake of convenience, all the records in this chapter are integers stored in arrays.

3.2 Insertion sort

We have an already sorted data sequence, and want to insert one number in it. We want the sequence to still be sorted after this insertion. In this occasion, we need to use a sorting method – insertion sort. The basic operation of insertion sort is to insert a piece of data into already sorted data to obtain new sorted data that contains one more element. Since the time complexity is relatively high, normally insertion sort algorithm is suitable for sorting with relatively small amount of data. This section introduces two main insertion sort algorithms: direct insertion sort algorithm and shell sort algorithm.

3.2.1 Direct insertion sort

Direct insertion sort is the simplest sorting method. The basic idea is to take a number that is not taken from the array previously, and insert it to the already taken numbers, so that the already taken numbers are still sorted. For example, when playing poker, we will want the cards in our hands after being obtained to be ordered: Three 8s should be together, followed immediately by two 9s. In this case, we will use insertion sort to insert the new card into an appropriate position among the cards in hand. Now, using the idea of insertion sort, we will perform incremental sorting on a set of unordered records:

$$R = \{49, 38, 65, 97, 76, 13, 27, 49, 55, 04\}$$

According to the idea of direct insertion sort, we first view record {49} as an ordered sequence, that is, the initial ordered sequence $\overline{R} = \{49\}$. Afterward, we need to insert all the remaining records into this ordered sequence $\overline{R}$, and ensure the orderliness of $\overline{R}$. Therefore, we can implement the algorithm following the below steps:
1. Construct the ordered sequence $\overline{R}$, and make $\overline{R} = \{49\}$.
2. Insert the second record 38 into the ordered sequence $\overline{R}$. Since the keyword 38 < 49, we need to insert record 38 before record 49, that is, $\overline{R} = \{38, 49\}$.
3. Insert the third record 65 into the ordered sequence $\overline{R}$. Since at this time the keyword 65 > 49, we insert record 65 after record 49, that is, $\overline{R} = \{38, 49, 65\}$.
4. Insert the fourth record 97 into the ordered sequence $\overline{R}$. Since at this time the keyword 97 > 65, we insert record 97 after record 65, that is, $\overline{R} = \{38, 49, 65, 97\}$.
5. Repeat the above steps.
6. Eventually, output the sorted sequence $\overline{R} = \{04, 13, 27, 38, 49, 49, 55, 65, 76, 97\}$.

Figure 3.3 describes the above process of direct insertion sort.

From the algorithm process we can see that the insertion step is the key to direct insertion sort algorithm. Each time the insertion step executed is one run of direct insertion sort. After we understand the idea of the algorithm, we can obtain the formal algorithm description.

The ith run of the direct insertion operation is to insert record i into an ordered sequence [1,2, …, i–1] with i–1 records, so it becomes an ordered sequence [1,2, …, i] with i records. At the same time, use sequential lookup algorithm to lookup the appropriate position of record i, and execute record insertion operation.

To better understand the code in C, we first give the pseudocode description of this algorithm:

Initial sequence	49	38	65	97	76	13	27	49	55	04

First selection: 49

Second selection: 38 49

Third selection: 38 49 65

Fourth selection: 38 49 65 97

Fifth selection: 38 49 65 76 97

Sixth selection: 13 38 49 65 76 97

Seventh selection: 13 27 38 49 65 76 97

Eighth selection: 13 27 38 49 49 65 76 97

Ninth selection: 13 27 38 49 49 55 65 76 97

Tenth selection: 04 13 27 38 49 49 55 65 76 97

Fig. 3.3: Illustration of direct insertion sort.

```
for(Process all data in loops starting from the second data)
   if(The ith data item is smaller than the i-1 data items before it)
      {
         Copy the ith data to a spare space for temporary storage. This space is
         the "sentinel"
         Find the appropriate position among the previous i-1 data items, and
         move the data elements starting from this position backwards
         Insert the sentinel data to the appropriate position
      }
```

Following the idea of this pseudocode, let us analyze the sorting process of three numbers {18, 15, 16}, as shown in Fig. 3.4.

After two runs of sorting, the sorting on the input sequence is completed. During the whole sorting process, we used an extra storage space (temp) to store a variable. In the end, we give the C language description of the algorithm.

```
/*================================================================
Functionality: Direct insertion sort
Function input: Starting address of the array *a, length of the array n
Function output: None
=============================================================*/
```

```
void InsertSort(int *a, int n)
{
   int i,j;
   int temp;      //temp is used as the sentinel, an extra storage space
   for(i=1;i<n;++i)
   {
      if(a[i]<a[i-1])
      {
         temp=a[i];         // Sentinel
         for(j=i-1; temp<a[j] && j>=0; --j)
         a[j+1]=a[j];
         a[j+1]=temp;       // L.r[j+1] i.e. L.r[i-1]
      }
   }
}
```

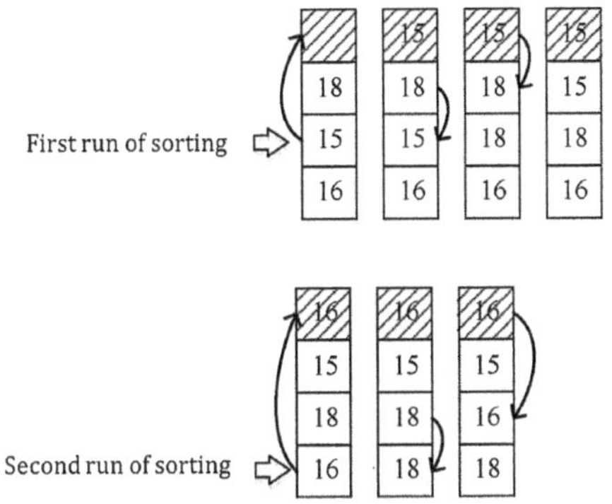

First run of sorting

Second run of sorting

Fig. 3.4: Implementation idea of direct insertion sort.

Algorithm efficiency: $O(n*n)$

The idea of direct insertion sort is clear, and the program is simple and easy to implement; hence, it has many advantages. What about the algorithm efficiency? Let us look at it. In terms of space, since we need to perform the exchange operation of two records, we only need the auxiliary space to store one extra record and the space complexity is relatively low. In terms of time complexity, the basic operations of direct insertion sort are: compare the keywords of two records and move the records. First, let us look at one run of direct insertion sort algorithm: the main loop cost would be spent on finding the position of record i in the ordered sequence $[1, i-1]$. Sequential lookup employs the method of comparing one by one to search; thus, there will be two potential extreme cases. Let us use the example of sorting a sequence in ascending order to illustrate.

1. When the original sequence was complete ascending ordered, rearrange it into descending order.

    Example 3.1 Sort the sequence R = {4,6,8,10,11,15,18} into a nondescending sequence.
    Since at this moment the sequence is an ascending sorted sequence, each new record to add during ordered search only needs to be compared with the first record of the ordered record, that is, only one comparison can determine that the position of the newly added record should be at the front of the whole sequence. In this occasion, the number of comparisons needed is the smallest:

    $$\sum_{i=2}^{n} 1 = n-1$$

2. When the original sequence was complete descending ordered, reorder it into ascending order.

    Example 3.2 Sort the sequence R = {18,15,11,10,8,6,4} into a nondescending sequence.

    Solution: Since in this occasion the sequence is descending ordered, when we perform the sequential lookup, each new record must be compared with all the elements in the ordered sequence before being added. The first record 4 does not need comparison, and the number of comparisons is 0; the second record 6 needs one comparison; the third record 8 needs 2 comparisons and so on, before we can finally ascertain that the position should be at the last in the whole sequence. Therefore, the number of comparisons in this occasion reaches the maximum value:

    $$\sum_{i=2}^{n} i = \frac{(n+2)(n-1)}{2}$$

If the sequence to be sorted is random, the number of comparisons needed in this occasion should be between these two extreme conditions; thus, we may use the average value of the maximum value and the minimum value listed above as the number of comparisons for direct insertion sort, which is approximately $n^2/4$. Therefore, the time complexity of direct insertion sort is $O(n^2)$. In cases where the data being sorted is huge, the time cost of direct insertion sort is relatively high. Under such situations, we may consider faster sorting methods such as binary search or quick search to improve the efficiency of the program.

3.2.2 Shell sort

After we analyze the direct insertion sort, we can discover that if the original sequence was basically "ordered," then the efficiency of executing direct insertion sort will be massively improved. On the other hand, if the original sequence has relatively few numbers of records that need to be sorted, then the number of records needing comparisons decreases, and the efficiency of direct insertion sort will be very high. Therefore, to improve the time efficiency of direct insertion sort, we can consider dividing one long sequence into several short sequences and perform highly efficient direct insertion sort on them, and then use direct insertion sort to perform an efficient sort on all records.

Based on the above idea, shell sort was proposed. Shell sort is also called decreasing incremental sort. It is also based on direct insertion sort, but it offers significant improvements in terms of time efficiency.

Let us use an example to illustrate the ideas of shell sort. Assume that we need to sort the array {49,38,65,97,76, 13,27,49,55,04}. According to the ideas of shell sort, we should divide this long sequence into several shorter sequences and perform sorting, respectively. We will divide the sequence with a gap of five numbers, then the long sequence will be divided into {49,13},{38,27},{65,49},{97,55},{76,04}. Performing sort on each short sequence, the result of one run of sorting obtained is shown in Fig. 3.5.

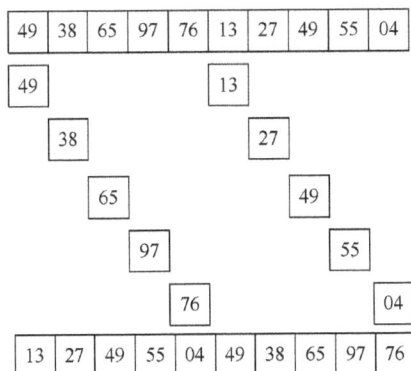

Fig. 3.5: Result of the first run of shell sort.

Afterward, we divide the sequence using a gap of three numbers. This time, the long sequence will be divided into {13,55,38,76}, {27,04,65} and {49,49,97}. Performing direct insertion sort on each short sequence again, we can obtain the result after two runs of sorting as shown in Fig. 3.6.

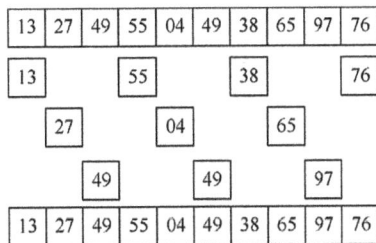

Fig. 3.6: Result of the second run of shell sort.

As we can see, after two runs, the whole sequence is already basically ordered. Then, we can divide the sequence with a gap size of one number, obtaining a "short" sequence that is actually just the original sequence itself. Therefore, in the third run we are just performing simple direct insertion sort, thus finishing the sorting of the whole sequence.

From the above example, we can see that shell sort is an improvement over direct insertion sort. A key is to compose new short subsequences every k numbers. In the above example, the first shell sort has the parameter $k = 5$; the second shell sort has parameter $k = 3$; the third shell sort has the parameter $k = 1$. According to the parameter n, the smaller records are not moved forward one place at a time, but in a striding manner. In the last run of shell sort we make the gap size 1. Therefore, the complexity of shell sort is lower than that of direct insertion sort.

After understanding the process of shell sort, we give the algorithmic expression of shell sort.

```
/*===============================================================
Functionality: Shell sort
Function input: Beginning address of the array *arr, length of the array n
Function output: None
===============================================================*/
void ShellSort(int *arr,int n)
{
    int dlta[3] = {5,3,1};
    //According to the delta sequence dlta[0..t-1],
    perform Shell Sort on the array
    for(int k=0;k<3;++k)
    ShellInsert(arr,n,dlta[k]);
}
```

Characteristics of shell sort:
- Repeatedly invoke one run of shell sort algorithm to realize sortings under different deltas.
- It can be applied to situations where n is relatively large.
- It divides the sequence into several smaller direct insertion sorts.
- It is not stable.
- Its time complexity is $O(n \log_2 n)$

```
/*===============================================================
Functionality: One run of Shell Sort algorithm
Function input: Beginning address of the array *arr, length of the array n,
gap dk
Function output: None
===============================================================*/
```

```
void ShellInsert(int *arr,int n,int dk)
{
    int i,j,k,temp;
    for (k=0;k < n-dk;k++)
    {
        for ( i=k+dk;i<n;i=i+dk)
        {
            if(arr[i]<arr[i-dk])
            // The data inserted is smaller than the previous data
            {
                temp=arr[i];          // Store the inserted data
                // Move until the position where to insert the data is found
                for( j=i-dk;j>=0&&temp<arr[j];j=j-dk)
                arr[j+dk]=arr[j];
                arr[j+dk]=temp;
            }
        }
    }
}
```

Shell sort is an efficient improved sorting algorithm. But the selection of the delta parameter k has always been a hard problem. Currently, there are only some local conclusions: under certain circumstances, there exist some good k sequences, to make the performance of shell sort relatively good. However, there has not been anybody who is able to prove that there exists an optimal sequence to maximize the performance of shell sort. Since differences in the k sequence will cause differences in the performance of shell sort, there has not been a real complete and rigorous complexity analysis on it. But normally speaking, it is considered that the time complexity of shell sort can reach $O(n \log_2 n)$ in best cases; thus shell sort is also one of the advanced sorting algorithms.

Knowledge ABC The developer of shell sort algorithm

Shell sort is named after its designer Donald Shell. The algorithm was published in 1959. Previously, it is stated that the selection of delta is the most important part of shell sort. Theoretically, as long as the final delta is 1, any delta sequence can work. Donald Shell suggested choosing $n/2$ as the initial delta and halving the value, until 1 is reached. However, there is still room for improvement for such delta selection. Currently, the known best delta sequence is designed by Marcin Ciura 1, 4, 10, 23, 57, 132, 301, 701, 1750, ...). Shell sort using such a delta sequence is faster than both insertion sort and heap sort, and is even faster than quick sort in the case of relatively small sequences. However, when a large amount of data is involved, shell sort is still slower than quick sort.

3.3 Exchange sort

The so-called exchange is to exchange the positions of the two records in the sequence according to the comparison result of the value of the keyword. The characteristic of switch sorting is to continuously move the records with relatively larger keyword value toward the rear of the sequence, so that records with relatively smaller keyword values are moved toward the front of the sequence, and thus eventually obtain an ordered sequence. This section introduces two switch sorting algorithms: bubble sort and quick sort. Quick sort is currently widely acknowledged to be the fastest sorting algorithm.

3.3.1 Bubble sort

Bubble sort is one of the simplest sorting algorithms. Its key idea is exchanging; thus, it belongs to exchange sorting. Here is an example from daily life: during a P.E. class, the teacher will need the students to line up according to height. After the initial lining up is complete, there will always be somebody out there to compare the heights of two adjacent students, saying: "You two, switch positions. You two, switch positions …" After a few times, the highest person will be at the end of the queue, and the shortest person will be at the front of the queue, which achieves the aim of sorting.

We continue using the previous example. The bubble sort on the sequence {49,38,65,97,76,13,27, 49,55,04} is shown in Fig. 3.7.

We can discover that, after one run of bubble sort, the largest record in the sequence {97} is already moved to the end of the sequence, while smaller records have all been moved toward the front. The larger records sink, while the smaller records float up; thus, this algorithm is called bubble sort algorithm. By executing bubble sort algorithm several times, we can eventually obtain an ordered sequence.

```
/*================================================================
Functionality: Bubble sort
Function input: Beginning address of the array *a, length of the array n
Function output: None
==============================================================*/
void BubbleSort(int *a, int n)
{
    int i,j,temp;
    int change=0;
    // Check whether there was any switch in the sequence. If there was no switch,
    // then the sequence is ordered, and we only need one run of the sorting.
    for(i=n-1; i>=0&&change==0; i--)
    //for loop. First check the condition, then see whether to execute
```

```
{
    change=1;              // The positions need to be switched
    for(j=0; j<i; j++)     // One run of sorting
    {
        if(a[j]>a[j+1])
        {
            change=0;
            temp=a[j];
            a[j]=a[j+1];
            a[j+1]=temp;
        }
    }
}
}
```

Initial Sequence

49	38	65	97	76	13	27	49	55	04

First run of bubble sort

49	38	65	97	76	13	27	49	55	04

38	49	65	97	76	13	27	49	55	04

38	49	65	97	76	13	27	49	55	04

38	49	65	97	76	13	27	49	55	04

38	49	65	76	97	13	27	49	55	04

38	49	65	76	13	97	27	49	55	04

38	49	65	76	13	27	97	49	55	04

38	49	65	76	13	27	49	97	55	04

38	49	65	76	13	27	49	55	97	04

38	49	65	76	13	27	49	55	04	97

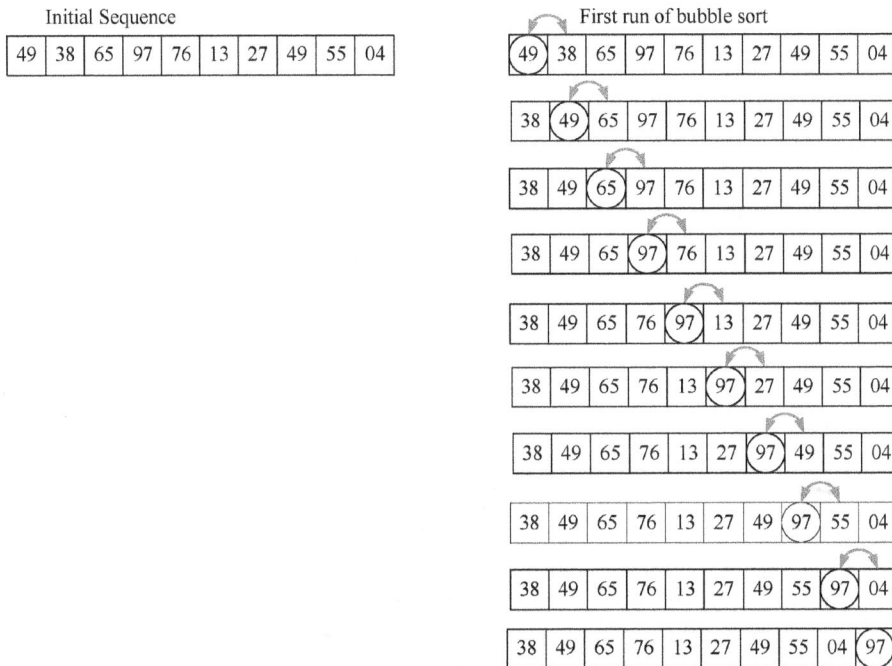

Fig. 3.7: One run of bubble sort.

Features of bubble sort: slow, relatively many numbers of movements.

Algorithm efficiency: $O(n^2)$

To analyze the efficiency of bubble sort, for a sequence with length n, we can discover that if the initial sequence to be sorted is already a "properly ordered" sequence, then we only need to run the sorting once, with $n-1$ record comparisons, and we do not need to move any records. However, if the initial sequence to be sorted is a "reversely ordered" sequence, then we need to perform $n-1$ sorts, and $\sum_{i=2}^{n}(i-1) = (n(n-1)/2)$ comparisons, with the same number of moves. Therefore, the complexity of bubble sort is $O(n^2)$. Compared with direct insertion sort, the time complexity of bubble sort is similar, but the number of movements is more. Bubble sort performs worse than the direct insertion sort; thus, it is rarely used in actual applications.

3.3.2 Quick sort

Quick sort is an improvement on bubble sort. Its basic idea is to divide the sequence to be sorted into two parts, the records of one part are smaller than the records of another part. Then, further divide these two parts, respectively, into two parts, so that the records of one part are all smaller than the records of another part. This process is repeated until the whole sequence is ordered.

To realize this aim, we need to select one special record, called pivot. The role of the pivot is to distinguish the two parts. If we can find one method, so that the pivot is always situated at an appropriate position, that is, all the records before the pivot are smaller than it, while all the records after the pivot are larger than it, then we have achieved the aim of quick sort.

Using the example of the sequence {49, 38, 65, 97, 76, 13, 27, 49, 55, 04}, let us illustrate the idea and method of quick sort, as shown in Fig. 3.8.

Figure 3.8 shows the illustration of the first run of quick sort algorithm. According to the ideas of quick sort, we need to first choose a number as pivot. Normally speaking, we choose the first number of the whole sequence as the pivot in the beginning, which is {49} in this example. In order to move the pivot to an appropriate position and ensure that all the numbers before the pivot are smaller than the pivot, and all the numbers after the pivot are larger than the pivot, we perform the movement of the pivot according to the following steps.

1. Set two pointers low and high, which point to the first record and last record in the sequence, respectively. At this time, low points to the first record in the sequence, that is, the location of the pivot. In the figure, we use black circle to represent pointer low, the record circled by the black circle represents the record pointed to by the pointer low; we use dashed circle to represent pointer high, and the record circled by the dashed circle represents the record pointed to the pointer high.

2. Compare the sizes of the records pointed by the two pointers: low and high. If the record low is smaller than the record high, then record high is larger than the pivot and should be positioned after the pivot; thus, there is no need to switch record low and record high. If the record low is larger than the record

Pivot

Initial sequence	49	38	65	97	76	13	27	49	55	04
First swap	04	38	65	97	76	13	27	49	55	49
Second swap	04	38	65	97	76	13	27	49	55	49
Third swap	04	38	49	97	76	13	27	49	55	65
Fourth swap	04	38	49	97	76	13	27	49	55	65
Fifth swap	04	38	49	97	76	13	27	49	55	65
Sixth swap	04	38	27	97	76	13	49	49	55	65
Seventh swap	04	38	27	49	76	13	97	49	55	65
Eighth swap	04	38	27	13	76	49	97	49	55	65
Ninth swap	04	38	27	13	49	76	97	49	55	65

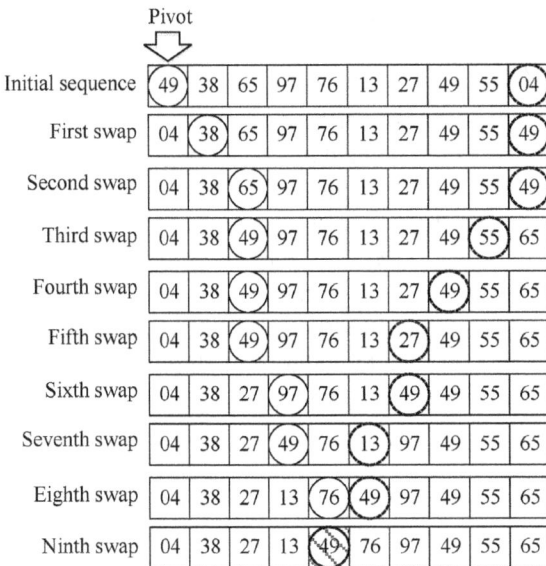

Fig. 3.8: Illustration of the first run of quick sort algorithm.

high, then it means the record high is smaller than the pivot, and should be positioned before the pivot; thus, we need to switch record low and record high. In this example, the value of the record low {49} is larger than the value of record high {04}; thus, we exchange the values of the record low and the record high. The value of the record low becomes {04}, while the value of the record high becomes {49}.

3. After we switched the values of the records low and high, we need to move the pointer. In order to move the pivot to an appropriate position, we need to select a record again to compare with pivot {49}. In this example, we move the pointer low by one position, reaching record {38}. (Think about why we did not move the pointer high instead?)

4. Now, the record low is {38}, the record high is {49}. The record low is smaller than the record high; thus, there is no need to switch the positions of the two.

5. We need to continue searching for appropriate records to be compared with the pivot. Now, the reasonable choice is to move the pointer low again to record {65}, and compare it with the pivot again. The value of the record low is larger than the value of the record high; thus, it should be after the record high. We exchange the records pointed by pointers low and high. The third switch is complete.

6. Now, the record pointed by the pointer low is changed to the pivot {49}. To find the next appropriate record to be compared with the pivot, the reasonable choice now is to move the pointer high, so that it points to the record {55}. Comparing the pivot pointed by the pointer low with the record pointed by the

pointer high, it is obvious that the pivot should be in front. Thus, we do not switch the records.

7. Continue to move the pointer high, select records to be compared with the pivot. Only in the fifth round, when the pointer high points to the record {27}, the pivot will be switched with the record pointed by pointer high.

8. Since a switch of record positions occurred, the pointer high now points to the pivot {49}; thus, we must move the pointer low to select an appropriate record to be compared with the pivot. Repeating the above algorithm, until the pointer low is pointing to the same record as the pointer high. Now, pivot {49} is moved to the middle of the sequence. We can ensure that all the records before the pivot {49} are smaller than the pivot, and all the records after the pivot are larger than the pivot. Now, we have finished the first run of the quick sort algorithm.

Obviously, we cannot finish the sorting on the whole sequence just after one run of the quick sort algorithm. Therefore, we need to continue invoking quick sort algorithm to sort the sequences before the pivot and after the pivot, as shown in Figs. 3.9 and 3.10. We can discover that the second and third sorting were essentially only to invoke the quick sort algorithm on shorter sequences. Thus, we do not explain them again.

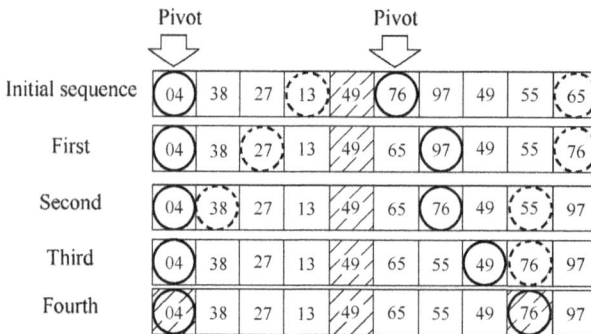

Fig. 3.9: The second run of the quick sort algorithm.

Such an idea to invoke the same algorithm within the algorithm itself is typical of the idea of using divide and conquer to solve practical problems. Divide and conquer is an effective idea to solve relatively hard problems with the computer. It is specifically illustrated in Volume 1 of this book. After understanding the flow of quick sort algorithm, we give the quick sort algorithm as follows:

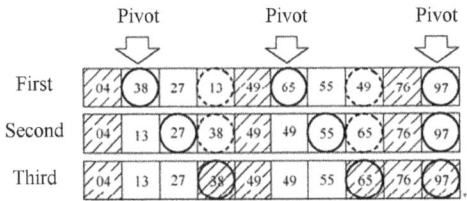

Fig. 3.10: The third run of the quick sort algorithm.

```
/*================================================================
Functionality: Quick Sort
Function input: Beginning address of the array *arr, length of the array n,
initial position of the pointer low, initial position of the pointer high
Function output: None
===============================================================*/
void QSort(int *arr,int n,int low,int high)
{ // Perform Quick Sort on the array arr[n], recursively invoke the function
   int pivotloc;
   if(low<high)
   {
      pivotloc=Partition_1(arr,low,high);
      QSort(arr,n,low,pivotloc-1);
      QSort(arr,n,pivotloc+1,high);
   }
}
/*================================================================
Functionality: Ensure that the input and output parameters of the sorting
algorithms are uniform
Function input: Beginning address of the array *arr, length of the array n
Function output: None
===============================================================*/
void QuickSort(int *arr,int n)
{ // Set initial value for the recursive function
   QSort(arr,n,0,n-1);
}
/*================================================================
Functionality: One round of Quick Sort, sort according to movement of the pivot
Search from both front and rear, One execution of Quick Sort
Function input: Beginning address of the array *arr, initial position of the
pointer low, initial position of the pointer high
Function output: None
===============================================================*/
```

```
int Partition_1(int arr[],int low,int high)
{  // Search from both front and rear
   int pivotloc=arr[low],temp;
   // Empty the position of pointer low,
   // use the first element as the value of the pivot
   while( low < high )
   {
      while(low<high && arr[high]>=pivotloc)
      --high;
      // Keep looping until a smaller value is found from the rear
      temp=arr[low];
      arr[low]=arr[high];
      arr[high]=temp;        // Position high is emptied
      while(low<high && arr[low]<=pivotloc )
      ++low;
      temp=arr[low];
      arr[low]=arr[high];
      arr[high]=temp;        // Position high is emptied
   }
   return low;
}
```

Characteristics of quick sort: repeatedly invoke one run of quick sort algorithm to perform quick sort on the array arr[n]; unstable.

Algorithm efficiency: $O(n \log n)$.

The time complexity of quick sort is $O(n \log n)$. Among all the sorting methods of the same order of magnitude, quick sort algorithm is the widely acknowledged sorting method with the best performance. However, when the initial sequence is basically ordered, since each comparison will involve the switch of pivot, quick sort will deteriorate into bubble sort, with time complexity deteriorating to $O(n^2)$. Currently, there are certain improved algorithms based on quick sort, which can improve the performance of quick sort to some extent. Interested readers can refer to related references.

3.4 Selection sort

Selection sort is an algorithm that sorts in a "brute-force" manner. Simply speaking, it can be described as: when sorting a sequence, select the smallest record from the remainder of the sequence each time, and put it to the beginning of the sequence, until all the numbers in the sequence have been put into their final positions. The idea of selection sort is relatively straightforward. The key problem is how to select the smallest record from the sequence. Different solutions to this key problem have generated two different sorting algorithms: simple selection sort and heap sort.

3.4.1 Simple selection sort

The basic idea of selection sort is to select the smallest record in all the remaining records in order to put it into an ordered sorted sequence. Repeating this process can achieve the ordering of an unordered sequence. As an example from our daily lives, let us imagine square pokers: there are 13 of them from A to K. ·In order to sort, we need to first find the smallest card among the 13 cards, A, put it aside and then find the second smallest card 2, put it aside... Repeating this process, we can realize the sorting operation from A to K. This idea is the basic idea of simple selection sort.

The implementation can be done in the following steps.

First step: Find the smallest number in the sequence $1-n$, switch it with the first number. Then the first number is sorted.

Second step: Find the smallest number in the sequence $2-n$, switch it with the second number. Then the first two numbers are sorted.

. . .

The $n-1$th step: Find the smallest number in the sequence $n-1-n$, and switch it with the $(n-1)$th number. The sorting ends.

Compared with bubble sort, simple selection sort is more effective. The reason is that bubble sort must immediately perform the data switch whenever it discovers that the previous number is bigger than the later number, after each comparison. The average number of data switch is larger than that in selection sort. On the contrary, there is at most one time of data switch with selection sort. Combined with the example {49,38,65,97,76,13, 27,49,55,04}, we illustrate the flow of simple selection sort, as shown in Fig. 3.11.

After understanding simple selection sort, we give the program of the algorithm in C language.

```
/*================================================================
Functionality: Selection Sort
Function input: Beginning address of the array *a, length of the array n
Function output: None
==============================================================*/
void SelectionSort (int *a, int n)
{   // Selection Sort
    int i,j,temp;
    for(i=0;i<n-1;i++)
```

```
for(j=i;j<n;j++)
{
    if(a[j]<a[i])
    {
        temp=a[j];
        a[j]=a[i];
        a[i]=temp;
    }
}
}
```

Initial sequence	49	38	65	97	76	13	27	49	55	04
First selection	04	38	65	97	76	13	27	49	55	49
Second selection	04	13	65	97	76	38	27	49	55	49
Third selection	04	13	27	97	76	38	65	49	55	49
Fourth selection	04	13	27	38	76	97	65	49	55	49
Fifth selection	04	13	27	38	49	97	65	76	55	97
Sixth selection	04	13	27	38	49	49	65	76	55	97
Seventh selection	04	13	27	38	49	49	55	76	65	97
Eighth selection	04	13	27	38	49	49	55	65	76	97
Ninth selection	04	13	27	38	49	49	55	65	76	97
Tenth selection	04	13	27	38	49	49	55	65	76	97

Fig. 3.11: The algorithm flow of simple selection sort.

Features of selection sort: The smallest element selected each time will be at the beginning; the algorithm is unstable.

Algorithm efficiency: $O(n \times n)$.

The main operations of simple selection sort are spent on the comparison operations of records. The first run is equivalent to finding the smallest record among all the records and retrieve it; the second run is equivalent to finding the smallest record in the $n-1$ remaining records and retrieve it ... until the last record. Therefore, no matter whether the initial sequence was basically ordered, the number of comparisons of simple selection sort will be $n(n-1)/2$. Therefore, its time complexity is also $O(n^2)$, and it is also a type of simple sorting algorithm.

3.4.2 Heap sort

Heap sort is an algorithm that sorts combining binary tree structure and an auxiliary record space. To understand the concept of "heap" in heap sort, we must combine it with the concept of binary tree.

Refer to the illustration of heap shown in Fig. 3.12. For a complete binary tree, if the values of all the nodes (except for the leaf nodes) are bigger than (smaller than) the values of the left and right child nodes, then this complete binary tree is called a heap. According to the definition of heap, we can discover that the heap top node (the root node of the binary tree) definitely corresponds to the biggest (smallest) record in the whole sequence. In this way, we can design a sorting idea that outputs the heap top record each time, as well as adjusts the remaining records so that they form a heap again. Repeating the above process, we can eventually obtain an ordered sequence and finish the sorting process. This sorting method is called heap sort.

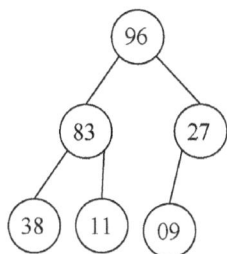

Fig. 3.12: Example of heap.

Therefore, for heap sort, there are two key questions:
- How to put all the records in an unordered sequence into a heap?
- After outputting the heap top record, how to again put the remaining records into a heap?

Let us first analyze the second question. We will use Fig. 3.13 as an example to illustrate the solution to the second question.

Figure 3.13 is already a heap. The heap top record {13} is the smallest record of the whole sequence. Therefore, we can first output the heap top record. Now, the structure of the remaining records is shown in Fig. 3.14.

We can see that now the left subtree and the right subtree are still heaps, which means record {38} is the smallest record in the heap formed by the left subtree, while record {27} is the smallest record in the heap formed by the right subtree. To proceed to find the smallest record and put it onto the top of the heap, we temporarily put the record that is stored in the last position according to the order of binary tree, {97}, to the top of the heap, as shown in Fig. 3.15.

However, now the whole binary tree no longer forms a heap. We need to adjust the order of the sequences in this binary tree, so that it is still a heap. In order to

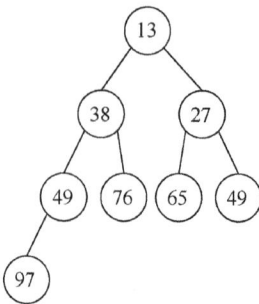

Fig. 3.13: Heap.

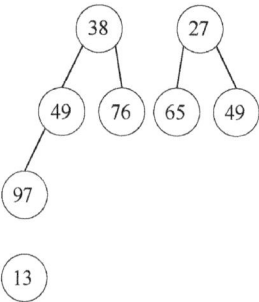

Fig. 3.14: The situation after outputting the heap top element.

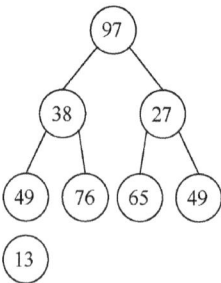

Fig. 3.15: The last record is put temporarily at the top of the heap.

achieve this goal, we consider the fact that {38} and {27} are, respectively, the heap top of the left and right subtrees. Therefore, we only need to select the smallest one among {97}, {38} and {27} and put it at the heap top in order to keep the heap top element smaller than the left and right children, as shown in Fig. 3.16.

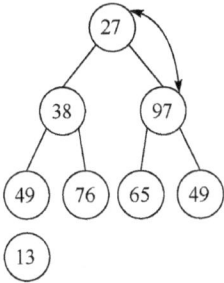

Fig. 3.16: The situation after one adjustment of the heap top element.

Now, although the heap top element satisfies the requirement for a heap, the heap structure of the right subtree is disrupted. To ensure that the right subtree still has the structure of a heap, we need to again compare the three records {97}, {65} and {49}, and select the smallest element to be put at the heap top of the right subtree, as shown in Fig. 3.17.

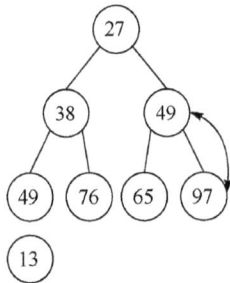

Fig. 3.17: The situation after the second adjustment on the heap top record of the right subtree.

Now, the remaining records are again arranged into a heap structure. We can output the heap top record {27}, after which the remaining records can still be arranged into a heap, as shown in Fig. 3.18.

Repeating this process, we can eventually generate an ordered sequence. We call this adjustment process from the heap top to the leaves as "filtering."

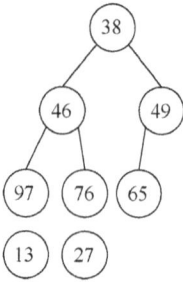

Fig. 3.18: After outputting the record {27}, the remaining records are again arranged into a heap.

Actually, the process of generating a heap from an unordered sequence in question 1 is also a process of continuously "filtering." Using sequence {49, 38, 65, 97, 76, 13, 27, 49, 55, 04} as the example, we illustrate how to establish a heap structure using "filtering."

First, according to the sequential storage principle, first construct a complete binary tree from all the records in the sequence, as shown in Fig. 3.19.

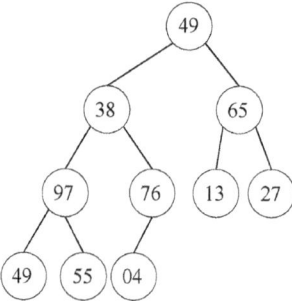

Fig. 3.19: The sequence directly constitutes a complete binary tree.

Now, the binary tree normally does not have the characteristics of a heap. Therefore, we need to construct it into a heap structure via "filtering." According to the "filtering" principle, first we know that the last nonterminal node in a complete binary tree is $\lfloor n/2 \rfloor$. Therefore, in this example, we use the fifth record {76} as the initial filtering node; first, compare {76} and its left and right children, and exchange according to the result of the filtering. We obtain the new complete binary tree, as shown in Fig. 3.20.

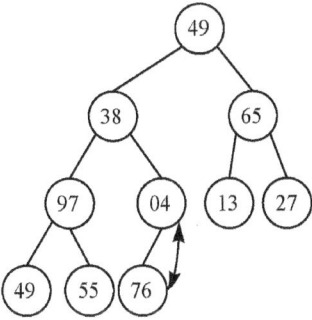

Fig. 3.20: The complete binary tree after one filtering.

Then, we process the second-to-last nonterminal node, that is, select the record {97}, and compare it with its left and right children and perform filtering operation. We can obtain an updated complete binary tree, as shown in Fig. 3.21.

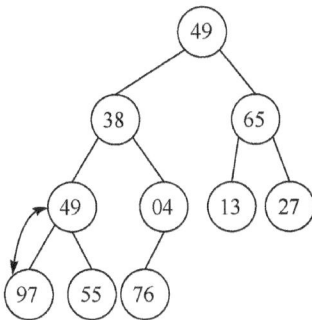

Fig. 3.21: The complete binary tree after the second processing.

Repeating the above process, we can eventually obtain a heap with complete structure, as shown in Fig. 3.22, and use it to record the heap sorting operation.

After understanding the basic ideas of heap sort, we now give the C-language implementation of heap sort.

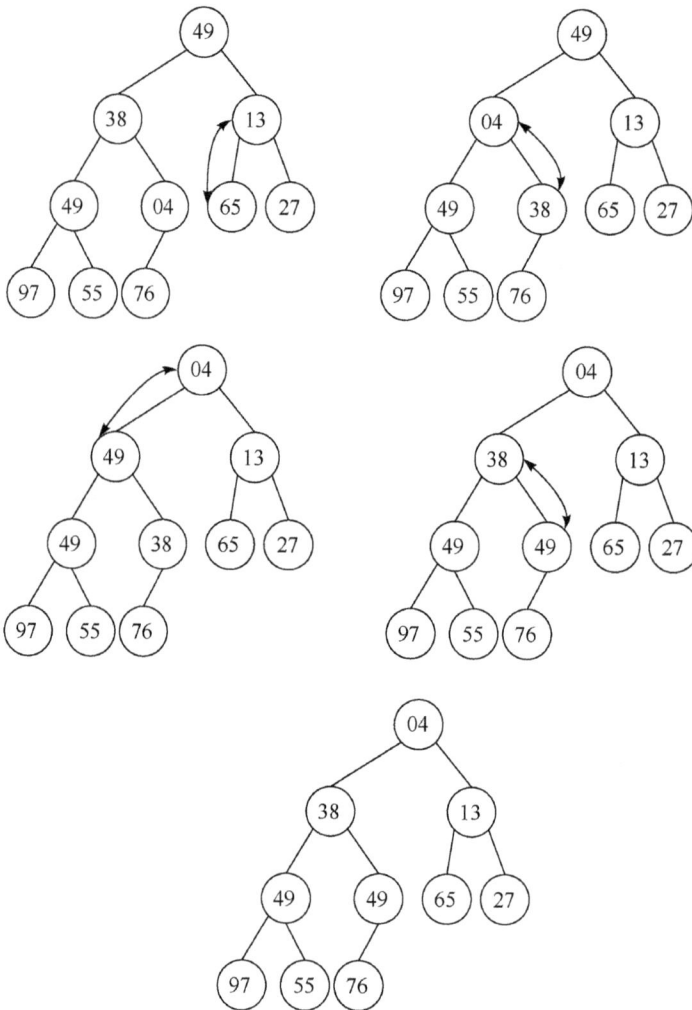

Fig. 3.22: The heap generated after processing of the sequence.

```
/*================================================================
Functionality: Heap sort
Function input: Beginning address of the array *H, Length of the array N
Function output: None
================================================================*/
void HeapSort(int *H, int N)
{
    int i, temp;
    for(i=N/2; i>=0; --i)
```

```
{
    MaxHeap(H,i,N);
    // Construct the max heap, find out the biggest element
}
for(i=N-1;i>=0;--i)
{
    temp=H[0];
    H[0]=H[i];
    H[i]=temp;
    // Swap, the maximum value is given to the last element in the array
    MaxHeap(H,0,i-1);
}
}
```

Characteristics of heap sort: uses binary tree for sorting. Only needs one extra slot to realize sorting.

```
/*=====================================================================
Functionality: Construct the max heap
Function input: Beginning address of the array *H, Position of the element in
the array n, Array length N
Function output: None
=====================================================================*/
void MaxHeap(int *H,int n,int N)
{   // Construct the max heap
    int l=2*n+1;
    //Arrays starting from index 0(l=2*n+1,r=2*(n+1))and starting from index 1
    //(l=2*n,r=2*n+1) have different indexes for left and right children
    int r=2*n+2;
    int max;         //l,r,max both represent array indexes
    int temp;        // Represents the temporary element to be used during swapping
    if (l<N&&H[l]>H[n])        max=l;        // The left node is bigger
    else max=n;        // The root is bigger
    if(r<N&&H[r]>H[max])       max=r;        // The right node is bigger
    if(max!=n)        // Swap
    {
        temp=H[n];
        H[n]=H[max];
        H[max]=temp;
        MaxHeap(H,max,N);        // Ensure the properties of the max heap.
    }
}
```

Normally speaking, when sorting sequences with relatively few records, heap sort is not recommended. However, when sorting sequences with a relatively large number of records, heap sort will be relatively effective, since the main time cost will be spent in the initialization of the heap and the "filtering" repeatedly done when adjusting the new heap. In the worst-case scenario, the time complexity of heap sort will still be $O(n \log_2 n)$, while in comparison quick sort's time complexity will deteriorate to $O(n^2)$. The fact that heap sort has relatively good time complexity even under the worst-case scenario is its biggest advantage.

3.5 Merge sort

Merge sort is a sorting method different from the ideas of the above methods. The so-called merging is to merge two or more than two ordered sequences to generate a new sequence and ensure that this new sequence is ordered. No matter with sequential structure or linked structure, it can be realized on $O(m + n)$ time complexity.

Let us first consider a simple question: how to merge two ordered sequences into a new ordered sequence. Suppose that there are two ordered sequences A and B, then we wish to merge A and B into a new ordered sequence C, where A = {1,3,5,7,9} and B = {1,2,6,8,10}. We can carry out the merging according to the below strategies.

The first step of the merging is shown in Fig. 3.23. First, since both A and B are ordered sequences, the first record of the ordered sequence C after the merge must come from the first record of either A or B, depending on which is smaller. Since they are both one, we arbitrarily choose the record 1 from sequence A, 1.

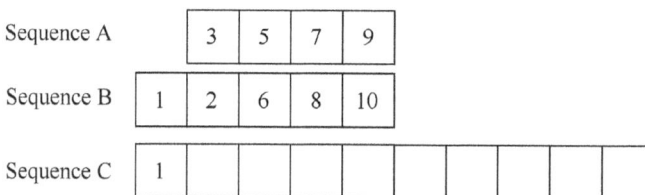

Sequence A		3	5	7	9					
Sequence B	1	2	6	8	10					
Sequence C	1									

Fig. 3.23: The first step of merging.

The second step of merging is shown in Fig. 3.24. The first record in sequence A has now become 3, and the first record in sequence B is still 1. Therefore, we still choose the smaller record among the first records of the sequences to output, and the second element in sequence C is record 1 from sequence B.

The next few steps are shown in Figs. 3.25 and 3.26.

Sequence A | 3 | 5 | 7 | 9 |

Sequence B | 2 | 6 | 8 | 10 |

Sequence C | 1 | 1 | | | | | | | | |

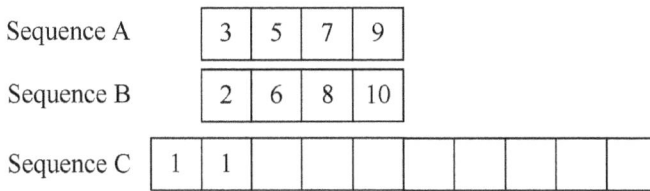

Fig. 3.24: The second step of merging.

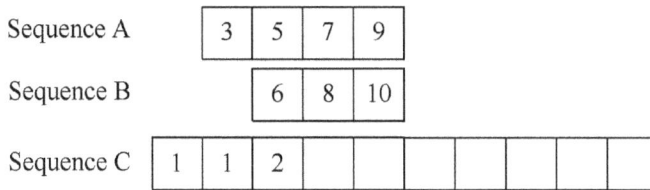

Sequence A | 3 | 5 | 7 | 9 |

Sequence B | 6 | 8 | 10 |

Sequence C | 1 | 1 | 2 | | | | | | | |

Fig. 3.25: The third step of merging.

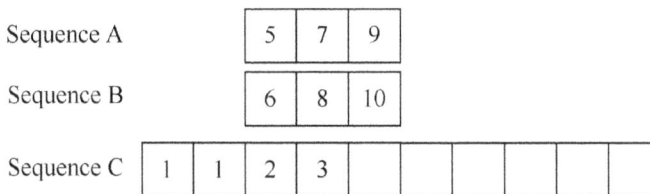

Sequence A | 5 | 7 | 9 |

Sequence B | 6 | 8 | 10 |

Sequence C | 1 | 1 | 2 | 3 | | | | | | |

Fig. 3.26: The fourth step of merging.

We skip several intermediate steps. Finally, we obtain the sequence C in Fig. 3.27, which is the result of merging.

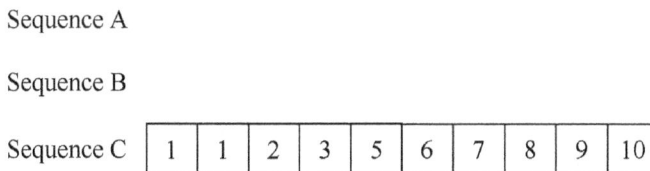

Sequence A

Sequence B

Sequence C | 1 | 1 | 2 | 3 | 5 | 6 | 7 | 8 | 9 | 10 |

Fig. 3.27: Merging ends.

Eventually, we obtain a new sequence C, which is ordered. Apparently, we have an efficient algorithm, which merges two ordered subsequences into a new ordered sequence. Merge sort repeatedly calls the above merging algorithm. It first views one single record as an ordered sequence, and continuously merges two adjacent ordered sequences to obtain a new ordered sequence. It does so repeatedly until eventually it obtains a complete ordered sequence. This algorithm is called two-way merge sort. Using the unordered sequence {49, 38, 65, 97, 76, 13, 27, 49, 55, 04} as an example to illustrate, the process of sorting is shown in Fig. 3.28.

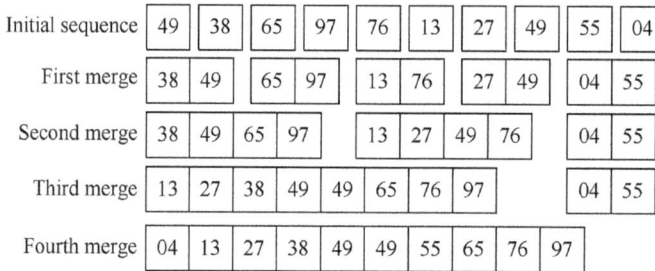

Initial sequence	49	38	65	97	76	13	27	49	55	04
First merge	38	49	65	97	13	76	27	49	04	55
Second merge	38	49	65	97	13	27	49	76	04	55
Third merge	13	27	38	49	49	65	76	97	04	55
Fourth merge	04	13	27	38	49	49	55	65	76	97

Fig. 3.28: Illustration of the execution of merge sort algorithm.

After understanding the idea of merge sort, we give the C-language program of merge sort as follows:

```
/*================================================================
Functionality: Merge sort
Function input: Beginning address of the array to be sorted *arr, Length of
array arr N
Function output: None
================================================================*/
void TMSort(int *arr,int N)
{
    int t=1;
    int *arr2=(int*)malloc(N*sizeof(int));
    while(t<N)
    {
        MSort(arr,arr2,N,t);
        //(Original array, array after sorting, total length, sub-array length)
        t*=2;
        MSort(arr2,arr,N,t);
        t*=2;
    }
    free(arr2);
}
```

```
/*==============================================================
Functionality: One run of merge sort algorithm
Function input: Beginning address of the array to be sorted *SR, Beginning
address of the array to store the sort result *TR, Length of the array n,
Length of sub-sequences for one merge t
Function output: None
==============================================================*/
void MSort(int *SR,int *TR,int n,int t)
{
    int i=0,j;
    while(n-i>=2*t)
    // Merge two adjacent ordered sub-sequences with length t into
    // a sub-sequence with length 2t
    {
        Merge(SR,TR,i,i+t-1,i+2*t-1);
        i=i+2*t;
    }
    if(n-i>t)
    // When the number of remaining elements is larger than
    // the length of one subsequence t
        Merge(SR,TR,i,i+t-1,n-1);
    else
    // When n-i <= t, it's simply equivalent giving the values from the
    // sequence X[i..n-1] to the sequence Y[i..n-1]
        for( j=i; j<n; ++j )
            TR[j]=SR[j];
}
```

Characteristics of one run of merge sort algorithm: merge two ordered sequences into one ordered sequence via merge function.

```
/*==============================================================
Functionality: Merge algorithm, merge two ordered sequences into one ordered
sequence
Function input: Beginning address of the array to be sorted *SR, Beginning
address of the array to store the sort result *TR, m divides SR into two ordered
sequences, Length of SR n, i is the address of the first data of array SR.
Function output: None
==============================================================*/
void Merge(int *SR,int *TR,int i,int m,int n)
{ // Merge sorted SR[i...m] and SR[m+1...n] into ordered TR[i...n]
    int j,k;
```

```
for(j=m+1,k=i;i<=m&&j<=n;++k)
{
    if(SR[i]<SR[j])        // Select the smallest value
        TR[k]=SR[i++];
    else
        TR[k]=SR[j++];
}
while(i<=m)          // Insert the rest
    TR[k++]=SR[i++];
while(j<=n)          // Insert the rest
    TR[k++]=SR[j++];
}
```

The program analysis of merge sort is as follows. The initial sequence that participates in the sorting is divided into subsequences with length 1 and are sorted for the first time using MSort function, which results in $n/2$ ordered subsequences with length 2 (if n is an odd number, there would still exist one subsequence with only one element). The second sorting is carried out by invoking MSort function again, which obtains $n/4$ ordered subsequences with length 4. In general, the ith sorting merges subsequences with length 2^{i-1} to obtain $n/(2^i)$ subsequences with length 2^i, until eventually there is only one subsequence with length n.

We can see that there need to be $\log_2 n$ sortings in total, each of which has a time complexity of $O(n)$. Therefore, the total time complexity of this algorithm is $O(n \log_2 n)$, but this algorithm requires $O(n)$ auxiliary memory space.

Characteristics of merge sort: repeatedly invoke one run of merge sort algorithm; stable.

Algorithm efficiency: complexity $O(n \log_2 n)$.

The best, worst and average time complexities of merge sort are all $O(n \log_2 n)$, while its space complexity is $O(n)$. Therefore, merge sort requires relatively much memory, but it is a relatively efficient sorting algorithm.

Using recursion to implement merge sort is concise in its implementation, but since it requires huge memory, it is often impractical. There are also nonrecursive implementation forms for merge sort. Interested readers can refer to the related references by themselves. Compared with quick sort and heap sort, the biggest characteristic of merge sort is that it is a stable sorting method.

3.6 Distribution sort

Different from the sorting algorithms mentioned earlier, distribution sort is a sorting method that trades space for time. Its basic idea is that, during the sorting process, it is unnecessary to compare the keywords. It instead utilizes additional space to carry out "distribution" and "collection" in order to realize the sorting. When the additional space is relatively large, the time complexity of distribution sort can reach linear $O(n)$. Simply speaking, distribution sort trades space for time; therefore, its time performance increases drastically compared with sorting methods based on comparison.

Common distribution sort algorithms include counting sort, bucket sort and radix sort. This section describes about bucket sort and radix sort.

3.6.1 Bucket sort

Normally speaking, when we look at the sorting algorithms mentioned earlier in this chapter, there is a lower bound $O(n \log n)$ for their best case time complexity. However, this does not mean that there are no better algorithms. Faster algorithms avoid most "comparison" operations with some restricting assumptions on unordered sequences. Bucket sort is proposed based on such a principle.

If we assume the input data are distributed evenly, the average case time complexity of bucket sort will be $O(n)$. Because the data are distributed evenly and independently in the whole range, bucket sort divides the sequence to be sorted into several ranges, each of which is viewed as a bucket. Afterward, based on a certain mapping function, it maps all the records in the sequence to be sorted into the corresponding bucket. Eventually, it performs comparison sort on all the records in all the buckets (e.g., effective quick sort), and outputs them according to the order, so that an eventual ordered sequence is obtained.

Using the sequence {49, 38, 65, 97, 76, 13, 27, 49, 55, 04} as an example, according to the idea of bucket sort, we must first ascertain the ranges. Since the values of all these records are smaller than 100, we can specify 10 buckets, with a mapping function of $f(k) = k/10$, and take the integer value of the result of $f(k)$. In this way, the conditions of all buckets after the mapping are shown in Fig. 3.29.

04	13	27	38	49 49	55	65	72		97
Bucket 0	Bucket 1	Bucket 2	Bucket 3	Bucket 4	Bucket 5	Bucket 6	Bucket 7	Bucket 8	Bucket 9

Fig. 3.29: Illustration of bucket sort.

Afterward, perform quick sort on each nonempty bucket, and we can eventually obtain the ordered sequence and finish the sorting. Of course, we can also customize the buckets according to the other rules (e.g., in our code we specified the buckets according to the number of digits of the data to be sorted). Afterward, we put the data according to the rules into the corresponding bucket, and then sort each bucket. Eventually, we combine all the ordered buckets into one ordered sequence.

Bucket sort utilizes the mapping relations of the functions and reduces almost all comparison work. Actually, the calculation of the mapping function for bucket sort is equivalent to dividing the huge amount of records to be sorted into basically sorted chunks of data (buckets). Then, we only need to perform efficient comparison and sorting on the small amount of records in each bucket. Therefore, the time complexity of bucket sort can be divided into two parts.

1. Calculate the result of bucket mapping function on each keyword. Its time complexity is $O(n)$.
2. Use advanced comparison sorting algorithm to sort all the data in each bucket. Its time complexity is $\sum n_i \log_2(n_i)$, where n_i is the number of data in the ith bucket.

Since the time complexity in part (1) is linear, part (2) is the key factor for deciding the performance of bucket sort, to improve the efficiency, we can only reduce the amount of data in each bucket as much as possible, so that the amount of data during in-bucket sorting is minimized. Therefore, when performing bucket sort, we need to pay attention to the following two points as much as possible.

– Mapping function f(k) can evenly divide the n data into m buckets. In this way, each bucket has an even amount of $[n/m]$ data. Therefore, bucket sort is normally used to sorting of data, which is independently and evenly distributed.
– Increase the number of buckets as much as possible. In the ideal situation, each bucket has only one data, in which case we completely avoid the "comparison" sorting operation for in-bucket data.

The average time complexity for bucket sort is linear $O(n + c)$, where $c = n \times (\log_2 n - \log_2 m)$. If for the same n, the number of buckets m is larger, then the efficiency will be higher. The best time complexity reaches $O(n)$. Of course, the space complexity of bucket sort is $O(n + m)$. If the input data is very large, and the number of buckets is also very large, then undoubtedly the space cost will be expensive.

After understanding the basic idea of bucket sort, we give the C-language implementation of bucket sort.

```
/*================================================================
Functionality: Bucket sort
Function input: Beginning address of the array to be sorted A, Length of array N
Function output: None
==============================================================*/
void BucketSort_1(int *A, int N)
{
    int i,j,MaxBit,number;
    //MaxBit represents the maximum number of bits of the elements in the array
```

```
MaxBit=AMaxBit(A,N);
//Bucket B[MaxBit][N+1],B[i][0] stores the number of elements in each bucket
int **B=(int**)malloc(MaxBit*sizeof(int*));
int temp;
if(B==NULL) return 0;
for(i=0;i<MaxBit;i++)
{
   //Add 1 because the first address doesn't store data elements
   B[i]=(int*)malloc((N+1)*sizeof(int));
   //Initially the number of elements in each bucket is 0
   B[i][0]=0;
}

for(i=0;i<N;i++)
{
   number=BitX(A[i]);
   //The row of data elements in each bucket is number-1,
   //since the array subscript starts from 0.
   ++B[number-1][0];
   temp=B[number-1][0];
   B[number-1][temp]=A[i];
   // Put it in the bucket
}

int k;
//Perform direct insertion sort on the elements in each bucket.
//Note that the first element in the bucket represents the
//size of the bucket element and doesn't participate in the sorting.
for(i=0;i<MaxBit;i++)
{
   for(k=2;k<B[i][0]+1;++k)
   {
      if(B[i][k]<B[i][k-1])
      {
         temp=B[i][k];      // Sentinel
         for( j=k-1; temp<B[i][j] && j>=1; --j)
            B[i][j+1]=B[i][j];
         B[i][j+1]=temp;       //L.r[j+1], which is also L.r[i-1]
      }
   }
}
}
```

```
k=0;
for( i=0; i<MaxBit; i++ )
// Give the value of the bucket elements to the array
for( j=1; j<=B[i][0]; j++ )
   A[k++]=B[i][j];
for( i=0; i<MaxBit; i++ )    // Release memory
   free(B[i]);
free(B);
}
```

Characteristic of bucket sort: effective on evenly distributed data; when the data type is integer, it normally distributes the data into buckets according to number of digits of the element.

```
/*=========================================================================
Functionality: Find out the maximum number of digits of the elements in the
array, and ascertain how many buckets are needed for the sorting.
Function input: Beginning address of the array to be sorted A, Length of array N
Function output: Maximum number of digits of the elements in the sequence to be
sorted, which is equivalent to the number of buckets needed
=========================================================================*/
int AMaxBit(int A[],int N)
// The number of digits of the maximum number to be sorted
{
   int largest=A[0],digits=0;
   for(int i=1;i<N;++i)
   {
      if(A[i]>largest)
      largest=A;
   }
   while(largest!=0)
   {
      ++digits;
      largest/=10;
   }
   return digits;
}
```

Bucket sort is an effective sorting algorithm. In the most under circumstances, it can reach linear time complexity. Of course, when the amount of data is huge, increasing the number of buckets without discretion can cause severe space wastage. Therefore, the complexity of bucket sort is essentially a trade-off problem between time cost and space cost. In addition, another advantage of bucket sort is that it is a stable sorting method.

3.6.2 Radix sort

Radix sort is a completely different algorithm when compared with the comparison sorting algorithms mentioned earlier. Its main idea is to realize sorting of single logical records via multikeyword sorting. The concrete implementation can be seen at the chapter stack and queue.

For radix sort, if there are n integers with d digits, each of which has k possible values, for example, sorting of natural numbers, k can take the values from 0 to 9, then the value of k is 10. Therefore, for n d-digit numbers, if the stable sorting method on any of the digit costs $O(n + k)$, then for the radix sort on d-digit integers, we can finish the sorting of all these integers in $O(d(n + k))$ time. Radix sort is an efficient sorting algorithm, but is not as widely applicable as other sorting methods.

3.7 Comparison of various sorting algorithms

Let us comprehensively review the several algorithms introduced in this chapter.

3.7.1 Quick sort

Quick sort essentially is a recursive algorithm using divide-and-conquer idea. Under normal circumstances, quick sort is faster than most sorting algorithms, and is currently the publicly acknowledged fastest sorting method. However, when the sequence is basically ordered, it will deteriorate into bubble sort, which affects the performance of the sort. In addition, quick sort is based on recursion and involves a huge amount of stack operations in the memory. For machines with very limited memory, it would not be a good choice.

3.7.2 Merge sort

Merge sort first views each individual record as an ordered sequence, and then obtains new sequences via two-way merges. In this way, it sorts all the records. Merge sort is faster than heap sort, though normally it is not as fast as quick sort.

3.7.3 Heap sort

Heap sort is suitable for situations where the amount of data is huge, for example, millions of records. Heap sort does not require a lot of recursion or multidimensional temporary arrays, which makes it suitable for sequences with a huge amount of data. When there are more than millions of records, since quick sort and merge sort essentially base their ideas on recursion, stack overflow error might occur.

Heap sort will construct a heap out of all the data, with the maximum (minimum) data at the top of the heap. Then it exchanges the top data with the last data in the sequence, then it rebuilds the heap and swaps data. In this manner, it sorts all the data.

3.7.4 Shell sort

Shell sort divides the data into different groups. It first sorts each group and then performs one insertion sort on all the elements, in order to reduce the swapping and moving of data. Shell sort is very efficient, but the quality of grouping will hugely impact the algorithm performance. Shell sort is faster than bubble sort and insertion sort, but it is slower than quick sort, merge sort and heap sort. Shell sort is suitable for situations where the number of data is lower than 5,000 and speed is not very important. It is very good for sorting arrays with relatively small amount of data.

3.7.5 Insertion sort

Insertion sort keeps inserting the values in the sequence into an already sorted sequence, until the end of this sequence. Insertion sort is an improvement on bubble sort. Now it is not widely used. However, since the algorithm is relatively simple, it is still effective for the sorting of some relatively small sequences.

3.7.6 Bubble sort

Bubble sort is the slowest sorting algorithm. Since it needs to compare each record in the sequence again and again, the number of comparisons is huge. It is the most ineffective algorithm, with a complexity of $O(n^2)$.

3.7.7 Selection sort

This sorting method is a sorting algorithm with swapping, with a complexity of $O(n^2)$. In actual application it is basically at the same place as bubble sort. They are just the initial stages in the development of sorting algorithm and are rarely used in actual applications.

3.7.8 Bucket sort

Bucket sort is a sorting strategy that trades space for time. Theoretically, if we use the same number of buckets as the number of records in the sequence, we can achieve linear time complexity. However, when the number of records is huge, it might cause huge space usage, and might even make the sorting impossible. Normally speaking, bucket sort is in practice slower than high-efficiency sorting algorithms such as quick sort. However, it is faster than traditional sorting methods.

3.7.9 Radix sort

Radix sort does not follow the same route as the normal sorting algorithms. It is a relatively new algorithm. Radix sort is best used against integers. If we are to apply it to the sorting of floating point numbers, we must be able to map the floating point number into comparisons of multiple keywords, which is really inconvenient.

The comparison of the performances of various sorting algorithms is given in Table 3.1.

As explained earlier, in all the algorithms discussed in this chapter, none has "the best performance." Some apply to situations where n is relatively large, some apply to situations where n is relatively small, some apply to situations where the sequence is basically ordered, some apply to situations where the sequence is completely unordered and so on. Since the best-case scenario and worst-case scenario of different sorting algorithms are different, we need to select different sorting algorithms under different situations, and might even combine multiple methods together.

The testing code for sorting algorithms

Table 3.1: Performance comparison of various sorting algorithms.

Sorting method	Average time	Worst-case scenario	Auxiliary storage
Bubble sort	$O(n^2)$	$O(n^2)$	$O(1)$
Selection sort	$O(n^2)$	$O(n^2)$	$O(1)$
Shell sort	$O(n \log n)$	$O(n \log n)$	$O(1)$
Quick sort	$O(n \log n)$	$O(n^2)$	$O(\log_2 n)$
Heap sort	$O(n \log n)$	$O(n \log n)$	$O(1)$
Merge sort	$O(n \log n)$	$O(n \log n)$	$O(n)$
Insertion sort	$O(n^2)$	$O(n^2)$	$O(1)$
Bucket sort	$O(n)$	$O(n^2)$	$O(n)$
Radix sort	$O(d(n+k))$	$O(d(n+k))$	$O(kd)$

Note: The n in the table refers to the number of elements to be sorted, d refers to the number of keywords and k refers to the range of possible values of keywords.

```
/********Testing of Sorting Algorithms********************/
#include<stdio.h>
#include<stdlib.h>
#include<malloc.h>
#define length 11    // Define the length of the sequence
void InsertSort(int *a,int n);    // Direct insertion sort
void ShellSort(int *a,int n);    // Shell sort
void BubbleSort(int *a, int n);    // Bubble sort
void QuickSort(int *a, int n);    // Quick sort
void SelectionSort(int *a, int n);    // Selection sort
void HeapSort(int *a,int n);    // Heap sort
void TMSort(int *a,int n);    // Two-way merge sort
void BucketSort(int *a,int n);    // Bucket sort
void RadixSort(int *a,int n);    // Radix sort
// Algorithm code for each sorting. For ease of understanding,
// the input elements are arrays. In actual applications
// they can be adjusted into structs according to needs.
int main()
{
    int arr[length]={15,124,2,19,14,321,415,1,62,58,55};
    int method;
    int i;
    printf("Array before sorting:\n");
```

```
for(i=0;i<length;i++)
printf("%3d ",arr[i]);
printf("\n");
printf("**************** Please select sorting method **************\n");
printf("Insertion sort    1: Direct insertion sort    2: Shell sort \n");
printf("Swap sorting      3: Bubble sort              4: Quick sort \n");
printf("Selection sort    5: Simple selection sort    6: Heap sort  \n");
printf("Merge sort        7: Two-way merge sort                     \n");
printf("Distribution sort 8: Bucket sort               9: Radix sort \n");
printf("**********************************************************\n");
scanf("%d",&method);
switch(method)
{
    case 1:  InsertSort(arr,length);  break;      //Insertion sort
    case 2:  ShellSort(arr,length);  break;      //Shell sort
    case 3:  BubbleSort(arr,length);  break;      //Bubble sort
    case 4:  QuickSort(arr,length);   break;      //Quick sort
    case 5:  SelectionSort(arr,length);  break;    //Normal selection sort
    case 6:  HeapSort(arr,length);  break;      //Heap sort
    case 7:  TMSort(arr,length);    break;      //Two-way merge sort
    case 8:  BucketSort(arr,length); break;      //Bucket sort
    case 9:  RadixSort(arr,length); break;      //Radix sort
    default:  InsertSort(arr,length); ;      //If the selection errs, just
    proceed with Insertion sort
}
printf("Array after sorting\n");
for(i=0;i<length;i++)
printf("%3d ",arr[i]);
printf("\n");
}
```

3.8 Chapter summary

The connections between the main contents of this chapter are shown in Fig. 3.30.

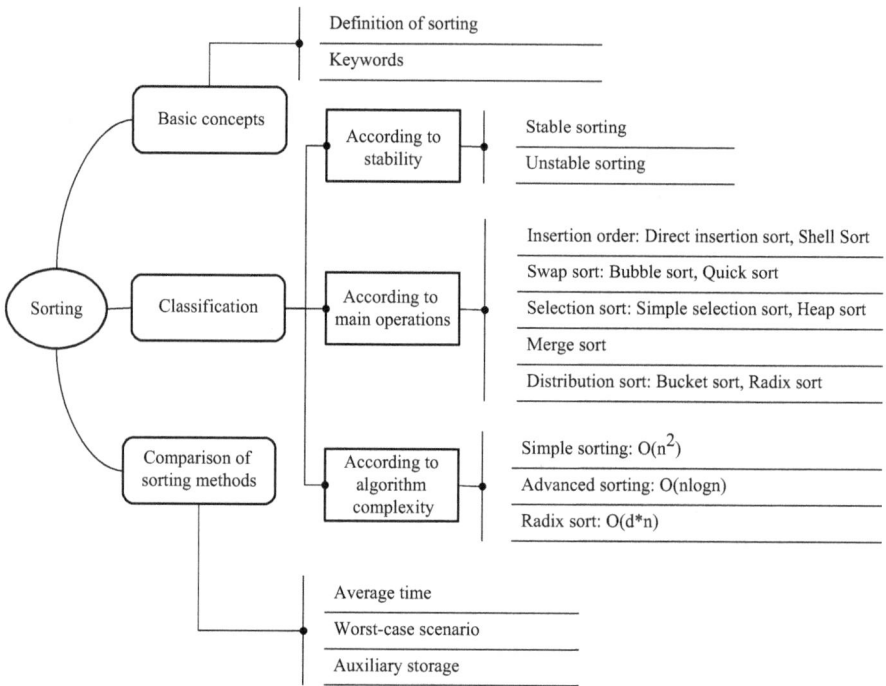

Fig. 3.30: Connections between various concepts of sorting.

3.9 Exercises

3.9.1 Multiple-choice questions

1. Among the internal sorting algorithms below:
   (A) quick sort   (B) direct insertion sort   (C) two-way merge sort
   (D) simple selection sort   (E) bubble sort   (F) heap sort
   (1) The algorithms whose number of comparisons is not related to the initial status of the sequence are ( ).
   (2) The unstable sorting algorithms are ( ).
   (3) When the initial sequence is basically ordered (i.e., after removing k elements from the n elements, it will be completely ordered, $k \ll n$), the algorithm with the highest efficiency is ( ).
   (4) The algorithms with average time complexity of $O(n\log_2 n)$ are ( ), of $O(n^2)$ are ( ).

2. The sorting method whose number of comparisons is not related to the initial status is ( ).

(A) direct insertion sort   (B) bubble sort
(C) quick sort   (D) simple selection sort

3. When sorting a set of data (84, 47, 25, 15, 21), the changes of data sequence during the sorting process are:
(1) 84 47 25 15 21   (2) 15 47 25 84 21   (3) 15 21 25 84 47   (4) 15 21 25 47 84

Then the sorting method used is ( ).
(A) selection sort   (B) bubble sort   (C) quick sort   (D) insertion sort

4. Which of the following sorting algorithm would not necessarily be able to select an element and put it into its final position after one run? ( )
(A) Selection sort   (B) Bubble sort   (C) Merge sort   (D) Heap sort

5. The keywords of a set of records are (46, 79, 56, 38, 40, 84), then using quick sort, the result of one run of division using the first record as the pivot is ( ).
(A) (38, 40, 46, 56, 79, 84)   (B) (40, 38, 46, 79, 56, 84)
(C) (40, 38, 46, 56, 79, 84)   (D) (40, 38, 46, 84, 56, 79)

6. Which of the following sorting algorithms costs the most time when the data is already sorted? ( )
(A) Bubble sort   (B) Shell sort   (C) Quick sort   (D) Heap sort

7. In terms of average performance, the best internal sorting method currently is ( ).
(A) bubble sort   (B) shell sort   (C) swapping sort   (D) quick sort

8. Among the below algorithms, which one uses the most auxiliary storage ( ).
(A) merge sort   (B) quick sort   (C) shell sort   (D) heap sort

9. If we use bubble sort to sort the sequence {10, 14, 26, 29, 41, 52} in descending order, we need to perform ( ) comparisons.
(A) 3   (B) 10   (C) 15   (D) 25

10. Quick sort is most ineffective under situation ( ).
(A) There is too much data to be sorted
(B) There are multiple identical values in the data
(C) There are odd number of data to be sorted
(D) The data to be sorted is already basically ordered

11. Which one of the following four sequences could be a heap? ( )
(A) 75, 65, 30, 15, 25, 45, 20, 10   (B) 75, 65, 45, 10, 30, 25, 20, 15
(C) 75, 45, 65, 30, 15, 25, 20, 10   (D) 75, 45, 65, 10, 25, 30, 20, 15

12. There is a group of data (15, 7, 20, 7, 9, 8, −1, 4). The initial heap constructed with heap sort filtering is ( ).
(A) −1, 4, 8, 9, 20, 7, 15, 7   (B) −1, 7, 15, 7, 4, 8, 20, 9
(C) −1, 4, 7, 8, 20, 15, 7, 9   (D) A, B, C are all wrong.

3.9.2 Cloze test

1. Most sorting algorithms have two basic operations: __ and __.
2. Performing direct insertion sort on a set of records (54, 38, 96, 23, 15, 72, 60, 45, 83) when inserting the seventh record 60 into an already ordered sequence, to find the insertion position, there need to be at least __ comparisons (the convention here is to perform the comparisons from back to front).
3. Between insertion and selection sort, if the initial data is basically ordered, then we select __ (until the end); if the initial data is basically reverse-ordered, then we choose __.
4. Between heap sort and quick sort, if the initial record is nearly ordered or reversely ordered, then we choose __; if the initial record is basically unordered, then it is better to choose __.
5. To perform bubble sort on a collection of n records, the time needed in the worst case would be __. If we then perform quick sort on it, then the time needed in the worst case would be __.
6. To perform merge sort on a collection of n records, the average time needed would be __ and the additional space needed would be __.
7. To perform two-way merge sorting on a list with n records, the whole merge sort needs to be run __ times, moving records __ times in total.
8. Suppose that we want to sort the keywords in the sequence (Q, H, C, Y, P, A, M, S, R, D, F, X) in ascending order, then:
    The result of one scan of bubble sort would be __.
    The result of one run of shell sort with initial step size of 4 would be __.
    The result of one scan of two-way merge sort would be __.
    The result of one scan of quick sort would be __.
    The result of initial heap construction from heap sort would be __.
9. Among heap sort, quick sort and merge sort, if we only consider storage space, then we should choose __, then __, at last __. If we only consider the stability of the sorting result, then we should choose __. If we only consider the fastest speed under average case, then we should choose __. If we only consider the fastest speed under worst case and also want to save memory, then we should choose __.
10. If there are multiple records with the same key value in the sequence to be sorted, and after the sorting the relative order of these records is not changed, then we call such a sorting method __, otherwise we call it __.
11. According to the differences in the storage devices involved in the sorting process, sorting can be divided into __ sorting and __ sorting.

3.9.3 Open-ended questions

1. We know that a sequence is basically ordered. What is the quickest sorting method under such a scenario? What is the average time complexity?

2. For a given set of keywords: 83, 40, 63, 13, 84, 35, 96, 57, 39, 79, 61, 15, draw the results of each run under direct insertion sort, direct selection sort, quick sort, heap sort and merge sort.

3. Decide whether the following sequences are heaps (they can be min heaps or max heaps; if it is not a heap, then adjust it to be a heap).

   (1) 100, 85, 98, 77, 80, 60, 82, 40, 20, 10, 66
   (2) 100, 98, 85, 82, 80, 77, 66, 60, 40, 20, 10
   (3) 100, 85, 40, 77, 80, 60, 66, 98, 82, 10, 20
   (4) 10, 20, 40, 60, 66, 77, 80, 82, 85, 98, 100

4. Design a direct selection sort using only linked lists.

5. Bubble sort moves large elements above (floating of the bubble) or moves small element below (sinking of the bubble). Please give a bubble sort algorithm that alternately carries out the floating and sinking processes (i.e., bidirectional bubble sort algorithm).

6. Output records of 50 students (each record includes name and grade), compose an array of records and output them in descending order of grades (each row should contain 10 records. The sorting method should be selection sort.

7. We already know that $(k_1,k_2,...,k_n)$ is a heap. Try to write an algorithm to adjust $(k_1,k_2,...,k_n,k_{n+1})$ into a heap. Do it with the following idea: an algorithm that fills in the elements one by one from an empty heap. (Hint: after adding k_{n+1}, the adjustment direction should be from leaf to root.)

4 Data processing – index and search technologies

Several decades ago, it was still a very arduous and time-costing task for people to retrieve some certain useful information. Even if you walk into the most advanced library at that time, it would be very difficult to find a needed book. The library front desks at that era had a lot of small cabinets as shown in Fig. 4.1, which are filled with index cards. One index card corresponds to the information of one book, and has a certain index – book index (e.g., in China, the Chinese Library Classification Number is used). Such an index is attached by hand to the spine of every book. The books are listed on the bookshelves with this as the basis for sorting. Whenever a reader looks up the book information, he/she needs to first look up the cards in the cabinet, and use specific information such as book name, publication date or author name as the basis for the lookup, and find the book index of the corresponding book. The librarian finds the location of the book in the library according to this book index, before delivering the requested book into the hands of the reader.

Nowadays, the amount of books stored in libraries is much more than several decades again. The lookup of books is via electronic index, and the feedback provided by search engines is fast and convenient. Figure 4.2 shows the book searching interface of Xidian University Library and the related book information reached by the book name *Introduction to Algorithms*. The operations that previously required search by hand are now all done instead by the computers.

In the current era of information explosion, there are multiple ways for people to get information. Modern people no longer worry about the inaccessibility of information, but the overload of information. How to filter and find useful information has now become a more urgent issue.

Facing the massive webpage information on the Internet, when we use the most popular search engine to perform information retrieval, click the search button and

https://doi.org/10.1515/9783110676075-004

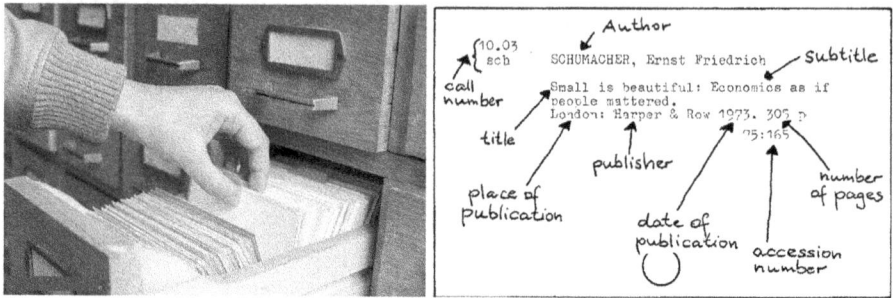

Fig. 4.1: Ancient index card cabinet and the book catalog cards.

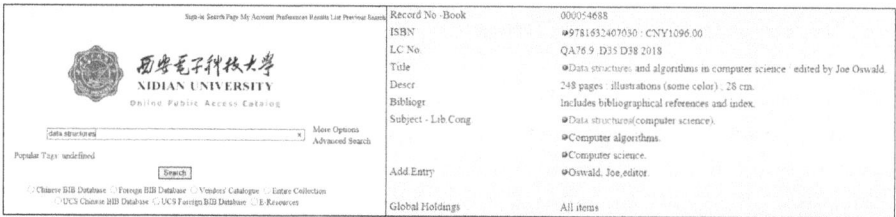

Fig. 4.2: Book lookup webpage.

miraculously obtain search results that are reasonably ordered in less than half a second, have we wondered how it is achieved?

This process actually has similarities to the book lookup in libraries, though it is on a larger scale and includes more details. The work performed by the search engine can be basically divided into the following three aspects:

- Collect information: use technologies such as web crawler, traverse the various websites on the Internet and collect the information there.
- Construct index: clean up the collected information, and construct an index list that is easy to search.
- Provide querying service: provide querying interface faced to the users, find the relevant information from background index and give feedback.

Actually, indexing technologies and searching algorithms are both the most crucial technologies in search engines. Using indexing technologies, the search engine company can put all the present webpage information on the Internet into its database. Using search technologies, whenever a user issues search commands, the search engine can immediately retrieve useful information to be used in information feedback.

Term Explanation
Index: Index is a clue-like guide. It is a kind of logical list between the keyword and the corresponding physical address.

Lookup: Lookup is the process of finding the data element corresponding to the given keyword with a certain method in a collection of data elements. The target collection of such a lookup can be linear or nonlinear, ordered or nonordered. The classical search algorithms include sequential search, binary search, binary ordered tree search and hash table search technology.

Index is born for quick lookup – usually the index itself is a linear or nonlinear data structure arranged in order based on the keywords.

4.1 Basic concepts of index

In our daily lives, the indexes that we usually use – for example, the directory in a dictionary or an index at the end of a technological publication – are all sources of inspiration for the index technology in computers.

Knowledge ABC The difference between index and directory
Index is commonly mentioned in the same breath with directory. In general, they have obvious differences in their usage – normally, the table of contents given in the beginning of the books we read are normally listed in the order of the concrete contents. Here, the "physical address" for books means page numbers, which are usually listed from small to large. On the contrary, index is sorted based on the keywords, while its corresponding "physical addresses" are not necessarily ordered. The corresponding example is shown in Fig. 4.3.

I Foundations

Introduction *3*
1 **The Role of Algorithms in Computing** *5*
1.1 Algorithms *5*
1.2 Algorithms as a technology *11*
2 **Getting Started** *16*
2.1 Insertion sort *16*
2.2 Analyzing algorithms *23*
2.3 Designing algorithms *29*
3 **Growth of Functions** *43*
3.1 Asymptotic notation *43*
3.2 Standard notations and common functions *53*
4 **Divide-and-Conquer** *65*
4.1 The maximum-subarray problem *68*
4.2 Strassen's algorithm for matrix multiplication *75*
4.3 The substitution method for solving recurrences *83*
4.4 The recursion-tree method for solving recurrences *88*
4.5 The master method for solving recurrences *93*
★ 4.6 Proof of the master theorem *97*
5 **Probabilistic Analysis and Randomized Algorithms** *114*

(a) Table of contents

AA-tree, 338
abelian group, 940
ABOVE, 1024
above relation ($\succeq_x$), 1022
absent child, 1178
absolutely convergent series, 1146
absorption laws for sets, 1160
abstract problem, 1054
acceptable pair of integers, 972
acceptance
  by an algorithm, 1058
  by a finite automaton, 996
accepting state, 995
accounting method, 456–459
  for binary counters, 458
  for dynamic tables, 465–466
  for stack operations, 457–458, 458 ex.
Ackermann's function, 585
activity-selection problem, 415–422, 450

(b) Index

Fig. 4.3: Examples of table of contents and index.

Even if it is catalogue and index generated about books, there exists a similar difference. The catalogue is only a macroscopic record of the documents on a whole, while index can provide microscopic revealing and lookup of a certain part, a certain perspective or a certain knowledge unit of the document from different angles. Of course, the general discipline of catalogue is a science that also covers index construction. Please be aware of the differences.

In general, understanding the rule of "indexing performs sorting according to keywords" would be very helpful in understanding the implementation essence of the indexing technology in computers.

4.1.1 Definition of index

Index technology creates an index list for the purpose of looking up the data in the data list. Each index list is composed of multiple index items. Each index item is a tuple of (keyword, address), where keyword is a data item that can uniquely mark the node to be looked up in the data list. The index items in the index list are usually sorted according to the keywords. For example, Fig. 4.4 shows a data list and an index list for student information.

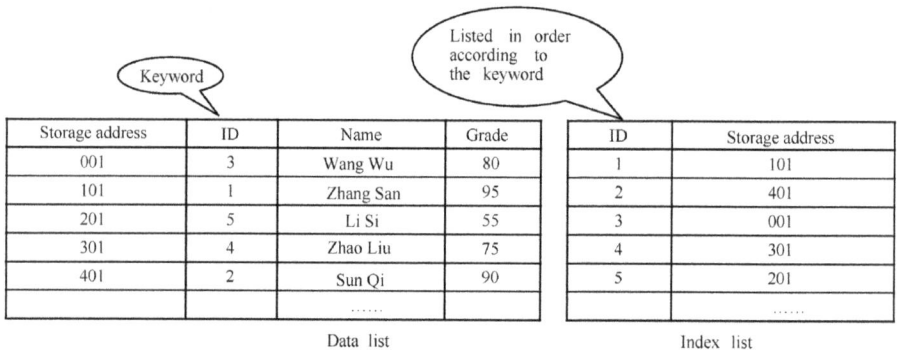

Storage address	ID	Name	Grade		ID	Storage address
001	3	Wang Wu	80		1	101
101	1	Zhang San	95		2	401
201	5	Li Si	55		3	001
301	4	Zhao Liu	75		4	301
401	2	Sun Qi	90		5	201
		......				......

Keyword · Listed in order according to the keyword

Data list   Index list

Fig. 4.4: Student information list and index list.

Using C language, an index item can be expressed as follows:

```
typedef struct
{
    KeyType key;          // Keyword
    DSType *pAddress;     // Address of data item corresponding to this keyword
} IndexEntry;
```

4.1.2 The logical characteristics of index

As long as it is a sortable logical structure, in theory it can be used in the construction of index list. Linear list and tree discussed in previous chapters can both be ordered; therefore, index list can be stored with linear data structures or with tree data structures.

Depending on whether the index is easy to modify, it is divided into static index and dynamic index. According to the logical structure used by the index, it is divided into linear index and tree structure index.

Based on the uniqueness in the data list of the item corresponding to the keyword in the index, keywords in the index can be divided into primary keyword and secondary keyword. In general, the index we speak of refers to index targeting primary keywords. However, a data element usually has multiple features. These features outside of primary keyword can be used as secondary keywords, and we can also construct the corresponding indexes on them.

The linear index constructed for primary keywords can be divided into dense index and sparse index (also called partitioned index) according to the quantity relationship between its index items and the data elements in the data list.

Indices constructed for secondary keywords have different forms such as multilist (also called multikeyword table) and inverted table.

If the processing on linear index with a huge amount of data is still time costly, we can again construct an index for it, which is called second-level index. Upon second-level index, we can still construct third-level index, fourth-level index and so on. Multilevel indexes will form tree-like structures, which is convenient for lookup tasks in practice. However, since this index structure is usually stored in the form of sequential list, its index items are not easy to be modified with the addition/deletion/changes of the data items. We call this type of index static index, which is suitable for scenarios where the data items are basically fixed.

Tree indexes are mostly used as dynamic indexes. The forms of these trees can conveniently change according to the changes to data items, while keeping the ordering properties of the index unchanged. They can be further divided into binary tree and nonbinary tree based on the form of a tree. Binary sorting tree (or called binary search tree) is a classic example of binary tree indexing, of which there also exists balanced binary tree, which is more suitable for indexing. The classic example of nonbinary tree indexing is B-tree, that is, balanced multiway search tree. It is a balanced search tree designed for hard disk or other direct access auxiliary storage devices (i.e., external storage compared with memory). Compared with binary tree, the advantage of B-tree is that it can lower the number of input/output operations on the disk, so that computational efficiency can be improved.

This chapter involves all the index types mentioned earlier, which will be illustrated in the way shown in Fig. 4.5. Note that this figure does not cover all types of indexes in practice, but only includes the most representative part of them.

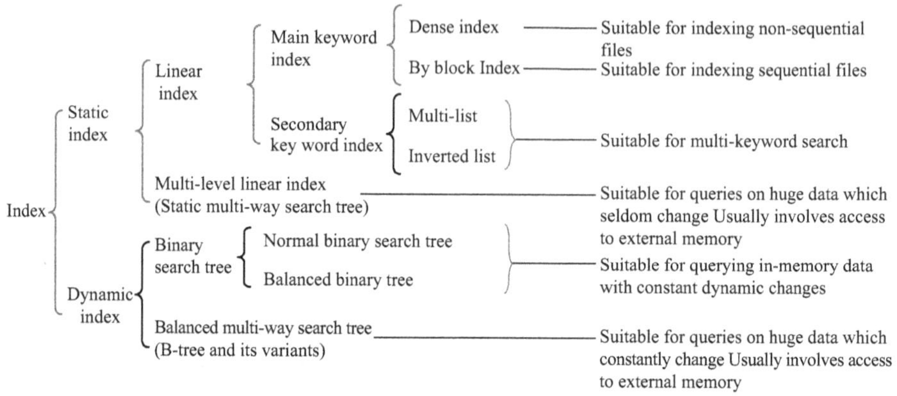

Index
- Static index
  - Linear index
    - Main keyword index
      - Dense index ——— Suitable for indexing non-sequential files
      - By block Index ——— Suitable for indexing sequential files
    - Secondary key word index
      - Multi-list
      - Inverted list } Suitable for multi-keyword search
  - Multi-level linear index (Static multi-way search tree) ——— Suitable for queries on huge data which seldom change Usually involves access to external memory
- Dynamic index
  - Binary search tree
    - Normal binary search tree
    - Balanced binary tree } Suitable for querying in-memory data with constant dynamic changes
  - Balanced multi-way search tree (B-tree and its variants) ——— Suitable for queries on huge data which constantly change Usually involves access to external memory

Fig. 4.5: Classification of index and the respective applicable scenario.

4.1.3 Major operations on index

1. Insertion: when content is added to the data list, insert the corresponding node into the index list.
2. Deletion: when content is deleted from data list, delete the corresponding node in the index list.
3. Lookup: look up the position of the corresponding data elements in the data list depending on the keywords.

As mentioned earlier, index list can be expressed with either linear or nonlinear data structure. Sections 4.2 and 4.3 illustrate, respectively, these two methods.

4.2 Linear indexing technology

Linear index is the most fundamental organization way of index list. In this case, index items are usually stored in the order of keywords.

4.2.1 Dense index

To organize events, a certain personnel of the student union, L, wants to investigate the interest of students in the class. He/she initiates an investigation activity. He/she carries a booklet to ask the extracurricular interest of each student he/she meets. However, when he/she looks back to organize the multitude of investigation records on his/her notebook, he/she realizes that he/she only reserved four pages of paper for each student before the investigation, but did not sort the students based on their student ID. Then, whenever he wants to view the information about one particular student, he/she needs to start the lookup from the beginning, which is very inconvenient. For more convenient lookup, he/she stuck a piece of paper at the beginning of the notebook, which organizes his records based on the student ID. The index list on the left in Fig. 4.6 is the content of this paper, and the data on the right is the contents recorded on each page of the notebook. Since the investigation process is completely random, this index list must exactly list the page where each record is. This index list is a dense index.

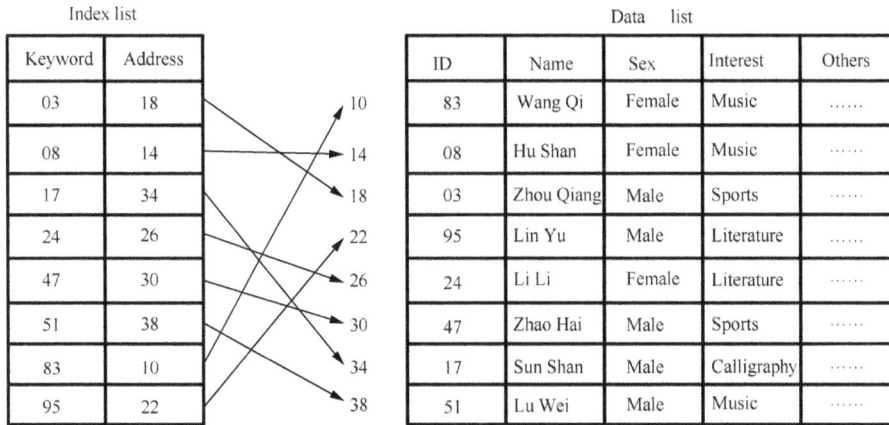

Index list

Keyword	Address
03	18
08	14
17	34
24	26
47	30
51	38
83	10
95	22

Data list

ID	Name	Sex	Interest	Others
83	Wang Qi	Female	Music	......
08	Hu Shan	Female	Music	......
03	Zhou Qiang	Male	Sports	......
95	Lin Yu	Male	Literature	......
24	Li Li	Female	Literature	......
47	Zhao Hai	Male	Sports	......
17	Sun Shan	Male	Calligraphy	......
51	Lu Wei	Male	Music	......

Fig. 4.6: Example of dense index – page number indexing of investigation records.

Term Explanation Dense index
Dense index refers to the practice that corresponds each record in the data set to one index item in the linear list.

Dense index is suitable for all kinds of data lists. The indexing, lookup and update of such type of files are relatively convenient. However, since there are many index items, which occupy a large number of space, whenever the records in the data list themselves are already ordered (indexing sequential files), we can choose to not use dense index. However, when the records are randomly stored in the data list (indexing nonsequential files), we must use dense index.

4.2.2 Block index

Actually, in many cases, the data itself is ordered. If we construct dense index in this situation, the index will occupy a lot of space, and the lookup will be difficult. Many people usually write down 26 English letters at the side of the dictionary, and lookup the word at the region where the first letter of this word is. These 26 English characters serve as a block index. The type of dictionary shown in Fig. 4.7 carefully directly marks the region of each English character at the side of the page, forming a very convenient index of characters at the side of the dictionary.

Fig. 4.7: Example of block indexing – the character index at the side of the dictionary.

Of course, if the data itself is not that orderly, but is clearly classified. Then the classification itself can serve as an index. This can also be viewed as a kind of block index. We might have the following experience when finding books in the library. Depending on whether the book is about social science and engineering or if it is a magazine, then we can select different rooms. According to the classification number of the book, we can quickly find the shelf on which the book rests. However, on which row, which position that book rests exactly, we need to search in more detail. In this scenario, we call the data to be ordered in sections, that is, the data list is divided into multiple sections; there is a certain order among the sections, but there might not be an order within each section. The keyword Chinese Library Classification Number on bookshelves has this feature. Note that in this case the classification information of books is also included in the Chinese Library Classification Number. Therefore, books of different categories are also ordered in sections according to the Chinese Library Classification Number as the keyword.

In summary, when indexing sequential files or constructing index for data list ordered in sections, we can use block index to reduce the number of index items.

Term Explanation Block index (sparse index)

i

To construct a data list ordered in sections, we divide n data elements into m blocks ($m \leq n$) in order based on the sections. It is not necessary for the nodes within each section to be ordered, but the sections must be ordered among themselves. For example, the keyword of any element in block 1 must be smaller than the keyword of any element in block 2, the keyword of any element in block 2 must be smaller than the keyword of any element in block 3 and so on. Selecting the largest keyword of each block of this data list ordered in sections to construct an index list is to construct a block index. Since we did not establish index items for each individual item, in contrast to the concept of dense index, block index is also called sparse index.

In computers, when performing lookups on databases, we normally load the index directly into the memory to perform the lookup. If there are too many index items, the index list might get too big to fit into the memory, which causes frequent data exchanges between the memory and the external storage (e.g., hard disk), which affects the lookup speed. Block storage and index are the most reasonable solutions under such a scenario.

When using block index to perform data lookup, two lookups are needed. The first is to find the corresponding region in the index, and the second is the lookup within this region.

Knowledge ABC Leveled index

i

When there are many index items, we can also construct index of indexes. Multilevel structure is thus formed.

The book classification problem mentioned earlier is actually a multilevel structure. The viewing rooms of many school libraries are classified according to the categories, for example, viewing room for social science books and viewing room for natural science books. Then, the process of finding the corresponding viewing room is equivalent to looking up the first-level index. Finding the bookshelf within the corresponding viewing is the process of looking up the second-level index.

Another example is the file organization method ISAM (Indexed Sequential Access Method) designed for file I/O on hard disks. It is also a typical multilevel index structure. Since a hard disk performs I/O according to the three levels of addresses "disk section, cylinder and track," we can establish multilevel index on the data files in the hard disk based on the disk section/cylinder/track, as shown in Fig. 4.8. In this case, not only the data list is very big, the index list is also very big. The reason for having multiple levels is that we cannot load the whole index list into the memory immediately to perform the lookup. We can only allow the index of the index (second-level index), even higher level index to reside in the memory constantly, while lower level index has to be put into external storage. In this case, the number of visits of the external storage is equivalent to the number of index loadings plus one reading of the object. The number of visits of external storage becomes the performance bottleneck of the whole system under such a scenario, since the response speed of external speed is magnitudes slower than that of memory.

Of course, multilevel index is no longer simple linear index structure. Later, we will continue the introduction of such leveled index structure in Section 4.3.

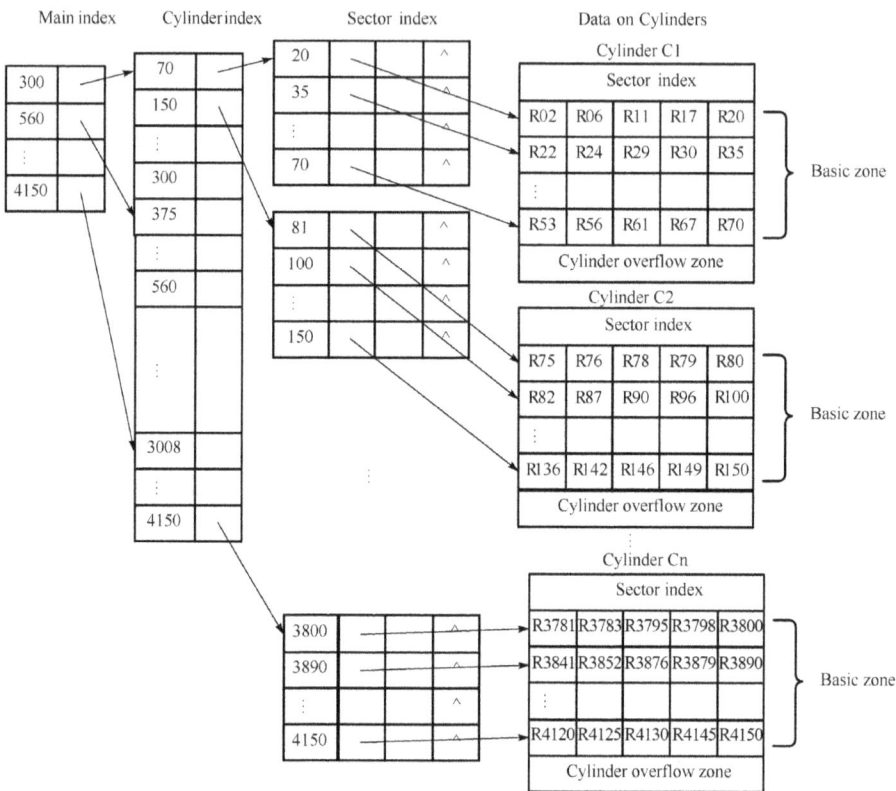

Fig. 4.8: Example of multilevel index – file organization method on hard disk ISAM.

4.2.3 Multilist

Student L from the student union finally was able to organize his/her interview records depending on the order of the student IDs. After reorganizing, the records
can be archived in the student union to serve as the reference for planning later
activities. However, during his work, he realized that it was not that convenient to
perform summarization and lookup on the sex and interest of each student, respectively. Therefore, he/she studied how to make the records easier to look up with
indexes. He/she came up with the idea to establish new indexes for sex and interest individually. In this case, each keyword in the index corresponds to not only
one record, that is to say, these "keywords" are actually secondary keywords.
When constructing secondary keyword indexes, he/she lets each keyword in the
index list to correspond to the page number of the first record corresponding to
this keyword. At the same time, each record contains one piece of information to

note what other record has the same value of this keyword. The result is the structure shown in Fig. 4.9, that is, a multilist structure.

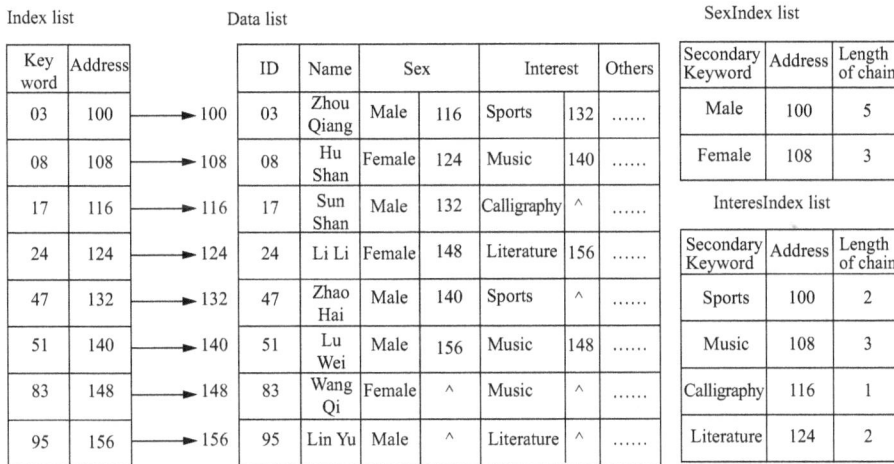

Index list   Data list   SexIndex list

Key word	Address
03	100
08	108
17	116
24	124
47	132
51	140
83	148
95	156

ID	Name	Sex	Interest		Others
→ 100	03	Zhou Qiang	Male	116	Sports	132	......
→ 108	08	Hu Shan	Female	124	Music	140	......
→ 116	17	Sun Shan	Male	132	Calligraphy	^	......
→ 124	24	Li Li	Female	148	Literature	156	......
→ 132	47	Zhao Hai	Male	140	Sports	^	......
→ 140	51	Lu Wei	Male	156	Music	148	......
→ 148	83	Wang Qi	Female	^	Music	^	......
→ 156	95	Lin Yu	Male	^	Literature	^	......

Secondary Keyword	Address	Length of chain
Male	100	5
Female	108	3

InteresIndex list

Secondary Keyword	Address	Length of chain
Sports	100	2
Music	108	3
Calligraphy	116	1
Literature	124	2

Fig. 4.9: Example of multilist – index on secondary keyword for the investigation records.

Term Explanation Multilist

Multilist file is an organizational method that combines indexing with linking. Its concrete organizational method is to construct an index for each secondary keyword that needs to be queried. At the same time, it connects the records with the same secondary keyword into a linked list, and store the head pointer, length and secondary keyword as an index item of the index list.

Normally, the main file in the multilist file is a sequential file.

To describe the data under such a scenario with C language, besides representing the index list in the form of normal index items, we also need some modifications to the data list.

```
typedef struct MLNode
{
    KeyType key;                     // Main/Primary keyword of the index
    DataType data;                   // Other data item
    SubKeyTpye1 subKey1;             // Value of the first secondary keyword
    struct MLNode * pNextBySubKey1;  // Linked list pointer related to the value
                                     // of the first secondary keyword
    SubKeyTpye2 subKey2;             // Value of the second secondary keyword
```

```
struct MLNode * pNextBySubKey2;      // Linked list pointer related to the value
                                     // of the second secondary keyword

    ...

} MultiList;
MultiList aMultiList[N];             // Secondary index list corresponding to
                                     // various types of secondary keywords
```

Then, construct the specific secondary keyword index. A secondary keyword index item can be defined as follows:

```
typedef struct
{
    SubKeyType subKey;           // Secondary keyword
    MultiList *pAddress;         // Address of the beginning item
    int linkLen;                 // The length of the linked list corresponding to
                                 // the secondary keyword
} SubIndexEntry;
```

In the example of investigation records, the main keyword is student ID, the secondary keywords are gender and interest. Two link bytes are set at the adjacent positions of these two keywords in each data element, which link together records with the same gender/interest. They together with the gender index list and interest table form the gender index and the interest index. With these indexes, it is easy to process various queries related to secondary keywords.

When performing information lookup according to the secondary keywords in a multilist, we need to find the corresponding index item according to the given value in the corresponding index list of secondary keywords. Then, we start from the head pointer pointed to this index item and list all records in this linked list.

4.2.4 Inverted list

After constructing multilist the last time, student L found it hard to find some male students who love music. Besides the need to lookup multiple index lists, he/she also needed to perform tedious search inside the data list to find the desired information. Now, the clever student L, when doing mathematics exercises on sets, realized that the method of obtaining intersection in sets is very suitable for this type of search. Therefore, he/she had a brilliant idea and came up with another method to index the secondary keywords. This method is to give all the values of the primary keyword of all relevant records at once. In this way, there is no need to add the pointer item in each record. The result is shown in Fig. 4.10. The data list obtained via this idea is not different from the ordinary data list. It only added an index list exclusively used for secondary keywords. From these two index lists, we can easily know how many students out there have identical values on two different conditions. If we want to locate who they are exactly, we can look them up in combination with the main index list. This secondary index structure is called inverted list.

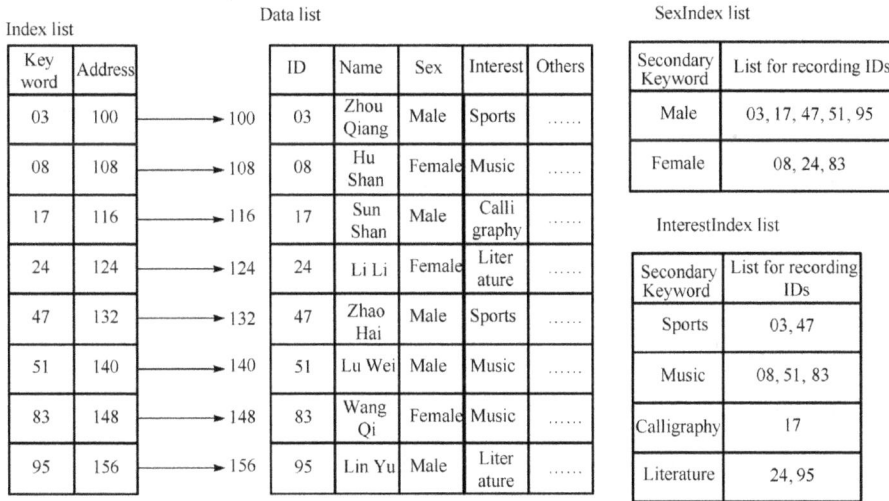

Index list				Data list						SexIndex list	
Key word	Address			ID	Name	Sex	Interest	Others		Secondary Keyword	List for recording IDs
03	100	→ 100		03	Zhou Qiang	Male	Sports	......		Male	03, 17, 47, 51, 95
08	108	→ 108		08	Hu Shan	Female	Music	......		Female	08, 24, 83
17	116	→ 116		17	Sun Shan	Male	Calligraphy	......			
24	124	→ 124		24	Li Li	Female	Literature	......		InterestIndex list	
47	132	→ 132		47	Zhao Hai	Male	Sports	......		Secondary Keyword	List for recording IDs
51	140	→ 140		51	Lu Wei	Male	Music	......		Sports	03, 47
83	148	→ 148		83	Wang Qi	Female	Music	......		Music	08, 51, 83
95	156	→ 156		95	Lin Yu	Male	Literature	......		Calligraphy	17
										Literature	24, 95

Fig. 4.10: Example of inverted list – the secondary keyword index of investigation records.

Term Explanation Inverted list

Inverted list is also called inverted index. It uses the nonprimary attribute value (secondary keyword) to lookup the secondary index of the record. It includes all the secondary keywords, and lists the primary keywords of all the related records. It is mainly used for complex queries.

There are several additional names for inverted list: inverted index and inverted document. The origin of the word "inverted" is because it does not ascertain attribute value according to the record, but uses the attribute value to ascertain the position of the record. Actually, the current mainstream search engines deal with complex queries based on inverted lists. They can quickly fetch the list of documents containing a word from that word.

If we describe data under such conditions with C language, we do not need to perform any modifications on the index list and data list. We only need to add a secondary index list related to the secondary indexes. Since the number of records in the inverted list is uncertain, we need to store the records with tools such as linked list. Therefore, we can consider the following form of definition:

```
typedef struct InvertedLisNodet
{
    KeyType key;    // The value of the primary keyword corresponding to this
                    // item which has the particular value of the secondary keyword
```

```
    struct InvertedLisNodet *pNext;    // The linked list pointer corresponding
                                       // to the secondary keyword
} InvertedList;

typedef struct
{
    SubKeyType subKey;        // The value of the secondary keyword
    InvertedList *pAddress;  // The address of the beginning item of the linked list
                             // corresponding to the value of the secondary keyword
} SubIndexEntry;
SubIndexEntry subIndList[M];       // A sort of inverted list corresponding to
                                   // secondary keywords.
```

Example 4.1 Simple inverted index for search engines.
There are three documents for which we construct indexes (in actual applications, there are countless documents):
- Document 1 (D1): mobile Internet of China develops rapidly.
- Document 2 (D2): mobile Internet has huge potential in the future.
- Document 3 (D3): the Chinese nation is industrious.

The dictionary set of the words in the document is: {China, mobile, Internet, develops, rapidly, future, of, potential, huge, Chinese, nation, is, has, industrious} (note: "in" and "the" are omitted in this example).

   To facilitate the logical computations of the computer, we can use one bit to represent whether each keyword is present in the document or not. We can obtain the result as shown in Table 4.1. When we search documents containing both "Internet" and "potential," the search engine will extract the two binary numbers "110" and "010" to perform Boolean operation, and obtain the value "010," which indicates that the set of documents containing both keywords has only document 2.

Table 4.1: Examples of inverted index list for search.

Keyword	China	Mobile	Internet	Develops	Rapidly	Future	Of
	100	110	110	100	100	010	011

Keyword	Potential	Huge	Chinese	Nation	Is	A	Industrious
	010	010	001	001	001	001	001

This simple indexing method can be used on small-scale data, such as indexing thousands of documents. There are two restrictions on this method.

1. There has to be enough memory to store the inverted index. For search engines, the data are counted in GB level. When the data size increases incessantly, it would be impossible to provide so much memory.
2. The algorithm is executed sequentially, and it would be hard to carry out parallel processing.

From this we can see that the implementation of search engines in reality is much more complicated.

Since sequential structures have the features of random access, easy access and easy lookup, linear indexes are usually stored in sequential storage structures. However, the insertion and deletion operations on sequential storage structures have obvious shortcomings – they need a huge amount of movements, which leads to high computational complexity. If the data list is a basically immutable collection, that is, dictionary, then it would be appropriate to use such type of linear index. If the collection of the data list itself frequently changes, for example, the Internet nowadays, where a lot of new contents emerge every day, then its indexes would involve constant changes. Linear index would no longer be appropriate.

Actually, linear index is suitable for collections whose contents barely change. We normally call it a type of static index. When targeting situations where the contents change a lot and rapidly, such as that of the Internet, we normally can use dynamic index to realize fast lookup. Considering lookup efficiency, dynamic index is usually constructed in the form of a tree, that is, a tree-like index.

4.3 Tree-like index

The example of multilevel index in Fig. 4.8 – disk file organization method ISAM's form of index is presented in the form of a tree. We can view it as a tree form index. The m-nary tree formed by it is also called m-way search tree. As shown in Fig. 4.11, when the corresponding data list (i.e., the corresponding file) structure was first created, the index was ascertained. During the execution, the structure of the tree will not be changed. This is effective in the situation where the files are not constantly modified. But it is not suitable for data files that constantly grow or shrink. If the data files (or the data list corresponding to the index) constantly change, we need to construct dynamic index to adapt to such a situation.

Dynamic tree index is introduced into the realm of indexes since it is convenient for dynamic construction. The average number of searches in tree index is related to the depth of the tree. If the number of nodes of the tree is N, then when the heights of the various branches of the tree differ by most 1, the height would be the smallest. Such a tree is called a balanced tree. In this occasion, the height of the tree is at most $\log_2 N$, and the time complexity of various operations on the tree is

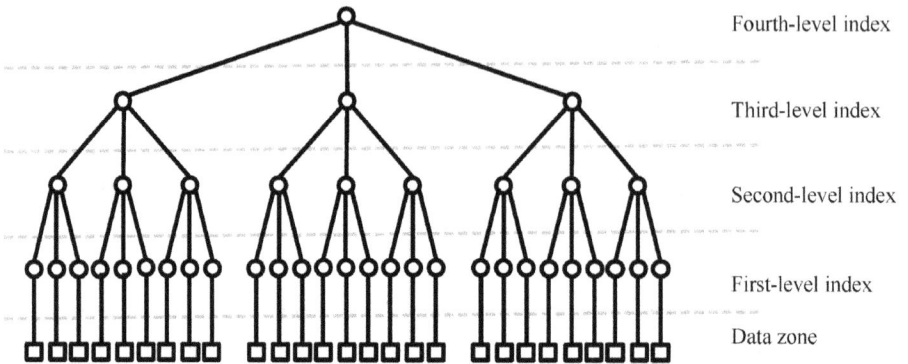

Fig. 4.11: Multilevel index structure forms *m*-way search tree.

$O(\log_2 N)$. The height of the tree is influenced by the form of the tree. Therefore, whether the form of the tree is balanced becomes a key feature for efficient search.

In the following, let us introduce indexes based on binary trees and nonbinary trees.

4.3.1 Binary search tree

The binary tree introduced in Chapter 1 can be used for search and sorting. It is called binary search tree. When it randomly encounters a piece of data, it can perform insertion on it at a very fast speed, and keep the ordered status of this binary tree under traversal. This characteristic makes it very suitable for dynamic indexes. Figure 4.12 gives an example of binary search tree.

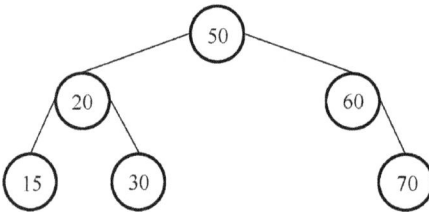

Fig. 4.12: A binary search tree.

Term Explanation Binary search tree
Binary search tree is either an empty tree or a binary tree with the following features:
1. If the left subtree is nonempty, then the values of all the nodes in the left subtree are smaller than the value of the root node.

2. If the right subtree is nonempty, then the values of all the nodes in the right subtree are smaller than the value of the root node.
3. The left subtree and the right subtree are both binary search trees, respectively.

Definition in C language

```
typedef struct BSTreeNode
{
    DataType data;
    KeyType key;
    struct BSTreeNode *lChild, *rChid;
} BSTree;
BSTree *bsTree;
```

We can build binary search trees with random construction, that is, treat the first data item encountered as the root node and construct a new binary search tree. The rest of the data will be merged into the tree by insertion into the appropriate position in the tree. Eventually, we form a binary search tree, which includes all the nonrepeating keyword values. Note that a binary search tree does not allow the existence of nodes with the same keyword value.

The most important operation in the construction of binary search tree is insertion. Insertion is based on search – we would first perform search, if we find the value, insertion will not be performed. If we cannot find it, we perform insertion at the eventual position reached. Steps for lookup.

- If the keyword value of the root node of the binary tree is equal to the keyword to be searched, the search succeeds.
- Otherwise, if it is smaller than the keyword value of the root node, recursively search the left subtree.
- If it is larger than the keyword value of the root node, recursively search the right subtree.
- If the subtree is empty, the search fails.

Example 4.2 Construct a binary search tree according to the given sequence of numbers: 50, 20, 15, 60, 30, 70.
Let us first construct a node with 50 as the keyword, and use it as the root node of the binary search tree. We obtain the single-node binary tree as shown in Fig. 4.13(a). Then, according to the steps of searching illustrated earlier, we recursively search the left subtree of the root node. We cannot find the corresponding node. Therefore, this position is where 20 should be inserted. After performing insertion there, we obtain the binary tree shown in Fig. 4.13(b). Continue to repeat the above steps, we obtain the binary tree shown in Fig. 4.13(c). We can see that it is completely identical to the one shown in Fig. 4.12.

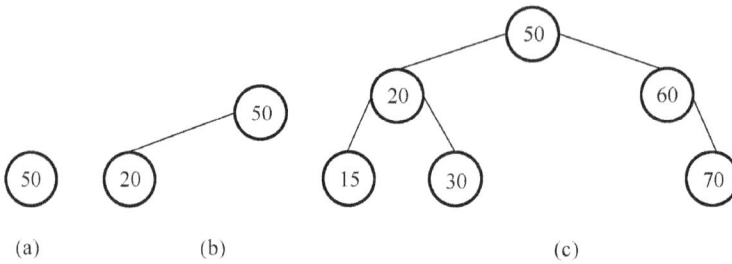

Fig. 4.13: Construction of binary search tree.

> **Think and Discuss** What kind of tree is constructed when we construct a binary search tree?
> **Discussion:** If we swap the order of data in this sequence, maybe we will obtain another binary
> search tree that contains the same key values but has a completely different form. For example,
> we can change the order to be 70, 50, 60, 30, 20, 15. Try to see what is the form of the tree
> constructed according to the above idea?
>   Observe the binary search tree obtained each time, and perform in-order depth-first traversal,
> what order will we obtain? It is a sequence in ascending order. This is also a typical feature of
> binary search tree. It is ordered under in-order depth-first traversal. This is the origin of the name
> binary sort tree. It is also the reason why we specifically discuss it in the index section.

The following is the function implementation of the insertion basic operation on binary search tree. There we can see how binary search tree preserves the feature of
being ordered under in-order traversal.

```
/*================================================================
Functionality: Insert the search keyword key in the binary search tree
Function input: Root pointer of the binary tree, keyword to insert
Function output: Whether the insertion succeeded
================================================================*/
bool InsertBST(BsTree *t, KeyType key)
{
    if(t == NULL)
    {
        if(!(t = (BsTree*)malloc(sizeof(BiTree))))
        {
            printf("Error in memory allocation!\n");
            return FALSE;
        }
        t->lChild=t->rChild=NULL;
        t->data=key;
        return TRUE;
    }
```

```
    if(key < t->data)
        return InsertBST(t->lchild,key);
    else if(key > t->data)
        return InsertBST(t->rchild,key);
    else
    {
        printf("This key is already exist in this tree!\n");
        return FALSE;
    }
}
/*=================================================================
Functionality: Construct a binary search tree from an array
Function input: Root pointer to the binary search tree, array storing the
keywords, number of keywords
Function output: Whether the creation of the binary search tree succeeded
=================================================================*/
bool CreateBsTree(BsTree *tree, KeyType d[], int n)
{
    int I;
    if(tree)
    {
        printf("The tree is not empty!");
        return FALSE;
    }
    for(i=0;i<n;i++)
    {
        if(!(InsertBST(tree,d[i])))
            return FALSE;
    }
    return TRUE;
}
```

Because we need to preserve the feature of being ordered under in-order depth-first traversal of the binary tree, the deletion of a node in binary search tree is much more complicated than the insertion operation. Interested readers can refer to the related materials themselves.

Under the condition of random construction, the ideal height of binary search tree is $O(\log_2 n)$. If the binary search tree is constructed based on an originally ordered sequence, as the example shown in the *Think and Discuss* after Example 4.1, the shape of binary tree would be a linear-like structure, where each level has only a left subtree or a right subtree (total height $O(n)$). In this scenario, the lookup operation is equivalent to lookup in a linked list, and has to start from the head node to perform the search sequentially, with a time complexity $O(n)$, which is not very efficient.

Then, is there any way to ensure that no matter we construct the binary search tree from a collection of whatsoever characteristics, it would not be such a "skewed" binary tree? The corresponding strategy would be to construct a balanced binary tree.

> **i** **Term Explanation** Balanced binary tree
> Balanced binary tree is also called AVL tree. It has the following characteristics: it is an empty tree, or the height difference between the left and right subtrees is not greater than 1. Also, its left and right subtrees are both balanced binary trees.

The main idea of constructing balanced binary tree is that if the insertion or deletion of a node would make the height difference greater than 1, we need to perform rotation between the nodes in order to again maintain the binary tree at a balanced state. The usual algorithms to construct and adjust balanced binary trees are red-black tree, extension tree, AVL, SBT, Treap and so on.

Performing direct insertion or deletion operation on any of the balanced binary trees mentioned earlier might cause the result to violate the definition of balanced binary tree. Therefore, we must modify the pointer structure of the tree accordingly. The modification is achieved via rotation.

Rotation is a local operation on the search tree that can preserve the properties of binary search trees. Figure 4.14 demonstrates two basic rotation operations: left rotation and right rotation. Via rotation, the height difference between left and right subtrees can be changed while keeping the precondition that its in-order traversal is ordered, thus achieving the aim of keeping the tree nearly balanced. The concrete operations to balance a binary tree will not be illustrated in this book.

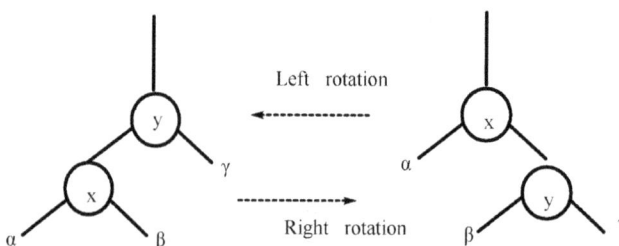

Fig. 4.14: Rotation operations on the binary search tree.

> **i** **Term Explanation** Red-black tree
> The binary search tree which satisfies the following properties is a red-black tree.
> Property 1: Each node is either red or black.
> Property 2: The root node is black.
> Property 3: Each leaf node (NIL node, i.e., empty node) is black.

Property 4: The two child nodes of each red node are black (i.e., there cannot be two consecu-
tive red nodes in any path from each leaf to the root).
Property 5: Any path from a node to its leaves must contain the same amount of black nodes.

Figure 4.15 is an example of red-black tree, where the black nodes are painted black, and the
black nodes are represented with shallow shadows. All the leaf nodes are omitted in the figure
(they are all black NIL nodes). Each path from the root node to a leaf contains four black nodes
(it also contains the empty node at the lowest level, which is black).

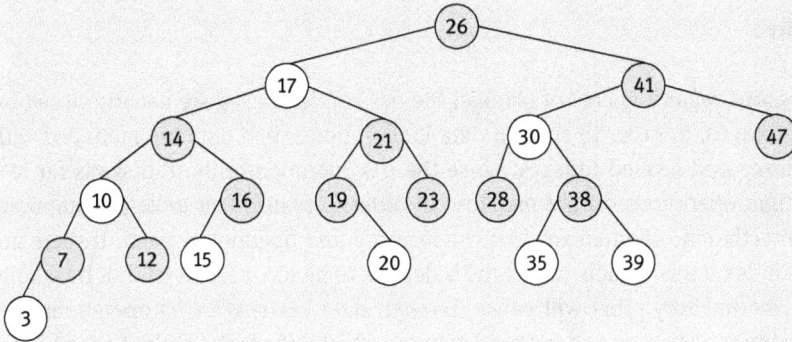

Fig. 4.15: Red-black tree.

The node definition of a red-black tree is as follows, where color can only take value red or
black:

```
typedef struct Node
{
    char color;        // Red/Black property
    KeyType key;       // Keyword value
    struct Node *left;
    struct Node *right;
    ...                //other data
} RBNode;
RBNode *rbTree;
```

To directly perform insertion or deletion operations on red-black trees might cause
the result to violate the definition of a red-black tree. Therefore, we must correspondingly change the colors of some nodes and pointer structures. The modification of pointer structures is achieved via rotation.

Knowledge ABC Why is a strict balanced binary tree not as widely used as a red-black tree?
A strict balanced binary tree needs more rotation operations (i.e., the left rotation and right rotation shown in Fig. 4.14) if it is to be kept balanced. Although strictly balanced binary trees are
more balanced than red-black trees (i.e., it has a lower height and is faster during lookups), the

time cost from too many rotation operations is usually greater than the time saving brought by faster lookup operations. To say it the other way round, red-black tree trades nonstrict balance for the reduction of rotation operations during insertion/deletion of nodes.

If the number of lookups in the application is far greater than those of insertion and deletion, then we should choose strict balanced binary tree. If the numbers of lookup, insertion and deletion are basically the same, especially if insertion/deletion operations are more, we should choose red-black tree. This situation is more common in practical applications.

4.3.2 B-tree

Due to reasons regarding cost of storage, the data of databases are usually stored on external storages. In order to shorten data lookup time, such data are managed with block indexes and leveled indexes. Since the I/O operation on hard disks is far less efficient than operations on the memory, in order to realize fast lookup, computers usually directly load the indexes into the memory and operate on them. If there are too many index items, which causes the index file to be too large, it cannot be wholly stored in the memory. This will cause frequent data exchange (I/O operations) between the internal storage and external storage, which affects the lookup speed.

The sorting algorithms introduced before mostly apply to in-memory sorting (also called internal sorting). Sorting on external storage (also called external sorting) needs to be differently designed due to restrictions from the hard disk I/O time.

Since the depth of the index tree directly impacts the number of I/O operations on the external hard disk (the total number of accesses is equal to the number of accesses to each bottom-level index plus one time of data access), it is better for the index tree to be as short as possible. Compared with a binary tree, where each node represents at most one piece of data, if each node can contain multiple keyword information and multiple branches according to the practical needs, the depth of the tree can be decreased. This would make the lookup of one element only load few nodes from the external hard disk into the memory, so that the data being looked up is quickly retrieved. The tree-like structure that realizes such a structural design is B-tree structure and the related variant structures: B+-tree structure and B*-tree structure. One claim asserts that B means "balanced," although there are multiple claims about the origin of this name.

B-tree is a leveled dynamic block index. Except for the leaf nodes, all the upper-level indexes are sparse. A node can have many children. It is a search tree similar to the balanced binary tree, which can agilely adjust the form of the tree according to the changes in the data to ensure that data are read with the minimum amount of accesses to the external storage. The main difference between B-tree and the example of multilevel indexes in Fig. 4.8 – hard disk file organization method ISAM – is that ISAM is a static multilevel index, while B-tree is a dynamic multilevel index, which is suitable for situations where the contents of the data change frequently.

In the following, we will start the introduction from the most simple B-tree – 2-3 tree. An actual example of the tree is shown in Fig. 4.16. The mean of "B-tree, 2-3" is that it is a balanced three-way search tree, with each node having at most two keywords and three branches.

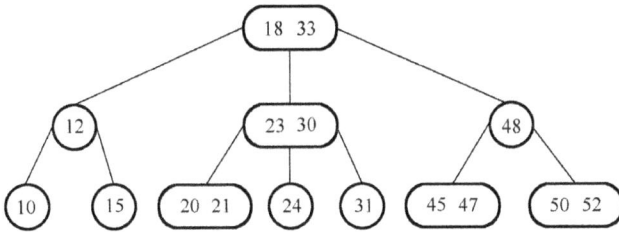

Fig. 4.16: Actual example of a 2-3 tree with height 3.

Term Explanation 2-3 Tree

A tree with the following properties is called a 2-3 tree.
1. A node contains one or two keywords.
2. Each internal node has two children or three children.
3. All the leaf nodes are at the same level of the tree. Thus, the tree is always balanced in height.
4. The keyword values of all the successor nodes in the left subtree of each node of the 2-3 tree are smaller than the value of the first keyword of the parent node.
5. The keyword values of all the successor nodes of the middle subtree are larger than or equal to the value of the first keyword of its parent node and smaller than the value of the second keyword.
6. If there is a right subtree, then the keyword values of all the successor nodes of the right subtree are greater than or equal to the value of the second keyword value of the parent node.

The C-language definition of 2-3 tree node can be expressed as follows:

```
typedef struct Node
{
    KeyType lkey;
    KeyType rkey;
    int Numkeys;
    struct Node *left,*center,*right;
} Tree23Node;
Tree23Node *type23Tree;
```

A node that contains two children in a 2-3 tree is called a two-node. A node that contains three children is called a three-node. The shape of a 2-3 tree is very similar to that of a full binary tree. If a 2-3 tree does not contain any three-nodes, it would look like a full binary tree – all of its internal nodes can have two children, and all leaves are at the same level. Therefore, a 2-3 tree with height h contains at least as many nodes as a full binary tree of the same height, which is 2^h-1 nodes. The height of a 2-3 tree with n nodes is not greater than the height of a binary tree containing the same amount of nodes $\lceil \log_2(n+1) \rceil$.

To maintain the balanced features, the insertion and deletion of nodes in a 2-3 tree usually also need special treatments, which are similar to the rotation operation used in a binary balanced tree. The insertion operation in a 2-3 tree requires special operations such as moving up, splintering, while during deletion the contrary merging and moving down operations are needed, as shown in Fig. 4.17.

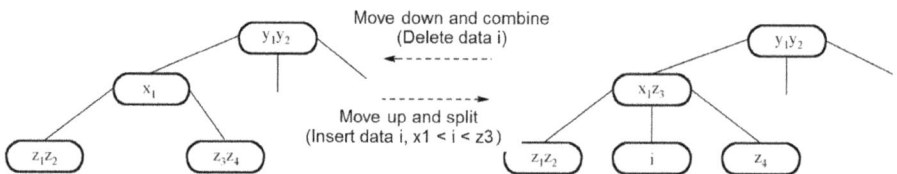

Fig. 4.17: The special treatments of insertion and deletion in B-tree.

More generally, a B-tree with order M is defined as follows:

> **Term Explanation** B-tree
> A B-tree of order m is a balanced M-way search tree. It is either an empty tree or a tree satisfying the following properties:
> 1. Each node of the tree contains at most M children ($M \geq 2$).
> 2. Each nonroot nonleaf node stores $\lceil \frac{M}{2} \rceil - 1$ to $M-1$ keywords, and these keywords are stored ascendingly.
> 3. Each nonleaf node has pointers pointing to its children, which are one more than the number of keywords it contains, that is, there are $\lceil \frac{M}{2} \rceil$ to M child nodes. Leaf nodes do not have any child pointers.
> 4. The keywords stored in nonleaf nodes delineate the range of the keywords stored in its various subtrees.
> 5. Every leaf node has the same depth, that is, the height of the tree.
> 6. All keywords appear only once in the whole tree.

The previously introduced 2-3 tree is a B-tree of order 3 according to this definition. The selection of M in practical problems should be related to the physical characteristics of the actual external storage.

Note that although the multilevel index example in Fig. 4.8 looks like a B-tree, its data are not stored in the middle nodes. Therefore, it is also not a B-tree. Also, the characteristics of the static index of multilevel indexes are fundamentally different from those of B-trees.

Since the number of I/O operations on the external storage depends on the height of B-tree, it is important to study the height characteristic of B-tree under the worst-case scenario. Given an order-M B-tree that contains N keywords and has height h^1, then:

$$h \le \log_{(\lceil \frac{M}{2} \rceil)} \frac{N+1}{2} + 1 \tag{4.1}$$

The proof of expression (4.1) is omitted. This expression for height illustrates that the lookup efficiency on B-tree is very high.

There are many variations of the B-tree, such as B+-tree, B*-tree, which are all widely applied in computer science.

4.3.2.1 B+-tree
On the basis of B-tree, it adds linked list pointers for the leaf nodes, and all keywords only appear in the leaf nodes, while the nonleaf nodes are only used as indexes for the leaf nodes. B+-tree only matches when it comes to the leaf nodes. The nonleaf nodes can then be viewed as the index part set up specifically for the leaf nodes. Figure 4.18 gives an example of B+-tree.

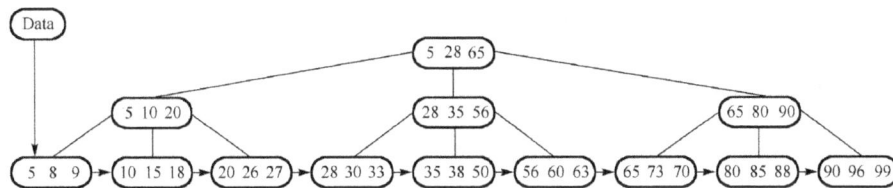

Fig. 4.18: An example of B+-tree.

The advantage of B+-tree is that a single scanning//lookup between data can be performed directly according to the linked list pointed to by the linked list pointers. This is more convenient than B-tree.

4.3.2.2 B*-tree
On the basis of B+-tree, linked list pointers are also added to nonleaf nodes. At the same time, it raises the minimum occupancy rate of nodes from 1/2 to 2/3, that is, each intermediate node (nonroot nonleaf node) has at least $\lceil \frac{2M}{3} \rceil$ nodes, compared

---

1 Note: Here the height of the B-tree does not count the level of the leaf nodes, since the leaf nodes in a B-tree do not store any data.

with the $\left\lceil \frac{M}{2} \right\rceil$ nodes before. Figure 4.19 shows the result of reforming the B+-tree in Fig. 4.18 into the shape of a B*-tree. Since the linked list pointers are added to the intermediate nodes at each level, it is relatively convenient to evenly distribute data among the levels of intermediate nodes, and reduce the construction operation of new nodes during the "moving up, splintering" process. In this way, the possibility of B*-trees allocating new nodes is lower than that of the B+-tree, which improves space usage efficiency.

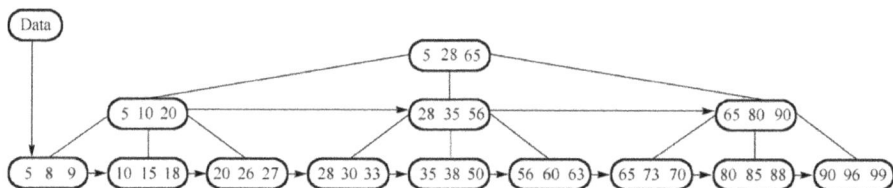

Fig. 4.19: An example of B*-tree.

4.4 Overview of searching

In the modern society of information explosion, to obtain the piece of information needed from the overwhelming amount of information, we must rely on searching technologies.

As mentioned earlier, the most important basic operation on index list, besides insertion and deletion, is searching – look up the position of the corresponding data element stored in the data list. Actually, index is born for searching. What are the performances of performing search on the various kinds of indexes mentioned earlier? Let us first review the basic concepts of searching. Note that the searching here is not limited to searching on indexes, but also includes searching on raw data.

4.4.1 Basic concept of searching

i **Term Explanation**
Lookup: the process of finding, among the data elements stored in the data list, the data element identical to the given keyword via certain methods is called lookup.

Keyword: keyword is a value of a certain item or composite item in the record. It can be used to mark the record. The record that can uniquely determine a record is called the main keyword. The keyword that cannot uniquely determine a record is called a secondary keyword.

Just like many questions can be solved with brute-force enumeration, no matter in what kind of index structure, or directly in the data list, is there any generally applicable lookup method? Actually, all the lookups can be performed with sequential lookup (if it is a tree shape, the lookup can be performed in the order of tree traversal). Sequential lookup is also the most time-costly method. In practice, especially among a huge amount of data, it is usually infeasible to perform sequential lookup. It can be said that lookup efficiency is an important indicator of the performance of the lookup algorithm.

4.4.2 Performance of the lookup algorithm

The time cost of the lookup algorithm is normally measured in terms of the number of comparisons performed on the keyword values. Then, what factors are related to the number of comparisons? Usually, the following aspects influence the number of comparisons during the lookup process:
1. Algorithm
2. Size of data
3. The position of the keyword to be looked up in the data list
4. Lookup frequency

among which algorithm is decisive. To evaluate an algorithm, we usually use the lookup length under the average-case scenario (or including the best-case scenario and the worst-case scenario) as the reference. For example, the average lookup length for a successful lookup is shown in eq. (4.2). Sometimes we also need to evaluate the average lookup length of lookup failures (there is no keyword to be looked up in the list).

$$\text{ASL} = \sum_{i=1}^{n} p_i c_i \tag{4.2}$$

where n is the size of the problem, p_i is the possibility of finding the ith record and c_i is the number of comparisons on the keyword needed of finding the ith record. In this case, the position of the keyword to be looked up is included in the concept of average value as a random variable. The lookup frequency is related to the actual application and not directly related to the algorithm itself. Therefore, the evaluation result of algorithm efficiency is normally a function related to the size of the data.

In some special cases, the time cost of the lookup algorithm is not measured in terms of number of comparisons of the keyword value. We have also already mentioned in Section 4.3.2 in some situations where the memory and the external storage need to exchange data, since the response time of the external storage already takes up the majority of the lookup time, the lookup performance will be determined completely by the number of data read operations from the external storage.

Just like other algorithms, the performance of the lookup algorithm will also differ based on the differences in the storage method and logical organization method of the data structure. In the following, we will analyze and discuss the performance of various lookup algorithms according to the different classifications of linear list and tree list.

4.5 The lookup technology of linear list

The linear list can be divided into sequential list and linked list according to the differences in storage methods.

In linked list and unordered sequential list, there is only sequential lookup, that is, one-by-one comparison from one end of the linear list to the other end.

4.5.1 Sequential lookup

> **Term Explanation** Sequential lookup
> Compare the keyword with the given lookup value one by one from the one end to the other end of the linear list. If the values are identical, then the lookup succeeds, and the position of this record in the data list is given. If after searching the whole list it still cannot find the keyword with the identical value as the given value, then the lookup fails and failure information is given.

The following is an example in C language about sequential lookup in sequential list:

```
/*=============================================================
Functionality: Sequential lookup
Function input: Pointer to the sequential list, length of the sequential list,
the value of the keyword to be looked up
Function output: The position of the found keyword, or return -1 if the keyword
is not found.
=============================================================*/
int SeqSearch (Node r[], int n, int iKey)
{
    int i=0;
    while (i<n && r[i].key != iKey) i++;
    if (i<n) return i;
    else return -1;
}
```

Suppose the possibility of finding the keyword is equal at each position, then, performing complexity analysis on this algorithm, the average length needed for a successful lookup is

$$\text{ASL} = \sum_{i=1}^{n} p_i c_i = \sum_{i=1}^{n} p_i (n - i + 1) = \frac{n+1}{2} = O(n) \tag{4.3}$$

where p_i is the possibility of looking up the data at position i, and c_i is the number of comparisons needed to find this data. The lookup length in the case of a failed lookup is always the length of the whole list, N; thus, this algorithm is also $O(N)$.

There is still room for improvement for the C code above. Since each time the while loop needs to check whether i exceeds the length of the list, which takes a fair amount of time, we can combine the two conditions into one by adding in a sentinel.

```
/*===============================================================
Functionality: Sequential lookup
Function input: Pointer to the sequential list, length of the sequential list,
the value of the keyword to be looked up
Function output: The position of the found keyword, or -1 if the keyword is not
found.
==============================================================*/
int SeqSearch (Node r[], int n, int iKey)
{
    int i=0;
    r[n].key=iKey;
    // Note: This can only be written this way if r allocated space for n+1
    // elements in the beginning.
    while (r[i].key != iKey) i++;
    if (i<n) return i;
    else return -1;
}
```

With this plan, although the calculation speed of the whole program is increased, the number of comparisons of the algorithm does not change. The time complexity is still $O(N)$.

The above analysis is about the lookup length for successful lookup. If the lookup fails, it would definitely be the case that the number of comparisons is equal to the length of the list, that is, the time complexity would be $O(N)$.

Through the above analysis, we can see that the complexity of sequential search is really high. Especially when the data list is long, this shortcoming will be fatal. However, since its calculation is simple, and does not require the linear list to be sorted, there is also no requirement on its storage method. It applies to a relatively wide range of scenarios.

4.5.2 Lookup on an ordered list

4.5.2.1 Binary search

We all have played the number-guessing game. In the game, you are asked to guess a number. If you are told that your guess is too low, you will increase the number to continue guessing. If you are told "too high," what will you do? The quicker way to reach the answer would be to lower the number by half to continue the guess. This is the general idea of binary search.

We can only perform binary search on ordered sequential lists.

Example 4.3 An ordered sequence of array elements is: 5, 10, 19, 21, 31, 37, 42, 48, 50, 55. Use binary search to search for elements whose k values are 19 and 66, respectively.

Binary search: we normally compare the key value with the element at the position mid = (low + high)/2. The result of the comparison can be divided into the following three situations:

1. Equal: the element at the location mid is what we want.
2. Greater than: we set low = mid + 1;.
3. Smaller than: we set low = mid + 1;.

The solution process is shown in Figure 4.20.

The following is the code in C language to perform binary search in a sequential list:

```
/*================================================================
Functionality: Binary search
Function input: Pointer to sequential list to be ordered; length of the
sequential list; the value of the keyword to be looked up
Function output: The position of the keyword to be looked up, or -1 when the
keyword is not found
================================================================*/
int BiSearch (Node r[], int n, int iKey)
{
   int low=0, high=n-1;
   while (low<=high)
   {
      mid = (low+high+1)/2;
      if (r[mid].key < iKey)
        low = mid+1;
      else if(r[mid].key > iKey)
          high = mid-1;
        else
          return mid;
   }
   return -1;
}
```

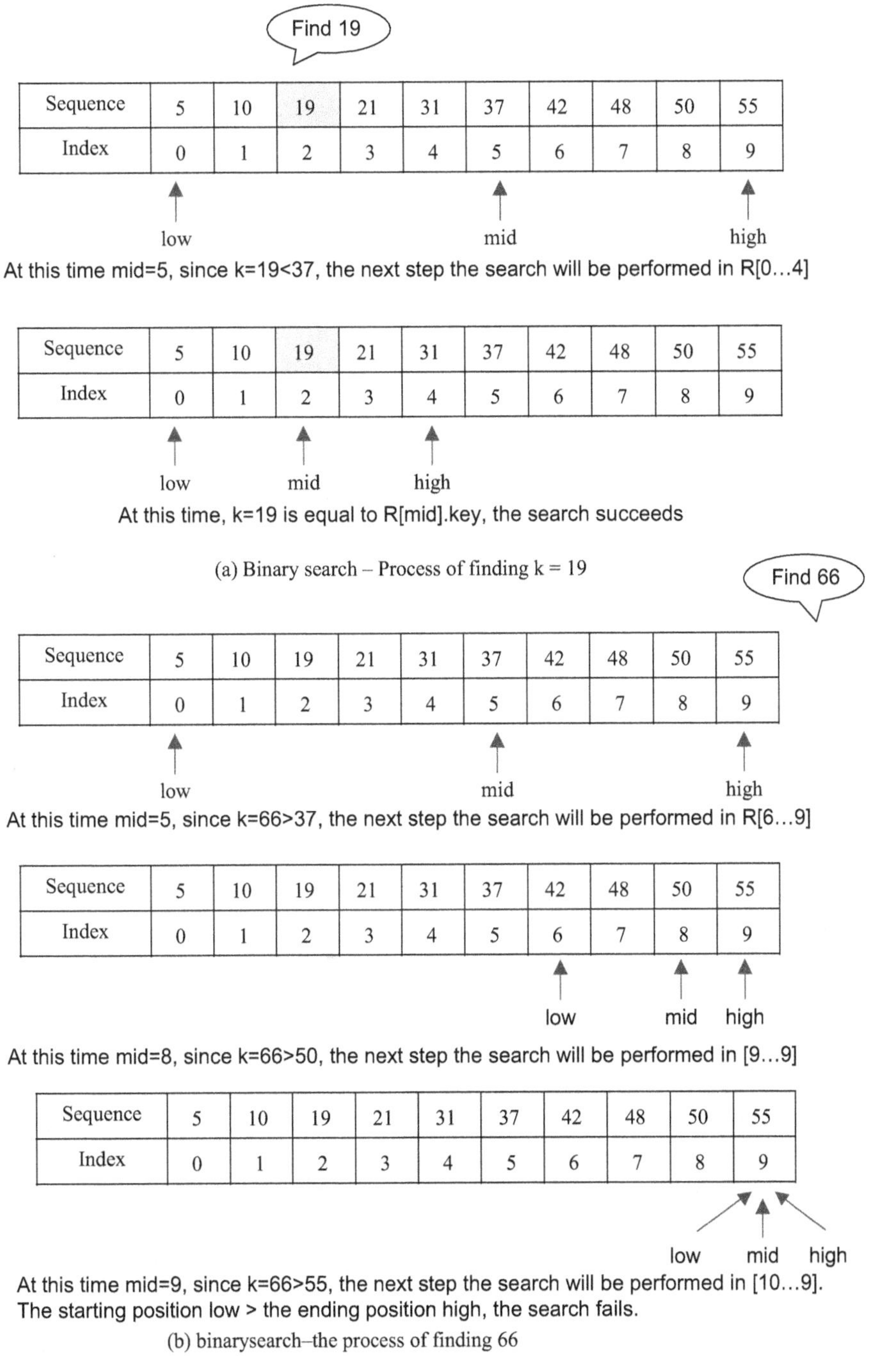

Find 19

Sequence	5	10	19	21	31	37	42	48	50	55
Index	0	1	2	3	4	5	6	7	8	9

low   mid   high

At this time mid=5, since k=19<37, the next step the search will be performed in R[0...4]

Sequence	5	10	19	21	31	37	42	48	50	55
Index	0	1	2	3	4	5	6	7	8	9

low   mid   high

At this time, k=19 is equal to R[mid].key, the search succeeds

(a) Binary search – Process of finding k = 19

Find 66

Sequence	5	10	19	21	31	37	42	48	50	55
Index	0	1	2	3	4	5	6	7	8	9

low   mid   high

At this time mid=5, since k=66>37, the next step the search will be performed in R[6...9]

Sequence	5	10	19	21	31	37	42	48	50	55
Index	0	1	2	3	4	5	6	7	8	9

low   mid   high

At this time mid=8, since k=66>50, the next step the search will be performed in [9...9]

Sequence	5	10	19	21	31	37	42	48	50	55
Index	0	1	2	3	4	5	6	7	8	9

low   mid   high

At this time mid=9, since k=66>55, the next step the search will be performed in [10...9].
The starting position low > the ending position high, the search fails.

(b) binarysearch–the process of finding 66

Fig. 4.20: Example of binary search.

This piece of code is a nonrecursive implementation of the algorithm. Actually, the idea of this algorithm is typically suitable for recursive algorithms. The reader can try to rewrite it into recursive form.

The time complexity analysis of this algorithm is as follows.

Suppose the time needed to perform binary search on n elements is $t(n)$, it is easy to know that:

if $n = 1$, then $t(n) = c_1$: if $n > 1$, then we have:

$$t(n) \le t\left(\frac{n}{2}\right) + c_2 \le t\left(\frac{n}{4}\right) + 2 \cdot c_2$$

(4.4)

$$\cdots$$

$$\le c_1 + (\log_2 n) \cdot c_2 = O(\log_2 n)$$

where c_1, c_2 are both constants. Therefore, the time complexity is $O(\log_2 n)$.

To better understand the above algorithm, we can imagine it as a binary decision tree, as shown in Fig. 4.21.

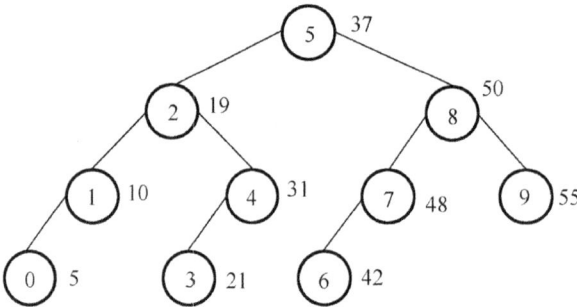

Fig. 4.21: The decision tree corresponding to binary search when $N = 10$.

4.5.2.2 Fibonacci search

The Fibonacci search is similar to the binary search. The division coefficient of binary search on the ordered list is 0.5, that is, the division point is calculated as (low + high) * 0.5, while the Fibonacci search has a division coefficient of 0.618.

For the Fibonacci sequence F[k] = {1, 1, 2, 3, 5, 8, 13, 21, 34, 55, 89}, the ratio of the two adjacent numbers approaches the golden ratio 0.618 as the array grows. For example for the 89 here, imagine it to be the number of elements in the whole ordered list, 89 = 55 + 34. Then, we divide the ordered list with 89 elements into 55 elements in the first part and 34 elements in the second part. If the element to be

looked up is in the first part, then we continue the division according to the numbers in the Fibonacci sequence: 55 = 34 + 21, with 34 elements in the first part and
21 elements in the second part. Continue this process until the search succeeds or
fails. In this way, the Fibonacci sequence is applied to the algorithm. Figure 4.22 is
an ordered list with 13 elements. We divide the ordered list according to the
Fibonacci numbers and obtain the order of the search for the positions of the various nodes. We can see that this tree is at a nearly balanced condition. Therefore,
the algorithm efficiency is close to that of the binary search.

order of lookup	5	4	3	4	2	4	3	1	4	3	2	4	3
Index	1	2	3	4	5	6	7	8	9	10	11	12	13

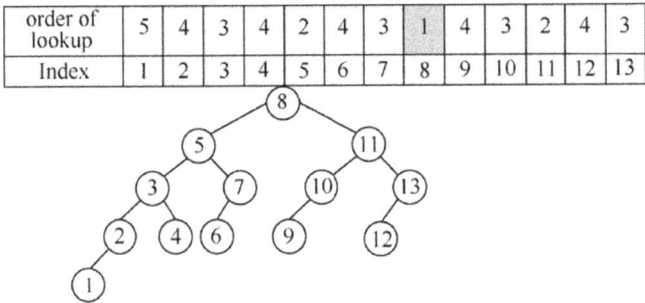

Fig. 4.22: A fifth-order Fibonacci tree.

When the number of elements in the ordered list n is not exactly a number in the
Fibonacci sequence, we need to patch up the number of elements in the ordered list
and make it a value in the Fibonacci sequence, as shown in Fig. 4.23.

Fibonacci sequence F[k]={0,1, 1, 2, 3, 5, 8, 13, 21, 34, 55, 89,....}

F[k]	0	1	1	2	3	5	8	13
k	0	1	2	3	4	5	6	7

Ordered sequence	5	10	19	21	31	37	42	48	50	55			
Index	0	1	2	3	4	5	6	7	8	9	10	11	12

F[k]

F[k-1]=8   F[k-2]=5

low   mid   high

Fig. 4.23: Dividing the sequence according to the Fibonacci numbers.

During the lookup, there are three possible results for the comparison between the key value and the mid position value. The rules for the various parameters that are to be computed are as follows:

1. Equal: The element at the midposition is what we are looking at.
2. Smaller than: high = mid – 1; k = k – 1.
3. Larger than: low = mid + 1; k = k – 2.

> **Example 4.4** An ordered sequence with 10 elements is 5, 10, 19, 21, 31, 37, 42, 48, 50, 55. Use the Fibonacci search method to lookup the elements with keys 19 and 66, respectively.
>
> **Solution:** the length of the array is 10, which is not a Fibonacci number. The smallest Fibonacci number larger than 9 is 13. Therefore, we need to patch up the length of the array to 13. Therefore, the Fibonacci division array is F[] = {0, 1, 1, 2, 3, 5, 8, 13}, with $k = 7$ at the beginning.
> In the first iteration, mid = F[k-1]-1 = 8-1 = 7. Note that mid is the subscript; thus, we need to subtract 1 from the length of the value F[k-1].

The solution process using the Fibonacci search is shown in Fig. 4.24. The related program implementation might be realized by the reader on him/herself.

4.5.3 Indexed search

The previously mentioned search methods refer to lookups targeting the data list. We can see that the binary search on an ordered list is much more efficient than sequential search. However, the massive amount of data would still be a burden during the search. In this occasion, a more efficient method would be to construct an index list and perform the search on the index list.

> **Example 4.5** Suppose the memory can only contain 64kb of data. At a certain time, the memory can at most store 64 objects for search. Now we have 14,400 objects in total, while each index item in the index list would occupy 4 bytes. Try to compare the efficiency of the various plans. The analysis is given in Table 4.2.

From this we can see the importance of indexed search. Actually this example corroborates our experience in daily life: when we get a dictionary, we usually use the index to find the word, which would be much more efficient than blindly opening the dictionary and trying to find the word directly.

To perform lookup on ordered index list data, besides the usually efficient binary search, there is another relatively quick search algorithm – interpolation search.

4.5.3.1 Interpolation search
We have previously mentioned that the edge of the dictionary usually provides indexes. When we lookup the dictionary, we normally do not strictly determine the

F[k]	0	1	1	2	3	5	8	13
k	0	1	2	3	4	5	6	7

Find 19

Makeup the length

Sequence	5	10	19	21	31	37	42	48	50	55	55	55	55
Index	0	1	2	3	4	5	6	7	8	9	10	11	12

low   mid   high

$k=7$, mid=F[k-1]-1=7, sincek=19<48, the next step the search will be performed in R[0...6].

Sequence	5	10	19	21	31	37	42	48	50	55	55	55	55
Index	0	1	2	3	4	5	6	7	8	9	10	11	12

low   mid   high

$k=k-1=6$, mid=low+F[k-1]-1=5-1=4, sincekey=19<31, the next step the search wil be performed in R[0...3]

Sequence	5	10	19	21	31	37	42	48	50	55	55	55	55
Index	0	1	2	3	4	5	6	7	8	9	10	11	12

low   mid high

$k=k-1=5$, mid=low+F[k-1]-1=3-1=2, sincekey=R[mid].key, the search succeeds

(a) Fibonacci search–The processing of finding key =19

Make up the length   Find 66

Sequence	5	10	19	21	31	37	42	48	50	55	55	55	55
Index	0	1	2	3	4	5	6	7	8	9	10	11	12

low   mid   high

$k=7$, mid=low+F[k-1]-1=7, sincekey=66>48, the next step the search will be performed in R[8...12]

Sequence	5	10	19	21	31	37	42	48	50	55	55	55	55
Index	0	1	2	3	4	5	6	7	8	9	10	11	12

low   mid   high

$k=k-2=5$, mid=low+F[k-1]-1=8+3-1=10, sincekey=66>55, the next step the search will be per formed in R[11...12]

Sequence	5	10	19	21	31	37	42	48	50	55	55	55	55
Index	0	1	2	3	4	5	6	7	8	9	10	11	12

low   mid high

$k=k-2=3$, mid=low+F[k-1]-1=11+2-1=12, sincekey=66>55, mid=high, the search goes through the data list but still hasn't found the number. Thus the search fails.

(b) Fibonacci search–the processing of finding key=66

Fig. 4.24: Concrete example of the Fibonacci search.

next page to read according to the binary search or the Fibonacci search. Instead, we go through the following thought process: if the word to be looked up is alphabetically much greater than the words on the current page, we would skip multiple pages. Otherwise, we would go through fewer pages.

In algorithm design, we can also use this idea. According to the interpolation method in mathematics, when we know that the keyword K is between K_1 and K_h, we will put the position for the next lookup at the point of proportion $\frac{K-K_1}{K_h-K_1}$.

Table 4.2: Efficiency analysis of the various storage and search methods.

Storage form		Processing method
Data list	Unordered list	It is impossible to read the data for all the objects into the memory at once. We need to read from the external storage multiple times. The number of reads on the external storage would be $O(n)$.
	Ordered list	It is impossible to read the data for all the objects into the memory at once. We need to read from the external storage multiple times. The number of reads on the external storage would be $O(\log_2 n)$.
	Index list	Number of bytes needed by 14,400 index items: 14,400*4/1,024 = 56.25 k Therefore, we can contain all the index items in the memory. First, search the index to ascertain the storage address of the data object. Then, with only one read operation on external storage, we can complete the search.

Note: The response time of reads on external storage is much longer than the response time of reads on memory. Therefore, we mainly use the number of reads on the external storage to evaluate the efficiency of the plan.

When the keywords basically increase at a steady pace, interpolation search can approach the position to be looked up more quickly than the binary search – binary search reduces the amount of work from n to $n/2$, while interpolation search reduces the amount of work from n to $\sqrt{n}$.

The following is an example in C language for interpolation search in sequential list:

```
/*=============================================================
Functionality: Interpolation search
Function input: Pointer to ordered sequential list, length of sequential
list, the value of the keyword to be looked up
Function output: Position of the keyword found, -1 if the keyword is not found
==============================================================*/
int BiSearch (Node r[], int n, int iKey)
{
    int low=0, high=n-1;
    while (low<=high)
    {
        mid = low+(high-low)*(iKey-r[low].key)/(r[high].key-r[low].key);
        if (r[mid].key < iKey)
            low = mid+1;
```

```
    else if(r[mid].key > iKey)
        high = mid-1;
    else
        return mid;
    }
    return -1;
}
```

We can see that when compared with binary search, this piece of code only changed the calculation method of mid. However, this small change can reduce the algorithm complexity from $O(\log_2 n)$ to $O(\log_2\log_2 n)$. In this case, it is impossible to draw a fixed-form search tree, since the next lookup position is determined by the concrete keyword to be looked up during each execution.

However, simulation experiments demonstrate that the complexity improvement of this algorithm is usually not that exciting, since the data distribution in reality is not necessarily so idealized. Also, the difference between $O(\log_2\log_2 n)$ and $O(\log_2 n)$ is not that great. The complex calculation when calculating the next lookup position will also cancel a part of the effect.

Nevertheless, interpolation search algorithm still has a wide range of successful applications. At the initial stage of external search, it can markedly reduce the number of accesses to external data, so that the response time is largely improved. The concept of external search can be referred to at the beginning of Section 4.3.2.

4.5.3.2 Block search

During searches with index, besides the methods mentioned earlier, there is also a situation of block search. This usually happens with block index, that is, sparse index. Since the data list corresponding to block index is at least ordered by block (although the data within a block is not necessarily ordered), we can quickly find the area containing the data using block index. However, that does not include the exact position of the data. The exact position of the data needs to be confirmed by sequential search within this area. From this we can see that the efficiency of block search is between sequential search and binary search.

4.6 Search techniques on tree lists

The tree list here refers to a tree list after ordering. The children in the tree are divided between left and right and are ordered based on size. The keyword values in the parent nodes divide the data boundaries in each child. Therefore, the lookup in the tree always starts from the root node and goes downward. The average search length is related to the number of data of tree nodes and the height of the tree.

4.6.1 Search on binary search tree

In a normal binary search tree, the amount of data n in a tree node is fixed, i.e. there is only one data element. Therefore, the search performance of binary search tree is only directly related to the tree height, that is, the number of lookups is proportional to the height of the tree. When the search succeeds, the complexity analysis of searches on different forms of binary search trees is as follows:

4.6.1.1 Worst-case scenario
The binary tree degenerates into a linear linked list. The complexity is therefore $O(n)$.

4.6.1.2 Average-case scenario
The search efficiency is proportional to the expected value of the height of the binary search tree. Therefore, the complexity is $O(\log_2 n)$.

4.6.1.3 Best-case scenario
The binary search tree under the best-case scenario is balanced. The property is similar to that of a complete binary tree, with height $\leq \log_2 n + 1$. Therefore, its complexity is $O(\log_2 n)$.

4.6.2 Search on B-tree

Search on B-tree starts from the root node. Binary search is performed on the ordered keyword sequence within a node. If the keyword is found, then the search ends. Otherwise, child node within the range to which the keyword belongs is searched. The above process is repeated until the corresponding child pointer becomes empty, or leaf node is reached.

4.6.3 Search on nonnumerical ordered list – dictionary tree

The keywords mentioned earlier are all numerical. However, searches in reality do not necessarily happen on lists composed of numbers. For example, the summary of word frequency performed by search engines on the pages is targeted at texts such as words. In this case, it would be inappropriate to copy the previous experience.

To perform word frequency summary on the given string, we can refer to the method of Huffman encoding. Figure 4.25 is a concrete example of a string search tree. The tree contains the set of strings ["Joe," "John," "Johnny," "Jane," "Jack"].

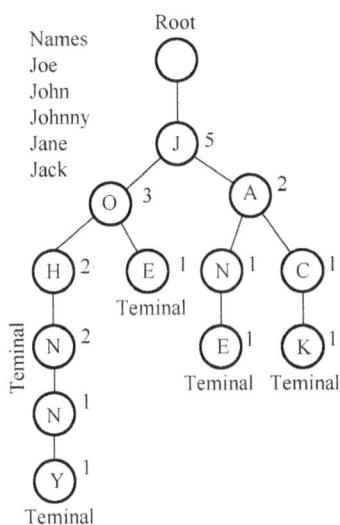

Fig. 4.25: A dictionary tree.

The process of search in this tree can be described as follows:
1. Perform one run of the search from the root node.
2. Obtain the first character of the keyword to be looked up. According to this character, select the corresponding subtree and continue the search in this subtree.
3. At the corresponding subtree, obtain the second character of the keyword to be looked up, and further select corresponding subtrees to perform searches on.
4. Repeat step 3 until all the characters of the keyword are searched (the search succeeds), or the corresponding character cannot be found (the search fails).
5. If at a certain node, all the characters of the keyword have been taken out, then read the information attached to this node, and finish the search.

The formal name for the above string search tree is a dictionary tree. It is also called word search tree. It is a type of sorted tree especially designed for searches aimed at strings. Dictionary tree has the following three basic properties:
-   The root node does not contain any character. All the nodes except for the root node contain one character.
-   If we connect all the characters on the path from the root node to a certain node, it forms the string corresponding to that node.
-   All the child nodes of a node contain different characters.

To perform information summary in dictionary tree is to add 1 to the count attached to the node when the search succeeds (which means one more instance of this word was discovered). When the search fails, a new corresponding node will be constructed and the count will be set to 1 (which represents we have discovered the current keyword once by now). The advantage of such a procedure is to utilize the common prefixes of strings to reduce search time, and reduce extraneous string comparisons as much as possible.

We can define a dictionary tree using C language data type:

```
#define MAX  26          // The size of the English alphabet
typedef struct TrieNode
{
    int nCount;              // Record the number of occurrences of this character
    struct TrieNode* next[MAX];   // Any character can follow each character
}TrieNode;
TrieNode* pRoot          // Root pointer of the dictionary tree
```

The implementation of the search operation in the dictionary tree is as follows:

```
/*================================================================
Functionality: Search on the dictionary tree
Function input: Root pointer to the dictionary tree, the value of the keyword
to be looked up
Function output: The information (word count here) corresponding to the
keyword found, or 0 when the keyword is not found.
================================================================*/
int SearchTrie(TrieNode* pRoot,char *s)
{
    TrieNode *p;
    int i,k;
    if(!(p=pRoot))
      return 0;
    i=0;
    while(s[i])
    {
      k=s[i++]-'a';
      if(p->next[k]==NULL) return 0;
      p=p->next[k];
    }
    return p->nCount;
}
```

The implementation of the function to insert newly encountered keyword into the dictionary tree is as follows:

```
/*========================================================
Functionality: Construct a new node in the dictionary tree
Function input: None
Function output: Address of the newly constructed node
====================================================*/
TrieNode* CreateTrieNode()
{
   int i;
   TrieNode *p;
   p=( TrieNode*)malloc(sizeof(TrieNode));
   p->nCount=1;
   for(i=0;i<MAX;i++)
     p->next[i]=NULL;
   return p;
}


/*============================================================
Functionality: Insert information of the newly encountered keyword into the
dictionary tree
Function input: Pointer to the root of the dictionary tree, the newly
encountered keyword
Function output: None
=========================================================*/
void InsertTrie(TrieNode* pRoot,char *s)
{
   inti,k;
   TrieNode*p;
   if(!(p=pRoot))
   {
     p=pRoot=CreateTrieNode();
   }
   i=0;
   while(s[i])
   {
     k=s[i++]-'a';        // Ascertain branch
     if(p->next[k]) p->next[k]->nCount++;
```

```
    // If found, increase count by 1
    else p->next[k]=CreateTrieNode();
    // If not found, create the new corresponding node
    p=p->next[k];
  }
  return;
}
```

4.7 Search techniques on hash table

Through the analysis of search techniques above, we can see that the search time of the search techniques mentioned up until now is all related to the data size. As the Internet expands in size, the data size increases as well, and the search becomes slower. It might be hard to imagine, as the data size on the Internet increases, a search engine using the search techniques listed above would be less and less efficient. The search time for a piece of data can degenerate from half a second to several seconds, several minutes and even several hours.

Is there any technology to help search engine to keep up the search feedback speed instead of losing itself amid the huge increase of data volume? The hard improvements such as the Moore law and cloud computation are all good, but can we still improve the search speed algorithmically?

4.7.1 Introductory example – idea derived from drawing lots and queuing

The participants of a certain competition need to go through a round of question and answer (Q&A). For fairness, the order of the Q&A is decided by drawing lots. Each participant can find her position in the whole Q&A queue just with the number information obtained from drawing lots. The reason of the rapid lookup is the direct correspondence between the number information and the position.

From the perspective of data structures, the problem of deciding the order by drawing lots is a lookup based on indices on a sequential list in which the values of data elements are the same as their indices. The sequential list is a random access storage structure. We can obtain the storage address directly via computation according to the index and the beginning address of the sequential list. The complexity of this lookup algorithm is $O(1)$. This is a highly efficient algorithm that has the lowest complexity among all the lookup algorithms that we have learnt so far.

The case where the values of the data elements are the same as the indices where they are stored at is only an exception. Can we extend this idea of directly computing addresses and apply it to more complex data structures in reality? For example, for the management of student registration, each student has information such as name, student ID and grades. The keyword "student ID" is an ordered integer. In this case, the keyword of the data element can correspond to its storage position. Then, we can still realize random access. The more general case is where neither the data element nor its keyword has direct correspondence with the storage position. In this case, if we can apply some transformation on the keyword via a function, so that the value obtained still corresponds directly with the storage position, then we can still achieve random access. Figure 4.26 describes this idea transformation process.

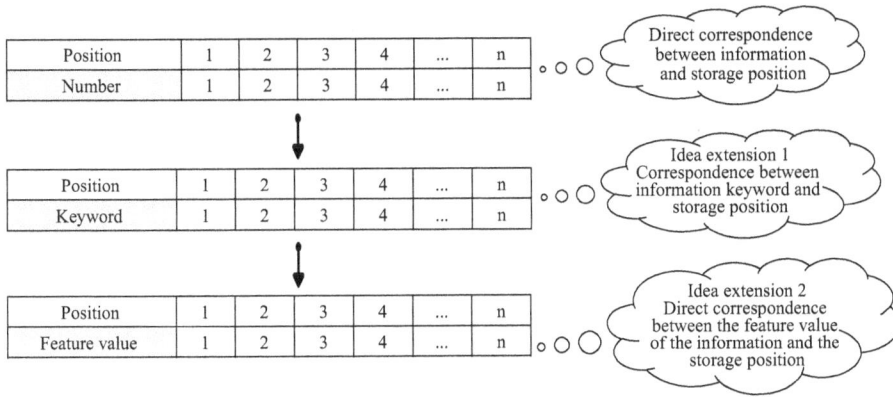

Fig. 4.26: Idea derived from drawing lots and queuing.

4.7.2 Introductory example – the search on words and pictures in the search engine

We are quite familiar with the search of related contents via inputting keywords of the traditional search engines such as Baidu and Google. However, if we want to search for the same picture with higher resolution, or want to know the origin and related contents of the picture, can a web search engine still satisfy the needs? Many years after the appearance of text search functionalities, this was still a dream for people. One reason is that pictures occupy huge amounts of data compared to texts. For example, the Chinese texts in an A4-sized page might occupy approximately 24K, while a photo might be 3M. It then corresponds to 3*1,024/ 24 = 128 pages of such an A4 document.

As the amount of data for online images increases, the needs for online image search from the users also increase. Image search engine was born because of such needs. To search for images with images is the process of search engines where they search identical or similar images depending on the image uploaded by the user.

The principle of traditional image search is that the search engine generates a "fingerprint" feature value according to the image uploaded by the user, and compares it with the existing image fingerprints in the database. The closer the result, the more similar the images. The key technology here is called "perceptual hash algorithm." The main steps of the algorithm are – minimize the image size to 8 * 8 pixels, simplify the colors, calculate the average grayscale value of each pixel, compare the grayscale of each pixel with the average value: if it is larger or equal to the average value, 1 is recorded; if it is smaller than the average value, 0 is recorded. Composing the comparison results, we obtain a 64-bit integer. This will be the fingerprint of this image.

The solution of quick search in the above introductory example is to calculate the storage address or feature value of the data element via a function mapping. This method is "hash method." The data table generated according to the data elements stored in the storage address is called a hash table. Apparently, compared with the various search technologies mentioned previously, to perform data search in the hash table, the idea situation only involves the time for calculating the hash function, that is, constant time, before we get the result. The time complexity for search is directly reduced to nearly $O(1)$, which massively improves the search efficiency.

4.7.3 Introduction to hash table

Term Explanation Hash
Hash is the technology to convert input of arbitrary length through a hash function (algorithm) into a fixed length output (hash value). It is both a storage method and a search method.

When applied to the storage of hash tables, the values of hash function are used as the addresses. Note that the addresses here refer to the relative addresses with reference to the starting storage address. The storage structure that uses relative address values to achieve quick access is sequential storage with the advantage of random access. Therefore, hash table is usually implemented based on linear storage structures.

Example 4.6 Design of hash table.
Question 1: there is a data table that includes the name, telephone and address of the users. Using the name of the user as the keyword, the set is:

S = {Anne,Ben,Davis,Eden,Frank, Greg, Isabel,Robert,Tony, Una, Will}

Design a linear storage structure that enables quick access according to the idea of hash.
Solution: the idea of hash is to find out the correspondence relationship between the feature value of the function and the storage address. By observing the keyword list, we can discover that the initial characters of each name are different. Therefore, we can use the position in the alphabet of the first character of the name as the address of the data element in the data list. In this way, we can set the storage space of the hash table as char HT[26][8]; .
  The hash function is: H(key) = key[0]–'a' .

The storage space of the constructed hash table is shown in Fig. 4.27.

Address of the hash	0	1	2	3	4	5	6	7
Keyword	Anne	Ben		Davis	Eden	Frank	Greg	
Address of the hash	8	...	17	18	19	20	22	...
Keyword	Isabel		Robert		Tony	Una	Will	

Fig. 4.27: Design of hash table 1.

Discussion: from this hash table, we can observe that the length of the hash table is larger than the number of nodes. This is a strategy to trade space for efficiency.

Question 2: if we add four keywords into the above set S to form a new set S1:

$$S1 = S + \{Alice, Elsa, Wilson, Uday\}$$

Obtain the new hash function H2(key).

Solution: from Fig. 4.28 we can see that if we keep using the hash function in question 1, then the keywords with the same initial character will share the same address. For example, there will be a conflict between "Anne" and "Alice." In this scenario, if we keep using the hash table from the above example S1, then we need to modify the hash function. We can define the new hash function in this way:

H2(key) = The average value of position in the alphabet of the initial and last characters of key.

Term Explanation Conflict

Whenever the situation occurs where different keywords are hashed to the same value, a conflict occurs. The different keywords that caused the conflict are called synonyms.

Address of the hash	0	1	2	3	4	5	6	7
Keyword	Anne	Ben		Davis	Eden	Frank	Greg	
	Alice				Elsa			
Address of the hash	8	...	17	18	19	20	22	...
Keyword	Isabel		Robert		Tony	Una	Will	
						Uday	Wilson	

Fig. 4.28: Design of hash table 2.

> **Think and Discuss** Can we ensure that no future conflicts occur with this new hash function H2?
> **Discussion:** Two different keywords might arrive at the same hash address using this hash function. In practical applications, hash functions that generate no conflicts rarely exist, that is, the occurrence of conflict is a high-probability event.

> **Knowledge ABC** The occurrence of conflict is a high-probability event in hashing
> Although we need to ensure that the size of the storage space is not smaller than the number of elements to be stored, we still have a compressed mapping relationship between the value space of the keywords and the address space. Although the conflicts caused by the random function based on such a property of compression are not many, they are actually highly probable to happen in terms of their existence.
> We can illustrate this topic with the birthday paradox. According to the pigeon's nest principle or drawer principle, there will be at least two people who have the same birthday among 367 people. That is to say, the possibility of some birthdays falling on the same day is 1. However, if we want to achieve a 97% probability of conflict, we actually only need 52 people. If we want to achieve a 50% probability of conflict, then we will only need 23 people. We can see that the possibility of conflict increases rapidly when the set of samples is still very small. When the number of samples in the set reaches half of the total address space, the existence of conflict is already a very high-possibility event.

Through discussions of practical examples, we have the below conclusions and solutions with regard to problems of hash table.

4.7.3.1 The configuration of the storage space of hash table

The value space of hash values is normally much smaller than the value space of the function input. The range of the hash values is smaller than the storage space, but the input value space of the samples is larger than the storage space, as shown in Fig. 4.29. To accommodate the data, the size of the storage space must be not smaller than the number of data elements to be saved. To reserve space for the design of the hash function, we normally set the size of the storage space as a value larger than the number of data elements to be stored.

The storage efficiency of hash table can be represented by the ratio between the space taken by the hash table, m, and the number of nodes filled into the table, n ($n < m$). We say $\alpha = n/m$ is the filling factor of the hash table. Its value is usually in the range $[0.65, 0.9]$.

4.7.3.2 Methods for resolving conflicts

Both the storage and searching using hash technique rely on using the hash function to compute the corresponding storage address. With a high probability, the hash function will cause different keywords to be mapped onto the same address, thus causing a conflict.

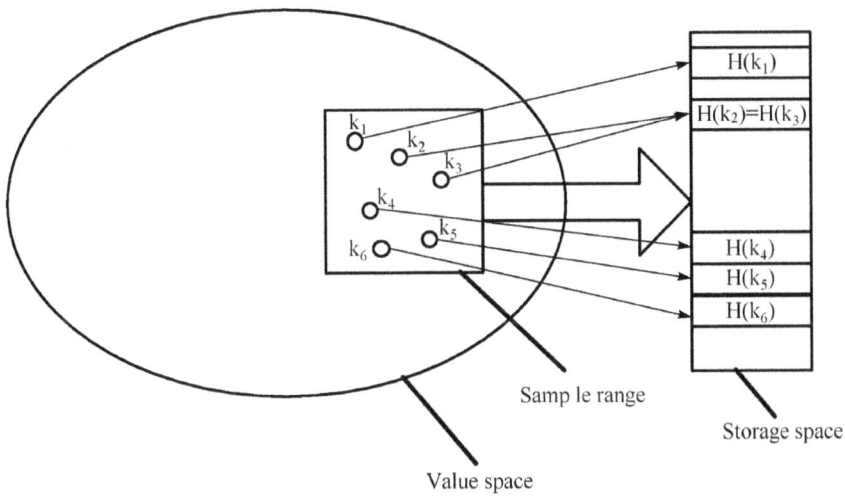

Fig. 4.29: The mapping relation of hash table space.

The synonyms conflict a need to find different addresses to store; when we encounter synonyms during a search, we need to be able to find the keyword really needed.

To reduce the bad impact of conflicts on search speed, the potential methods to consider are:
- reduce the possibility of conflicts via adequate design of the hash function;
- design a method to completely tell the storage positions of the synonyms apart when a conflict happens.

We can summarize the main design principles of hash table technique as follows: contain the data and resolve the conflicts.

4.7.4 The design of hash function

The design goal of hash function is to reduce the possibility of conflict as much as possible, and at the same time satisfy the need of hash values within the legal address space. Therefore, we can deduce some of the basic design principles for hash functions:
- the hash function values are within the allocated length space of the hash table;
- the output results of the hash function should be distributed throughout the whole address value space as evenly as possible;
- the computation of the hash function should be as simple as possible.

The hash functions written according to the above principles will normally also need to have the following four properties.

- Long to short: the hash algorithm can hash a piece of data of arbitrary length into a piece of fixed-length data.
- Fast: the hash algorithm basically never performs complex operations. Its speed is very fast.
- Irreversible: it would be difficult to find the prehashing string from the hashing result.
- Low collision: the situation where the hashing outputs are similar for different prehashing inputs exists. However, in most scenarios, different inputs lead to different outputs.

There are certain functions that satisfy the above conditions and properties. However, for a hash function to be really secure the mathematics involved will be complicated. We introduce some of the simpler and common ones below.

4.7.4.1 Square-and-take-middle-digits method

In some scenarios, the distribution of the various digits of the keyword is not very even. In this case, it would not be easy to get the appropriate address value with numerical analysis method. For example, the keyword in binary number, or the keyword represented with a text of similar characters. If we still want to use a numerical analysis method, we will need to break apart the number distribution at various digits. Squaring is an appropriate operation.

Example 4.7 We know a set of keywords (AO, AB, A1, AD, DA). Try to use square-and-take-middle-digits method to select a hash function for the storage spaces with 100 addresses and 1,000 addresses, respectively.

The operation on text usually needs to be first converted into numbers that represent such text in the computer. For example, we can use ASCII values to perform the conversion. We obtain the set of keyword values (6579, 6566, 6549, 6568, 6865). Then it is obvious that the distribution on all the digits is not very even. However, if we first perform the squaring, which produces (43283241, 43112356, 42889401, 43138624, 47128225), the distribution will be more apart. In this case, performing digit analysis and selecting relatively even digits to perform the combination would be easier.

For example, we can select the middle two digits (the fourth and fifth digits) to use as the value under the situation with 100 addresses. The address values will be (83, 12, 89, 38, 28). If we select the middle three digits (the 4th, 5th and 6th) as the value under the situation with 1,000 addresses, then we get address values of (832, 123, 894, 386, 282).

4.7.4.2 Remainder method

Remainder method is also called division hash method. This method maps k onto one of the m addresses by taking the remainder of k divided by a certain factor m. The hash function is

$$h(k) = k \bmod m$$

In general, it is advised to select a prime value m, which is near the length of the storage space. This can ensure the relative evenness of the function value in most cases.

> **Example 4.8** We already know a set of keywords (26, 36, 41, 38, 44, 15, 68, 12, 6, 51, 25), with a storage space size of 15. Try to select the appropriate hash function with the remainder method.
>   Because the storage space size is 15, we can choose the maximum prime number smaller than 15, 13 as the divisor. We obtain the hash function:
>
> $$h(k) = k \bmod 13$$
>
> Then the addresses of the various keywords will be (0, 10, 2, 12, 5, 2, 3, 12, 6, 12, 12) respectively. We can see that there are still conflicts. In this case, we need to rely on some conflict resolution method to store every value into the hash table.

4.7.4.3 Base conversion method

This method looks at the keyword as data in another base, converts it back to the original base and then take certain digits from it as the address. Normally, a number larger than the original base is used as the conversion base, and it would be good if the two bases are relatively prime. For example, a keyword 596238 in base 10 can be viewed as a number in base 11, which gives us a new number as follows:

$$(596238)_{11} = 5 \times 11^5 + 9 \times 11^4 + 6 \times 11^3 + 2 \times 11^2 + 3 \times 11 + 8 = 945293$$

If the storage space size is 1,000, then we can take the last three digits as the hash address.

4.7.4.4 Random number method

Since our target is to let the output distribute on the domain as randomly as possible, then any random function, after modification to let it satisfy the requirement of the domain, can be used as the hash function. This method is more effective when the keywords are of different lengths:

$$h(k) = \text{random}(k)$$

Besides, there are various kinds of hash functions, such as whole-domain hash method, shifting method, folding method, subtraction method and multiplication method. Those algorithms used a lot of "hashing" methods. This is how the hash function and hash table got the name.

4.7.5 Conflict resolution methods

Since conflicts happen with a large probability, the resolution is an essential part of hash table design. In real life, there are also many examples of conflicts. For example, when you go to the movies, you find that your favorite seat is already occupied. How do you resolve the conflict in this case? With the assumption that you still want to see this movie, there are usually two ways to solve it – spatial and temporal. The former is to have a list of backup seats around your favorite seat (second-best choice, third-best choice, etc.), and find a seat as comfortable as possible according to this list. The latter is to still choose your favorite seat, but then you might have to wait until the next or even the one after the next screening.

Similar to the above examples in daily life, the conflict resolution methods for hash table can normally be divided into two types: open addressing (find another position within the current space) and chaining method (still stick to the current position; it is just that there are various different keyword records at this position).

4.7.5.1 Open addressing method
Open addressing is also called closed hashing. This is a solution under purely linear storage space, where all the elements are stored in a linear hash table.

When a conflict occurs, performing data insertion would require continuously checking the hash table, until an empty position is found to store the data. This process is called probing. According to the differences in probing sequences, this method can be divided into linear probing, reprobing and double hash lists. Suppose the storage space is m, then the hash function obtained under these methods can be expressed uniformly as follows:

$$h''(k,i) = (h(k) + h'(k,i)) \mod m, \ i = 0, 1, \ldots, m-1$$

where $h(k)$ is the original hash function, $h'(k,i)$ is the function related to the probing sequence, which is normally related to i. i is the count of the number of probings.
- In linear probing method, we have $h'(k,i) = i$.
- In reprobing method, we have $h'(k,i) = c_1 i + c_2 i^2$.
- In double hash lists method, we have $h'(k,i) = f(i, h_2(k,i))$, where $h_2(k)$ is another hash function different to $h(k)$. $f(a,b)$ is a function to perform simple calculations on the values of a and b, for example, multiplication of a and b.

To perform the lookup in a rehashed hash table, we need to start the lookup from the position of the hash value according to the probing sequence it uses.

- If the position at hash value is empty, return fail.
- If the keyword is found at the position of the hash value, return success.
- If it is not found at the position of the hash value, then search the subsequent positions according to the probing sequence. If found, return success.
- If an empty position is encountered during the lookup process, then there is no such a piece of data in the table. The lookup fails.

Example 4.9 There is a set of keywords (26, 36, 41, 38, 44, 15, 68, 12, 6, 51, 25), try to construct the hash table for this set of keywords.

Take $\alpha = 0.75$, $m = \lceil \frac{n}{\alpha} \rceil = 15$, the hash table will be HT[15]. If we take the largest prime smaller than m, 13, as the divisor, then the hash function will be

$$h(k) = k \bmod 13$$

By processing the conflicts with linear probing method at the same time, we can obtain the results in Fig. 4.30. The results also list the number of comparisons. We can see that after the conflict occurred, some data only found the storage position after one or multiple new comparisons. For example, the hash position of 12 is 12, but 38 has already occupied this position. Therefore, we need to use linear probing formula to calculate a new hash address:

$$H''(key, i) = (H(key) + H'(key, i)) \% m = (12 + 1) \% 15 = 13$$

key	26	36	41	38	44	15	68	12	6	51	25
Key%13	0	10	2	12	5	2	3	12	6	12	12

Hash address	0	1	2	3	4	5	6	7	8	9	10	11	12	13	14
Keyword	26	25	41	15	68	44	6				36		38	12	51
Number of comparisons	1	5	1	2	2	1	1				1		1	2	3

Fig. 4.30: Result obtained via linear probing method.

We should note that for the comparisons with positions 3 and 15 at the insertion of 68, and the comparisons with positions 0 and 26 at the insertion of 25, those are both keywords with different hash function values, and should originally not cause any conflict. However, conflicts occurred during the process of looking up the insertion position. This phenomenon is called clustering.

Term Explanation Clustering

The phenomenon where nodes with different hash addresses via the same subsequent hash address is called clustering.

[i]

The occurrence of clustering is the main disadvantage of the rehashing method. There is another disadvantage of the rehashing method, that is, the storage unit cannot be completely freed after the deletion of a piece of data. This is actually also caused by clustering: since data with different hash values will perform comparisons between each other. This disadvantage during the insertion process will also happen during the lookup process. If we delete the keyword 15 at position 3 in the above example, then, when looking for the keyword 68, fail will be returned because the hash address 3 is empty. To avoid this situation, we need to specially mark the deleted position, which complicates the algorithm.

The data structure used by the rehashing method is very simple. It is just sequentially stored linear list. The data structure can be described with the below C language code.

```
typedef struct       // Node structure of hash table
{
   KeyType key;
   DataType other;
} HashTable;
HashTable HT[M];
/*===============================================================
Functionality: Data lookup with linear probing method
Function input: Pointer to the sequential hash table, the value of the keyword
to be looked up
Function output: The position of the keyword found, or -1 if the keyword is not
found
===============================================================*/
int H(KeyType k);
// This is the calculation function for the hash function value
int LinPrbSrch(HashTable HT[ ], KeyType k)
{
   int d,i=0;
   //i is the incremental value for the address when a conflict occurs
   d=H(k);     //d is the hash address
   while ((i<m) && (HT[d].key ! =k) && (HT[d].key!=Nil))
   // Nil here indicates that the keyword is empty
   {
       i++;
       d=(d+1) % M
   }
   if(HT[d].key==k) return d;
   return -1;
   // If HT[d].key = k, then the lookup succeeds. Otherwise it fails
}
```

```
/*====================================================================
Functionality: Data insertion with linear probing method
Function input: Pointer to the sequential hash table, the node to insert
Function output: None
======================================== =========================*/
void LinPrbInS (HashTable HT[ ], HashTable s)
{
    int d;
    d= LinPrbSrch (HT[ ], s.key)     // Lookup the insertion position of s
    if (HT[d].key= =Nil)
    HT[d]=s;  // d is an available address, insert s
    else
    printf("ERROR");    // The node already exists or the table is full
    return;
}
```

4.7.5.2 Separate chaining method

Open hashing is also called separate chaining. This method is a solution based on mixed storage space. This method puts all the elements that hash to the same address into a linked list, while the main linear structure of the hash table stores a pointer at each of its addresses. The pointer points to the list head of the linked list, which stores all the elements that hash to this address. If there is no such an element, this pointer will be empty.

Example 4.10 For the set of data keywords in Example 4.8 (26, 36, 41, 38, 44, 15, 68, 12, 6, 51, 25), we still take the hash function as follows:

$$h(k) = k \bmod 13$$

Using the separate chaining method to deal with the conflicts, we obtain the results as shown in Fig. 4.31.

Using C language we obtain the following description of the storage method:

```
typedef struct NodeType
{
    KeyType key;
    DataType other;
    struct NodeType *next;
} ChainHash;
ChainHash *HTC[m];
```

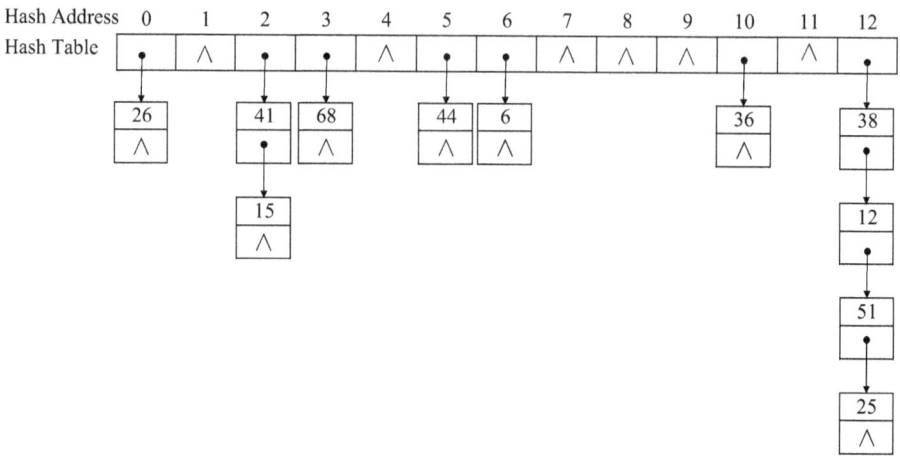

Fig. 4.31: Results obtained with separate chaining method.

```
/*================================================================
Functionality: Data lookup under separate chaining method
Function input: Beginning address of the sequential hash table, the value of
the keyword to look up
Function output: The position of the found keyword, or empty pointer if the
keyword is not found
================================================================*/
int H(KeyType k);
// This is the hash function
ChainHash *ChnSrch(ChainHash *HTC[ ], KeyType k)
{
    ChainHash *p;
    p=HTC[H(k)];      // Get the head pointer of the linked list that contains k
    while (p && (p->key ! =k)) p=p->next;      // Sequential lookup
    return p;
    // The lookup succeeds, return the node pointer.
    // Otherwise return empty pointer
}
/*================================================================
Functionality: Data insertion under separate chaining method
Function input: Beginning address of the sequential hash table; the node to
insert
Function output: None
================================================================*/
```

```
void ChnIns(ChainHash *HTC[ ],*s)
{
    int d ;
    ChainHash *p;
    p= ChnSrch (HTC,s->key);
    // See whether there is the node to insert in the table
    if (p) printf("ERROR");     // The table already has this node
    else { d=H(s->key); s->next=HTC[d]; HTC[d]=s; }    // Insert s
    return;
}
```

4.7.5.3 Extension of separate chaining method – hash tree

The chaining method can solve clustering relatively well. However, to perform in-formation lookup under the singly linked list under one hash address, we still need to proceed linearly, which is inefficient. Referring to the idea of rehashing in the open addressing method, we can extend this singly linked list in separate chaining method to a multibranching linked list defined with another or multiple other hash functions, that is, to a hash tree, in order to improve the search efficiency.

To improve the storage efficiency, we can store the hash tree in a purely linked manner.

Example 4.11 Still performing analysis on the set of keywords (26, 36, 41, 38, 44, 15, 68, 12, 6, 51, 25) from Example 4.8.

Since these 11 numbers have a range of 6–68, we can take two hash functions, respectively, as follows:

$$h(k) = k \bmod 7$$

$$h(k) = k \bmod 11$$

The prime numbers 7 and 11 are naturally prime to each other. The result of the mod-ulo operation can cover 77 different data. Combining the two operations, we can en-sure that the above data will cause absolutely no conflict. (Ensuring that no conflict occurs is not mandatory, if conflict occurs, we can still deal with it with conflict reso-lution methods.) The corresponding hash values of the above two hash functions are given in Table 4.3. From it we can obtain the hash tree shown in Fig. 4.32.

Table 4.3: Hash values of the hash tree.

Key	26	36	41	38	44	15	68	12	6	51	25
Key%7	5	1	6	3	2	1	5	5	6	2	4
Key%11	4	3	8	5	0	4	2	1	6	7	3

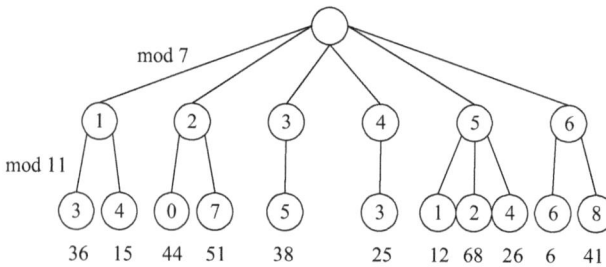

Fig. 4.32: Result obtained with hash tree.

When performing lookup and storage on the hash tree, the complexity will then be related to the height of the hash tree.

Readers might note that the shape characteristics and lookup method are very similar to the dictionary tree mentioned in Section 4.6.3. Actually, dictionary tree can be viewed as a special type of hash tree. Here, the data structure and algorithm implementation in C is omitted. Readers can write it referring to the section about dictionary tree.

4.7.6 Performance analysis of hash table lookup

To evaluate the performance of a lookup algorithm, we normally use the normal case scenario, or including the average lookup length under the best-case scenario and worst-case scenario as a reference, for example, the average lookup length for a successful lookup. Sometimes we also need to evaluate the average lookup length of failed lookups (there is no keyword being looked up in the table). Expression (4.3) is the unified expression to calculate the average lookup length, whether the lookup succeeds or fails.

Suppose the lookup on each keyword is equiprobable, then if we analyze the results in Examples 4.8, 4.9 and 4.10 with expression (4.3), then we can obtain the result as given in Table 4.4.

We can see that the differences in conflict resolution methods in the hash method will influence both the average lookup length and longest lookup length for the hash table obtained against the same set of data under the same hash function.

Between two conflict resolution methods, the worst case of the lookup performance (when the hash function did not evenly distribute the results) is the operation time that is proportional to the length of the table. However, in most cases the hash function controls the length of each singly linked list to be really short, and the lookup time is nearly $O(1)$.

Table 4.4: The analysis of the lookup times about the various conflict resolution methods in the examples.

		Linear probing method	Separate chaining method	Hash tree method
Average lookup length	Lookup success	1.82	1.64	2
	Lookup failure	4	0.85	2
Longest lookup length	Lookup success	5	4	2
	Lookup failure	5	4	2

If the insertion operation is performed by insertion to head, it can be completed in $O(1)$ time. The deletion operation's performance is similar to that of the lookup operation.

We can prove that the average lookup length of any conflict-resolving hash table is a function of the filling factor α. Therefore, choosing the filling factor appropriately and designing it well will have a huge impact on the application effect of the hash technique.

Although hash table has very good lookup performance, it has a fatal flaw viewing only from the lookup perspective, that is, it would be inconvenient to perform multikeyword complex queries or range queries. For example, if we have queries such as "find the female students from Shanxi province in the university" and "find the students over 24 years old in the university," these types of queries can be easily realized in an index table. However, for the hash table, which only determines the address according to a main keyword, while student origin, sex, age and others are all secondary keyword information, we need to perform by-item iteration to obtain the results. Its characteristic of quick lookup cannot be used here.

Therefore, in search engines, although some concrete problems might be accelerated with hash technique, the daily information queries still have to be supported by index technique. Index technique can be said to be the biggest foundation of search engine. Of course, hash can also be combined with index techniques, for example, hashed index.

Knowledge ABC The application of hash is not limited to storage and lookup
Besides storage, the application of hash technique also includes fields such as cryptography and copyright protection. For example, the famous MD5 value is a type of hash value. There are also techniques to use hashing to generate information fingerprints, which can be used for the lookup of audio and video, as well as the comparison of files. Google used hashing to produce information fingerprint while summarizing the information from various webpages, and will simultaneously store the information with hash table to prevent "repeated querying" – preventing its crawler from repeatedly visiting the same page.

4.8 Chapter summary

The connections between the main contents of this chapter are shown in Figs. 4.33–4.35.

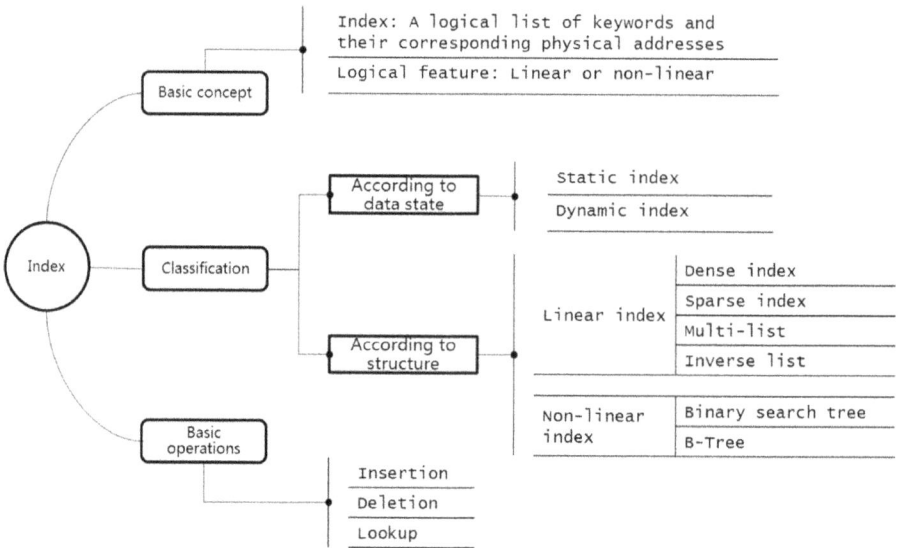

Fig. 4.33: Connections between the concepts of index.

The two major techniques for information lookup are index and hashing.

The data is huge and is either ordered and unordered. How do we quickly find what we need?

Indexing is the mapping list between the key value and the corresponding storage address after ordering the data based on a designated key value.

First index list, then data list, and this lookup strategy reduces the lookup time.

Data list is either static or dynamic. It has multiple forms. Choose to use static index or dynamic index depending on the situation you can handle.

The user can perform the queries based on keywords and secondary keywords. The index list and inverted list have the corresponding nice designs to handle those scenarios.

If the amount of data is huge, they are stored in the external storage and the response time when reading them would be huge.

The solution is to build multilevel tree index to store the memory for a long time.

Reduce hard disk reads as much as possible to limit the total search time.

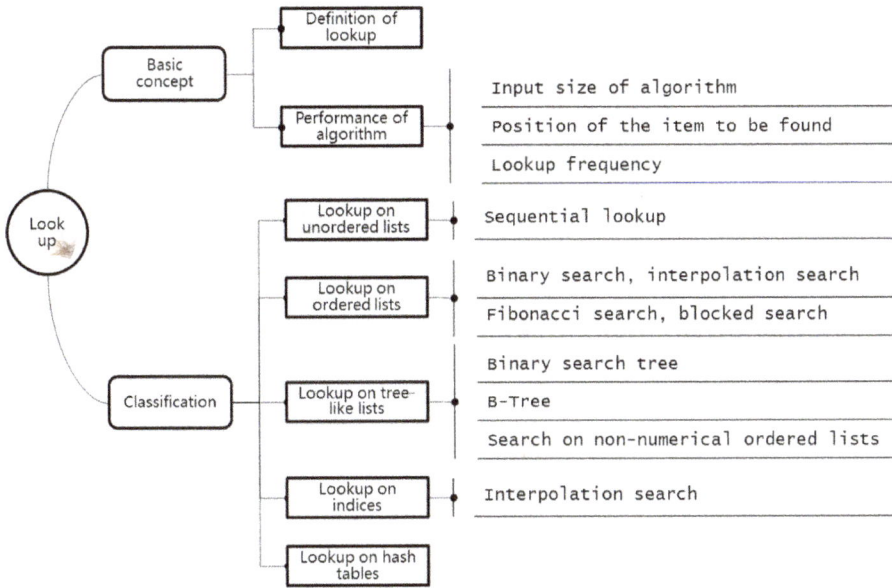

Fig. 4.34: Connections between the contents of lookup.

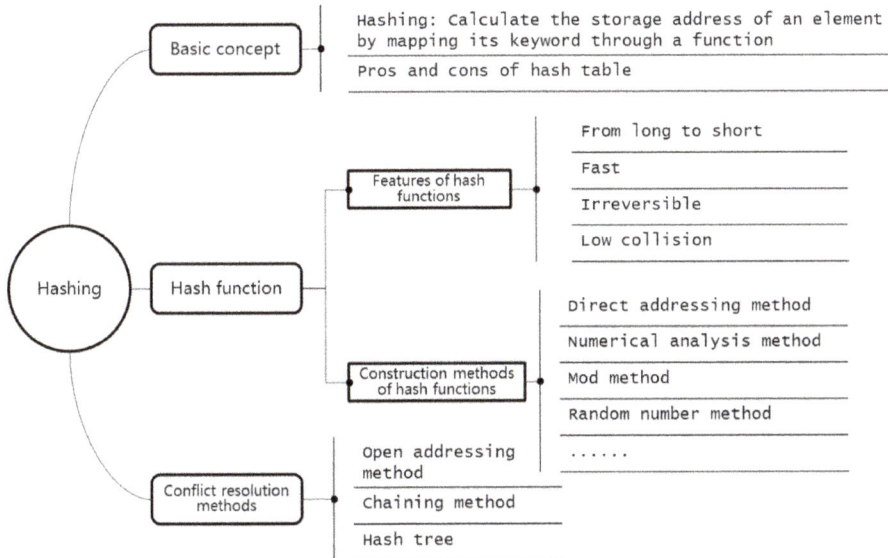

Fig. 4.35: The connections between the concepts of hashing.

Tree-like index is suitable for dynamic situations. The lookup performance will be good if the tree structure is as balanced as possible.

The most basic lookup technique is sequential lookup.

Although it is inefficient, it has a wide range of applicability.

Quick lookups can be based on ordered lists.

Classic algorithms for it include the Fibonacci search and binary search.

The lookup on tree structure is layer-by-layer scanning.

An example is B-tree, which offers multiway search.

For special lookup and storage techniques, the method of hashing is clever.

The storage address can be directly obtained via function mapping of the keyword.

The lookup time is nearly constant time in Big-O, which is wonderful.

The key to hashing is for the hashing function to be simple and effective. Conflict resolution needs careful thinking.

To improve the average lookup performance of hashing technique, the filling factor is the most important.

There are various advantages of hashing. However, for complex scenarios such as multikeyword lookup or range lookup, it would be unhelpful.

4.9 Exercises

4.9.1 Multiple-choice questions

1. In the following index structures, which index structure is most suitable for indexing a huge amount of data that constantly undergo changes? ( )
    (A) B-Tree   (B) Binary search tree
    (C) Multilevel index   (D) Linear index

2. Which of the following search methods is not suitable for multikeyword combined search? ( )
    (A) Multilist   (B) Inverted list   (C) Hash table

3. Which of the following search methods can be used to nonordered data list? ( )
    (A) Sequential search   (B) Binary search
    (C) Fibonacci search   (D) Interpolation search

4. Which of the following search methods is the most efficient on an ordered data list with sparse index? ( )
    (A) Search by block + binary search
    (B) Search by block + sequential search
    (B) Search by block + Fibonacci search
    (D) Directly perform binary search

5. When employing search by block, if there are 1,024 data elements in the linear list, on the assumption that each data element can be encountered with equal probability, if the data within each block is unordered, then which of the following block size can help achieve the highest search efficiency? ( )
(A) 1024   (B) 32   (C) 256   (D) 16

6. The basic idea of hashing storage is to determine the hash address with the keyword and hash function. In this method, "conflict" refers to: ( )
(A) Different keywords are mapped to the same address
(B) Data with different hash values vie for the same address
(C) Different keywords are stored to the same address
(D) Different keywords are linked after the same address

7. Suppose the address space for hashing is $0 \sim m - 1$, if we use the remainder method as the hash function, that is, $H(k) = k \bmod p$, then which would be the most appropriate value for p? ( )
(A) An odd number smaller than m
(B) The smallest prime number larger than m
(C) The largest prime number smaller than m
(D) Equal to m

8. Under which of the following situations would clustering occur? ( )
(A) Conflict resolution using separate chaining method in a hash table
(B) Conflict resolution using open addressing method in a hash table
(C) Index construction using B-tree
(D) Lookup using Fibonacci method

4.9.2 Closure

1. The basis upon which to decide whether to construct dynamic index or static index on a data list is __. While for a data list suitable for constructing static index, the basis upon which to decide whether to construct linear index or tree-like index is __.
2. When constructing a tree-like index, in order to reduce lookup complexity, we normally consider the __ feature of the shape of the tree.
3. Normally speaking, the time complexity of lookup operation on binary search tree is __, and the time complexity under the worst-case scenario is __.
4. The average algorithm efficiency of Fibonacci search is __ than that of binary search.
5. The time complexity of binary search is __.
6. When performing lookup on data in __, interpolation search is significantly better than the binary search.

7. When performing lookup on data completely stored in the memory, the algorithm efficiency of B-tree lookup is ___ than that of the binary search tree.
8. The two major problems in the application of hashing technique are ___ and ___.
9. The design of hashing function requires the computation complexity to be ___, the evenness of function values to be ___ and the range of values to not exceed the range of the addresses.
10. Filling factor α refers to the ratio between the data stored in the hash table and the length of the hash table. When the value of α is ___, there is a bigger possibility for conflicts.
11. According to the birthday paradox principle, the possibility of conflicts between hash addresses is ___.
12. The search efficiency of hash technique is mainly influenced by the choice of ___.
13. Is hash technique applicable to querying by range of a keyword? ___
14. There are two main methods for conflict resolution in hash tables: ___ and ___.

4.9.3 Practical questions

1. We already know that a data list is ordered. What index structures can we use to construct index on it? If the data within seldom change, and are large in number, what index structure is the most appropriate? What would be the highest lookup efficiency in this case?
2. For a set of given keywords (56, 89, 12, 58, 79, 23, 35, 62, 47, 15, 96, 24), draw the corresponding randomly generated binary search tree, red-black tree.
3. For a set of ordered keywords (14, 23, 25, 35, 42, 49, 56, 78, 84, 96, 135), draw the binary search decision tree and Fibonacci search decision tree for the lookup of keywords 96 and 87, respectively.
4. Use the remainder method to design hash function for the storage of the following set of keywords, and draw the status of the storage space obtained by resolving the conflict with open addressing method and separate chaining method (the filling factor can be taken as 0.75).
   Set of keywords: 85, 75, 27, 40, 65, 98, 74, 89, 12, 5, 46, 97, 13, 69, 52, 26, 19, 92.

4.9.4 Algorithm design questions

1. Please write executable code for binary search and interpolation search completely, and come up with two examples, where for one example the average lookup length of binary search is lower than that of the interpolation search, and for the other example the average lookup length is higher. Please illustrate the design idea when writing examples and verify them with the code.

2. Please write an executable code that implements hashing with base conversion method and deals with conflicts with separate chaining method, where construction and lookup operations can be performed.

3. Please write the construction and lookup algorithms for the data structure of hash tree in Example 4.11.

Appendix A   Answers to selected exercises

Chapter 1 Nonlinear structure with layered logical relations between nodes – tree

1. (C) data with branching leveled relations between elements
2. (C) both sequential and linked storage structures can be used
3. (B) Huffman tree
4. (C) is a tree and a binary tree
5. (A) unique
6. (C) 2^k-1
7. (A) 31
8. (B) R[2i+1]
9. (B) a is to the left of b
10. (B) cannot
11. (C) 4
12. (C) 69
13. (C) 5
14. (A) 98
15. (B) DFEBCA

Chapter 2 Nonlinear structure with arbitrary logical relations between nodes – graph

1. (C) 2
2. (B) 1
3. (C) $n(n-1)/2$
4. (C) 7
5. (D) n^2
6. (B) queue
7. (A) stack
8. (D) a tree
9. (B) sum of number of noninfinite elements at the ith column
10. (B) pre-order traversal
11. (A) symmetric matrix
12. (B) linked storage structure
13. (A) vertex sequence
14. (A) incoming edges
15. (A) G1 is a subgraph of G2
16. (B) set all elements in the ith row of the adjacency matrix to 0
17. (D) has one or multiple possibilities

https://doi.org/10.1515/9783110676075-005

18. (C) unconnected graph cannot be searched via DFS
19. (D) layered traversal
20. (A) the longest path from source vertex to sink vertex
21. (B) topological ordering
22. (B) 516234
23. (B) one or multiple

Chapter 3 Data processing methods – sorting technologies

3.9.1 Multiple-choice questions

1. (1) (E)   (2) (A)   (3) (B)   (4) (ACF), (BDE)
2. (B)
3. (A)
4. (C)
5. (C)
6. (C)
7. (D)
8. (A)
9. (C)
10. (D)
11. (C)
12. (D)

3.9.2 Cloze test

1. Compare the two keywords; save the pointer pointing to the record or change the saved pointer
2. 3
3. Insertion sort; selection sort
4. Heap sort; quick sort
5. $O(n^2)$; $O(n^2)$
6. $O(n \log_2 n)$, the extra space needed is $O(n)$
7. $\log_2 n$; $n \log_2 n$
8. H, C, Q, P, A, M, S, R, D, F, X, Y;
    P, A, C, S, Q, D, F, X, R, H, M, Y ;
    H, Q, C, Y, A, P, M, S, D, R, F, X ;
    F, H, C, D, P, A, M, Q, R, S, Y, X ;
    A, D, C, R, F, Q, M, S, Y, P, H, X 。
9. Heap sort, quick sort, merge sort; merge sort; quick sort; heap sort

10. Stable; unstable
11. Internal/in-memory; external

Chapter 4 Data processing – index and search technologies

4.9.1 Multiple-choice questions

1. (A)
2. (C)
3. (A)
4. (A)
5. (D)
6. (A)
7. (C)
8. (B)

4.9.2 Cloze test

1. Whether the data item changes frequently; the amount of data
2. Balancedness
3. $O(\log_2 N)$ $O(N)$
4. Lower
5. $O(\log_2 N)$
6. External storage
7. Lower
8. How to design the hashing function; how to handle conflicts
9. Low; high
10. High
11. Very high
12. Filling factor α
13. No
14. Open addressing/closed hashing; separate chaining/open hashing

References

[1]   Wang X. Fast binary tree traversal algorithm used in embedded systems. Computer
      Engineering and Design. 2013, 34(3): 873–877.
[2]   Wang X. Fast algorithm on binary trees for inherent properties from the indices of nodes.
      Computer Engineering and Application. 2011, 47(9): 16–20.

https://doi.org/10.1515/9783110676075-006

Index

https://doi.org/10.1515/9783110676075-007

www.ingramcontent.com/pod-product-compliance
Lightning Source LLC
Chambersburg PA
CBHW080714220326
41598CB00033B/5414